Matrix Metalloproteinase

Matrix Metalloproteinase

Editors

Magnus S. Ågren
Ulrich auf dem Keller

MDPI • Basel • Beijing • Wuhan • Barcelona • Belgrade • Manchester • Tokyo • Cluj • Tianjin

Editors
Magnus S. Ågren
University of Copenhagen
Denmark

Ulrich auf dem Keller
Technical University of Denmark
Denmark

Editorial Office
MDPI
St. Alban-Anlage 66
4052 Basel, Switzerland

This is a reprint of articles from the Special Issue published online in the open access journal *International Journal of Molecular Sciences* (ISSN 1422-0067) (available at: https://www.mdpi.com/journal/ijms/special_issues/matrix_metalloproteinase).

For citation purposes, cite each article independently as indicated on the article page online and as indicated below:

LastName, A.A.; LastName, B.B.; LastName, C.C. Article Title. *Journal Name* **Year**, *Article Number*, Page Range.

ISBN 978-3-03936-648-4 (Hbk)
ISBN 978-3-03936-649-1 (PDF)

Contents

About the Editors

Magnus S. Ågren is Affiliate Professor of Surgery at the Department of Clinical Medicine, Faculty of Health and Medical Sciences, University of Copenhagen, Denmark, and active at the Copenhagen Wound Healing Center and Digestive Disease Center, Bispebjerg Hospital. Professor Ågren received his doctoral degree (dr.med.sci.) in Experimental Pathology at the University of Linköping, Sweden 1990. He subsequently was a postdoctoral fellow at the University of Miami School of Medicine, Miami, Florida, USA from 1991–1993 with affiliation to the Departments of Dermatology (Professor William Eaglstein), Biochemistry (Professor Fred W. Woessner, Jr.), and Molecular and Cell Biology (Professor Gary Grotendorst). Since 1993, Dr. Ågren has worked as Senior Scientist and manager in both academia and industry. He was appointed Associate Professor at University of Linköping 1994 and became Professor at the University of Copenhagen 2011. Dr. Ågren' s research has focused on mechanisms and novel treatments for wound healing in the skin and the gastrointestinal tract. In these fields he has developed reliable and translatable cell and animal wound healing models, studied pathological wound healing, and worked on the identification of biomarkers of normal and aberrant wound healing. His specific scientific interests include the role of zinc, extracellular matrix metabolism and fibroblast phenotypes in wound healing. Worth mentioning among his scientific achievements are the demonstration of the physiological requirement of MMPs in epidermal repair, beneficial effect of MMP inhibitors in colonic wound repair, and the detrimental role of senescent fibroblasts in chronic cutaneous wound healing. Professor Ågren has supervised 12 basic scientists and medical doctors, and has hosted several visiting researchers from continental Europe. He has published 140 wound-healing articles in internationally renowned and peer-reviewed medical journals. His current H-index is 31 (Web of Science). Finally, Professor Ågren has edited a textbook series on Wound Healing Biomaterials (34 chapters), authored 4 book chapters and holds 3 patents. Dr. Ågren is Scientific Advisor for several Scandinavian and international companies in the health care sector. He is Section Editor of Acta Dermato-Venereologica, and acts on the Editorial Board of Wound Repair and Regeneration and Journal of Wound Care. Dr. Ågren organized the 7th Joint meeting of European Tissue Repair Society (ETRS) with the Wound Healing Society and the 25th Annual Meeting of ETRS in Copenhagen, 2015, and served as the President of ETRS from the year 2016 to 2017.

Ulrich auf dem Keller is Professor (with special responsibilities) and head of the Section for Protein Science and Biotherapeutics at the Department of Biotechnology and Biomedicine at Technical University of Denmark. Professor auf dem Keller received his diploma in Biochemistry in 2000 from the University of Tubingen, Germany and his PhD in 2005 from ETH Zurich, Switzerland, after completing his doctoral studies supported by a Boehringer Ingelheim Fonds PhD fellowship in the Institute of Cell Biology with Prof. S. Werner. After a short postdoctoral period in the same department he joined in 2006 Prof. C. M. Overall's laboratory at the University of British Columbia, Vancouver, Canada as a recipient of a Research Fellowship from the German Research Foundation (DFG) to work on protease proteomics. In 2009 he returned to ETH Zurich as a Senior Scientist and Junior Group Leader, where he applied newly developed advanced proteomics methods to elucidate the function of proteases in healthy and diseased skin. In 2017, Prof. auf dem Keller moved to Denmark to establish a research program on proteolytic networks in skin homeostasis, inflammation and repair with a focus on matrix metalloproteinases (MMP) and supported by a Young Investigator Award from the Novo Nordisk Foundation. Professor auf dem Keller is a pioneer in mass spectrometry-based proteomics technologies and has published several highly cited papers on MMP substrate discovery and global analysis of proteolysis in the healing skin wound. He served as president of the International Proteolysis Society from 2017 to 2019 and is the current president of the ETRS.

International Journal of *Molecular Sciences*

MDPI

Editorial

Matrix Metalloproteinases: How Much Can They Do?

Magnus S. Ågren [1,2,*] and Ulrich auf dem Keller [3]

1 Digestive Disease Center and Copenhagen Wound Healing Center, Bispebjerg Hospital, University of Copenhagen, 2400 Copenhagen, Denmark
2 Department of Clinical Medicine, Faculty of Health and Medical Sciences, University of Copenhagen, 2400 Copenhagen, Denmark
3 Department of Biotechnology and Biomedicine, Technical University of Denmark, 2800 Kongens Lyngby, Denmark; uadk@dtu.dk
* Correspondence: magnus.agren@mail.dk; Tel.: +45-3863-5954

Received: 16 March 2020; Accepted: 9 April 2020; Published: 12 April 2020

Abstract: Zinc-dependent matrix metalloproteinases (MMPs) belong to metzincins that comprise not only 23 human MMPs but also other metalloproteinases, such as 21 human ADAMs (a disintegrin and metalloproteinase domain) and 19 secreted ADAMTSs (a disintegrin and metalloproteinase thrombospondin domain). The many setbacks from the clinical trials of broad-spectrum MMP inhibitors for cancer indications in the late 1990s emphasized the extreme complexity of the participation of these proteolytic enzymes in biology. This editorial mini-review summarizes the Special Issue, which includes four review articles and 10 original articles that highlight the versatile roles of MMPs, ADAMs, and ADAMTSs, in normal physiology as well as in neoplastic and destructive processes in tissue. In addition, we briefly discuss the unambiguous involvement of MMPs in wound healing.

Keywords: extracellular matrix; inflammation; wound healing; cytokines; proteinases; interstitial collagens

---

More than half a century ago, Gross and Lapière discovered a true collagenase, which was the first vertebrate matrix metalloproteinase (MMP) responsible for the resorption of the tail in the metamorphosing tadpole [1]. We now know that vertebrate collagenase belongs to the metzincins, which is a clan of metalloendopeptidases found in all living organisms [2,3]. The metzincins, with their third ligand being histidine or aspartate in the active site, comprise not only the MMP family, which has 23 members in humans [4], but also other metalloproteinases, such as adamlysins or reprolysins, including ADAMs (a disintegrin and metalloproteinase domain; 21 members in humans) and ADAMTSs (a disintegrin and metalloproteinase thrombospondin domain), consisting of 19 secreted enzymes and at least 7 ADAMTS-like proteins that are devoid of catalytic activity [5,6], astacins (e.g., meprins, bone morphogenetic protein-1), leishmanolysins, serralysins, and snapalysins [2].

Research activities that followed the discovery by Gross and Lapière focused in the beginning on the critical role of these proteinases in extracellular matrix (ECM) remodeling in homeostatic balance and imbalance [7]. Thus, it turned out that MMPs are necessary for multiple and diverse physiological processes, such as reproduction, morphogenesis, embryonic development, bone remodeling, angiogenesis, and tissue repair, but they can also contribute to tissue destruction during cancer development and spreading and in arthritis/osteoarthritis and fibrotic diseases. The association of pathologies with MMP overexpression was also the impetus for the intense exploration of synthetic MMP inhibitors (MMPIs), especially those targeting cancer diseases, in the mid and late 1990s [8,9]. The results of randomized controlled trials were overwhelmingly disappointing for small-molecule MMPIs, due to their poor oral bioavailability, lack of efficacy, dose-limiting toxicities, and undesired musculoskeletal side effects [10]. These first-generation synthetic MMPIs targeted broadly by mimicking MMPs' natural substrate (usually collagen) [8], but later innumerable substrates were identified [11]. Unfortunately, these early MMPIs also inactivate proteinases unrelated to the disease but necessary for one or more physiological processes [8]; the importance of ADAMs and ADAMTSs was unknown at the time. Today, there is consensus that MMPs, ADAMs, and ADAMTSs function in many cell-signaling pathways, in which they are probably even more important than in ECM remodeling [11,12]. In parallel with our increased understanding of the diverse biological roles of these proteinases, the current goal is the development of highly selective inhibitors [8,13] although there is still no approved MMPI available. The only exception is low-dose oral doxycycline (Periostat®), which was approved in 1998 as an adjunct for the treatment of adult periodontitis. The mechanism of the beneficial doxycycline at subantimicrobial doses (20 mg b.i.d.) involves inhibiting collagenase (MMP-8) activity [14].

The content of this Special Issue highlights the multiple biological functions of these proteolytic enzymes and includes four review articles [4,15–17] and 10 original articles [18–27]. The articles address the role of MMPs, ADAMs, and ADAMTSs, in normal physiology as well as in neoplastic and tissue destructive processes.

Reflecting their pleiotropic activities, MMPs are closely associated with direct and indirect cell communication by modifying cell adhesion via integrin interactions and by activating or inactivating cytokines/chemokines or other signaling biomolecules and their cognate receptors. The specific roles of MMPs in these complicated signaling networks are highlighted by Young et al. [17].

A specific example of MMP-dependent cytokine signaling was elucidated by Sammel et al. [27]. They showed that not only ADAM-10 but also ADAM-17 and several other MMPs can shed the interleukin (IL)-11 receptor to induce transsignaling, a process where soluble receptor fragments interact with the ligand to act on cells not responsive to the ligand alone. Functional redundancy ensures a robust response, even in the absence of individual members of the shedding machinery. Although the current study remains at the level of cell culture and overexpression systems, it warrants further investigation of this elaborate proteinase network in animal models and in samples from patients with disturbances in individual components.

Another example of the importance of MMPs as proteinases with essential functions that have evolved within robust networks with redundant activities is presented by Rogerson et al. in their studies on ADAMTS [26]. In their research study in genetically manipulated mice, Rogerson et al. identified ADAMTS-9 as a novel aggrecanase that, in the absence of the aggrecanolytic ADAMTS-4 and ADAMTS-5, is highly increased in its abundance and might assume the critical functions of ADAMTS-4 and ADAMTS-5 in normal skeletal development [26]. ADAMTS-4 and ADAMTS-5 are thought to contribute to osteoarthritis by degrading the proteoglycan aggrecan in articular cartilage [28]. ADAMTSs are highly conserved in mice and humans, but it remains to be explored if similar redundancies also contribute to human pathologies.

The way in which membrane-tethered MMPs can interact with soluble members of the MMP family was demonstrated by Albrechtsen et al. at the University of Copenhagen [18]. By identifying basigin, an inducer of soluble MMPs, as a novel shed substrate of ADAM-12, they revealed a new function of the proteinase within the MMP network. This mechanistic insight has the potential to help devising novel strategies for inhibiting aberrant MMP activity in carcinogenesis, although in vivo validation is required.

Yip et al. [16] thoroughly reviewed the literature on MT4-MMP (MMP-17), another membrane-anchored MMP that has only been poorly studied [4]. MT4-MMP is one of the two members (the other one is MT6-MMP) of the family that is tethered to the membrane by a glycosylphosphatidylinositol anchor and shows specific activity in processing only a few ECM components. These appear to be important in many diseases, particularly in several types of cancer, which again demonstrates the strong need to better understand every member of the MMP family as a potential target for therapy.

Several studies in this Special Issue corroborate findings that show that MMPs have to be tightly controlled to prevent detrimental activity in disease. This might also be achieved by posttranslational modification (PTM) of the proteinase or the substrates. Many PTMs in MMPs and their target proteins have been identified, but their dynamic relationships are still poorly understood. In a review article, Madzharova et al. [4] summarize the current knowledge of the PTM (glycosylation, phosphorylation, and glycosaminoglycans)-mediated control of MMP activity and review the technologies used to study this topic.

As an additional layer of the control of MMP activity, soluble MMPs can also activate each other, e.g., by propeptide removal. For instance, the stromelysin MMP-3 is incapable of cleaving triple-helical fibrillar collagens but increases collagenolysis via the activation of collagenases [29]. Mirastschijski et al. [24] has convincingly demonstrated the profound effect of the lack of MMP-3 on tumor necrosis factor-α (TNF-α)-initiated collagenolysis in the skin of MMP-3-deficient mice, a finding that supports the indirect pathological role of MMP-3 in skin collagen catabolism through the activation of the human collagenase MMP-1 [30]. The authors did not demonstrate a direct link between murine MMP-13 and collagenase activity.

Oku et al. [25] have studied the effect of inflammation on cancer metastasis and the involvement of matrix metalloproteinase-9 (MMP-9). They used exogenous TNF-α to mimic the in vivo conditions and found that TNF-α upregulated MMP-9 at the transcriptional and translational levels in gastric cancer and mesothelial cell lines. The peritoneal barrier was modeled in vitro by mesothelial cells grown on a basement membrane matrix. TNF-α increased cancer cell invasion of the mesothelium via a process that was inhibited by MMP-9 gene silencing but not by MMP-2 gene silencing and was rescued by the addition of MMP-9 (Figure 1).

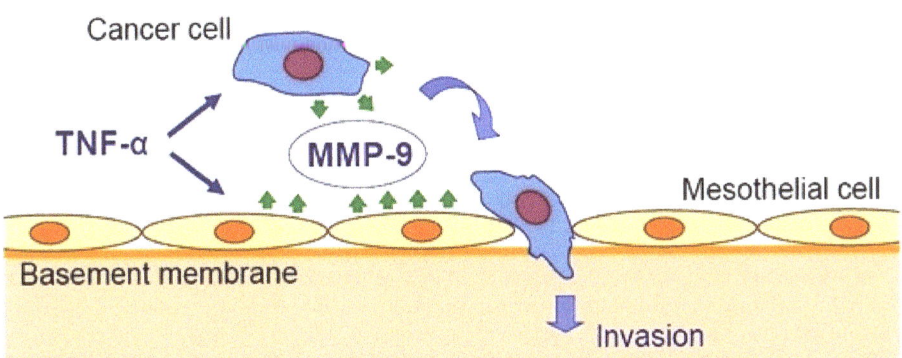

Figure 1. Tumor necrosis factor-α (TNF-α) stimulates gastric carcinoma cells and peritoneal mesothelial cells to secrete matrix metalloproteinase-9 (MMP-9), which promotes cancer cell invasion [25].

Huber et al. [22] set out to elucidate the cellular mechanisms involved in the differential effect of high-molecular-weight heparin (HMWH) compared with that of other anticoagulants on MMP-9 blood levels. For this purpose, monocytic and T and B lymphocytic cell lines were cocultured. The researchers demonstrated that HMWH increased IL-16 and sICAM-1 secretion by T lymphocytes, which in turn increased IL-8 and MMP-9 production by monocytes (Figure 2).

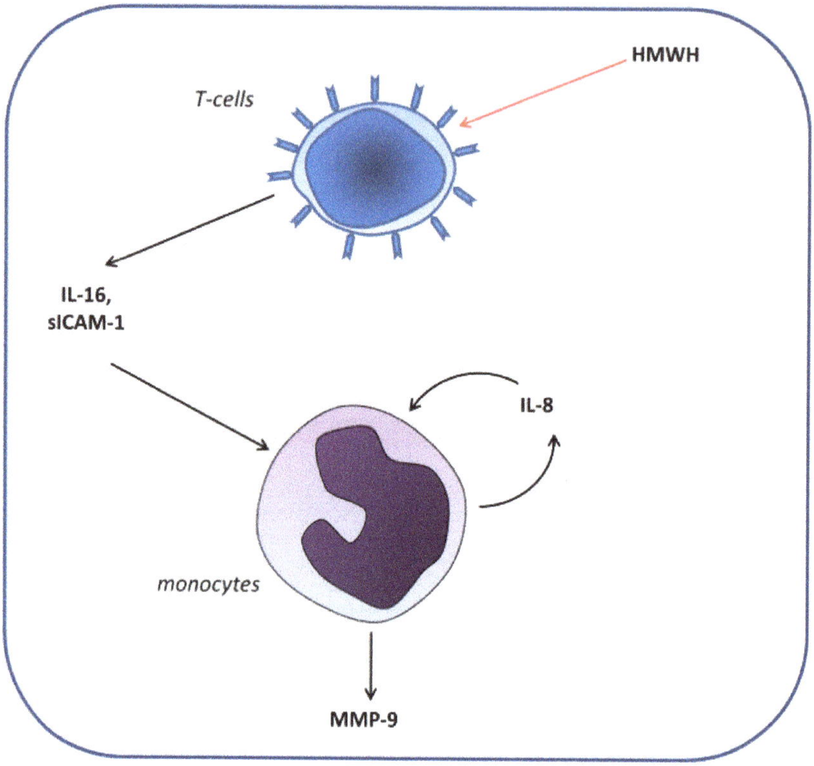

Figure 2. Proposed model of the induction of matrix metalloproteinase-9 (MMP-9) production in monocytes by high-molecular-weight heparin (HMWH)-treated T cells. T cells secrete the mediators IL-16 and sICAM-1, which induce monocytic IL-8 production. Together, these factors induce continuous IL-8 secretion as well as enhanced MMP-9 production by monocytes [22].

To investigate the emerging functions of MMPs as immune modulators, Bates et al. [19] mimicked the influence of inflammatory processes initiated by *Porphyromonas gingivalis* in vitro, causing periodontal destruction. They developed a 3-cell coculture model that includes monocyte-derived dendritic cells, CD4+ T lymphocytes, and primary gingival keratinocytes. MMP-7 and MMP-12 production differed significantly between the single cell cultures and the cocultures.

Dreschers et al. [20] addressed the clinical problem of premature delivery due to intrauterine infection. They hypothesized that detrimental persistence of inflammation is caused by reduced apoptosis of neonatal monocytes following phagocytosis of the infectious agent, e.g., *Escherichia coli*. The reduced apoptosis of infected neonatal monocytes compared with that of infected adult monocytes was attributed to increased shedding of CD95L by MMP-9. TACE (ADAM-17) was not involved in this process. One drawback of the study was the use of the general MMP inhibitor chlorhexidine; thus, the involvement of other MMPs could not be ruled out [31].

Tuberculosis remains a serious infectious disease, causing 2 million deaths every year at a global level. In a comprehensive review of MMP involvement in tuberculosis, Rohlwink et al. [15] summarize the current knowledge of the functions of MMPs in tuberculosis infections in both the lung and the brain. They outline the critical activities of MMPs in ECM degradation and pathogen release in the lung as well as in modulating immune responses. The use of various MMPIs in preclinical models of pulmonary and central nervous system tuberculosis is also reviewed. The ambiguous roles of MMPs in disease progression, particularly in childhood tuberculous meningitis, underscore the need for an increased understanding of how to balance MMP activity in treatment strategies.

Neuropathic pain is very difficult to treat, and there is an unmet medical need for effective therapeutic approaches. Kwan et al. [23] evaluated the effect of MMP-2/MMP-9 inhibition on neuropathic pain (Figure 3). They used an orally bioavailable MMP-2/MMP-9 inhibitor (AQU-118) in the spinal nerve ligation rodent model of neuropathic pain and discovered a novel relationship between elevated MMP-2 mRNA expression levels and caspase-3-mediated cell death, once again highlighting the complex interactions of MMPs with other proteinases within the proteinase network.

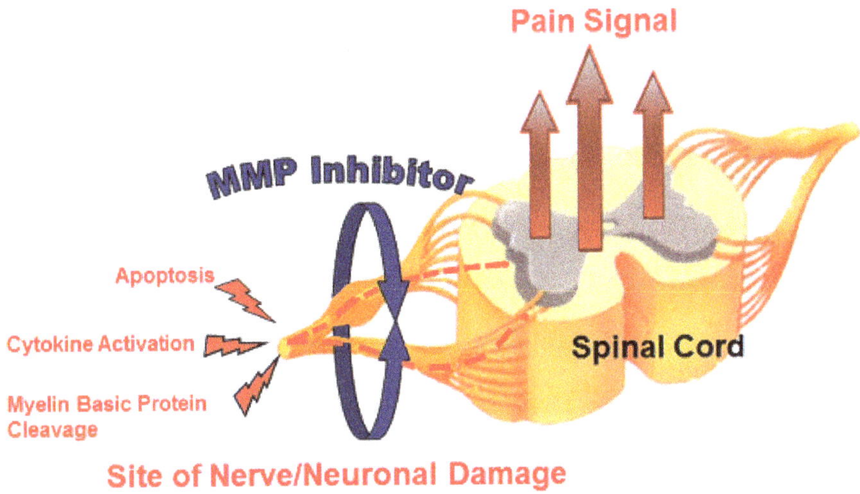

Figure 3. MMP inhibitor (MMPI) targets (increased apoptosis, increased cytokine activation, and decreased myelin basic protein levels) in neuropathic pain [23].

Snake venom metalloproteinases belong to the adamlysins, and their antithrombotic effects are well-known [32]. Huang et al. [21] tested the therapeutic effect of the metalloproteinase SP, which was isolated and purified from the venom of a moccasin snake (*Agkistrodon acutus*), in animal models of induced thrombosis and pulmonary embolism. The protein prolonged the coagulation time and inhibited platelet aggregation and thrombosis. Mechanistically, metalloproteinase SP cleaved the α, β, and γ chains of fibrinogen.

During wound healing, MMPs are involved in multiple cellular, molecular, and biochemical processes [33]. To decipher the role of MMPs, we used synthetic nondiscriminative MMPIs in animal and human acute injury wound-healing models [34–39]. The main conclusion of these studies is that neoepithelium formation is severely impaired by blocking MMP activity, while, paradoxically, increased collagen deposition in skin or peritoneal or intestinal wounds does not result from inhibiting MMP activities [34–37]. Without the protective epithelium, there is an increased susceptibility to infection. This presents a therapeutic challenge for nonhealing chronic cutaneous wounds, which often present with excessive MMP activity [33,40,41]: therapy should block the activity of pathogenic MMPs but not affect the activity of the MMPs required for reepithelialization [39]. The MMPs responsible for reepithelialization are unknown, but mouse models indicate that MMP-9 is one candidate [42], and in vitro studies suggest MMP-1 and MMP-7 as important candidates as well [43,44]. MMP-10 is highly upregulated in keratinocytes of the migrating tip in wounds with dermal involvement [39,45] and can cleave several cell adhesion and bioactive proteins [46]. Nonetheless, MMP-10-deficient mice did not show severe deficits in epithelial healing [47], indicating that the role of redundancy in the interconnected MMP network needs to be explored. Another clinical example is the regeneration of the mucosal epithelium of anastomotic wounds after resection of diseased colorectal tissue. While they are beneficial for normal anastomotic wound healing [9,48], broad-spectrum MMPI therapies are detrimental to anastomotic wound repair under complicated conditions [49]. Under these conditions, nonselective MMPIs severely delay epithelial coverage resulting in increased invasion by pathogenic microorganisms, abscess formation, and anastomosis insufficiencies [49]. Additionally, inhibition of the antimicrobial MMP-12 may contribute to weakening of the host defense [50]. The use of more selective MMPIs or local delivery of the MMPI to avoid systemic side effects may be a solution [37,51].

In conclusion, the present Special Issue has shed some light on the complex functions of MMPs, ADAMs, and ADAMTSs, in physiological and pathological processes. The new experimental and collected data that are provided here add to the current knowledge. The identification of novel substrates has been critical to the recent advances of the MMP research field [52], and novel high-throughput degradomics technologies have been instrumental in extending the classical view on MMPs as simple tissue degraders to precise signaling scissors modulating complex immune responses. The use of these promising technologies together with valid disease models and clinical studies has the potential to translate into effective therapeutic modulators inhibiting detrimental MMP activities and enhancing their beneficial effects as targets and antitargets in diseases [12,53].

Funding: Ulrich auf dem Keller acknowledges support by a Novo Nordisk Foundation Young Investigator Award (NNF16OC0020670).

Conflicts of Interest: The authors declare no conflict of interest.

Abbreviations

ADAM	A disintegrin and metalloproteinase domain
ADAMTS	A disintegrin and metalloproteinase thrombospondin domain
ECM	Extracellular matrix
HMWH	High-molecular-weight heparin
IL	Interleukin
MMP	Matrix metalloproteinase
MMPI	Matrix metalloproteinase inhibitor
PTM	Posttranslational modification
sICAM-1	Soluble intercellular adhesion molecule
TACE	Tumor necrosis factor-α converting enzyme
TNF-α	Tumor necrosis factor-α

References

1. Pardo, A.; Selman, M. MMP-1: The elder of the family. *Int. J. Biochem. Cell Biol.* **2005**, *37*, 283–288. [CrossRef] [PubMed]
2. Gomis-Rüth, F.X. Structural aspects of the metzincin clan of metalloendopeptidases. *Mol. Biotechnol.* **2003**, *24*, 157–202. [CrossRef]
3. Rawlings, N.D.; Barrett, A.J.; Woessner, J.F.; Salvesen, G. *Handbook of Proteolytic Enzymes*, 3rd ed.; Elsevier Science & Technology: San Diego, CA, USA, 2012.
4. Madzharova, E.; Kastl, P.; Sabino, F.; Auf dem Keller, U. Post-translational modification-dependent activity of matrix metalloproteinases. *Int. J. Mol. Sci.* **2019**, *20*, 3077. [CrossRef] [PubMed]
5. Le Goff, C.; Cormier-Daire, V. The ADAMTS(L) family and human genetic disorders. *Hum. Mol. Genet.* **2011**, *20*, R163-7. [CrossRef] [PubMed]
6. Brocker, C.N.; Vasiliou, V.; Nebert, D.W. Evolutionary divergence and functions of the ADAM and ADAMTS gene families. *Hum Genomics* **2009**, *4*, 43–55. [CrossRef] [PubMed]
7. Amar, S.; Smith, L.; Fields, G.B. Matrix metalloproteinase collagenolysis in health and disease. *Biochim. Biophys. Acta Mol. Cell Res.* **2017**, *1864*, 1940–1951. [CrossRef]
8. Fischer, T.; Senn, N.; Riedl, R. Design and structural evolution of matrix metalloproteinase inhibitors. *Chemistry* **2019**, *25*, 7960–7980. [CrossRef]

9. Ågren, M.S.; Jorgensen, L.N.; Delaissé, J.M. Matrix metalloproteinases and colon anastomosis repair: A new indication for pharmacological inhibition? *Mini Rev. Med. Chem.* **2004**, *4*, 769–778.
10. Coussens, L.M.; Fingleton, B.; Matrisian, L.M. Matrix metalloproteinase inhibitors and cancer: Trials and tribulations. *Science* **2002**, *295*, 2387–2392. [CrossRef]
11. Rodriguez, D.; Morrison, C.J.; Overall, C.M. Matrix metalloproteinases: What do they not do? New substrates and biological roles identified by murine models and proteomics. *Biochim. Biophys. Acta* **2010**, *1803*, 39–54. [CrossRef]
12. Kessenbrock, K.; Plaks, V.; Werb, Z. Matrix metalloproteinases: Regulators of the tumor microenvironment. *Cell* **2010**, *141*, 52–67. [CrossRef] [PubMed]
13. Levin, M.; Udi, Y.; Solomonov, I.; Sagi, I. Next generation matrix metalloproteinase inhibitors—Novel strategies bring new prospects. *Biochim. Biophys. Acta Mol. Cell Res.* **2017**, *1864*, 1927–1939. [CrossRef] [PubMed]
14. Golub, L.M.; McNamara, T.F.; Ryan, M.E.; Kohut, B.; Blieden, T.; Payonk, G.; Sipos, T.; Baron, H.J. Adjunctive treatment with subantimicrobial doses of doxycycline: Effects on gingival fluid collagenase activity and attachment loss in adult periodontitis. *J. Clin. Periodontol.* **2001**, *28*, 146–156. [CrossRef] [PubMed]
15. Rohlwink, U.K.; Walker, N.F.; Ordonez, A.A.; Li, Y.J.; Tucker, E.W.; Elkington, P.T.; Wilkinson, R.J.; Wilkinson, K.A. Matrix metalloproteinases in pulmonary and central nervous system tuberculosis-a review. *Int. J. Mol. Sci.* **2019**, *20*, 1350. [CrossRef] [PubMed]
16. Yip, C.; Foidart, P.; Noel, A.; Sounni, N.E. MT4-MMP: The GPI-anchored membrane-type matrix metalloprotease with multiple functions in diseases. *Int. J. Mol. Sci.* **2019**, *20*, 354. [CrossRef]
17. Young, D.; Das, N.; Anowai, A.; Dufour, A. Matrix metalloproteases as influencers of the cells' social media. *Int. J. Mol. Sci.* **2019**, *20*, 3847. [CrossRef]
18. Albrechtsen, R.; Wewer Albrechtsen, N.J.; Gnosa, S.; Schwarz, J.; Dyrskjot, L.; Kveiborg, M. Identification of ADAM12 as a novel basigin sheddase. *Int. J. Mol. Sci.* **2019**, *20*, 1957. [CrossRef]
19. Bates, A.M.; Fischer, C.L.; Abhyankar, V.P.; Johnson, G.K.; Guthmiller, J.M.; Progulske-Fox, A.; Brogden, K.A. Matrix metalloproteinase response of dendritic cell, gingival epithelial keratinocyte, and T-cell transwell co-cultures treated with *Porphyromonas gingivalis* hemagglutinin-B. *Int. J. Mol. Sci.* **2018**, *19*, 3923. [CrossRef]
20. Dreschers, S.; Platen, C.; Ludwig, A.; Gille, C.; Kostlin, N.; Orlikowsky, T.W. Metalloproteinases TACE and MMP-9 differentially regulate death factors on adult and neonatal monocytes after infection with *Escherichia coli*. *Int. J. Mol. Sci.* **2019**, *20*, 1399. [CrossRef]
21. Huang, J.; Fan, H.; Yin, X.; Huang, F. Isolation of a novel metalloproteinase from Agkistrodon venom and its antithrombotic activity analysis. *Int. J. Mol. Sci.* **2019**, *20*, 4088. [CrossRef]
22. Huber, R.; Attili/Abedalkhader, R.; Kuper, D.; Hauke, L.; Luns, B.; Brand, K.; Weissenborn, K.; Lichtinghagen, R. Cellular and molecular effects of high-molecular-weight heparin on matrix metalloproteinase 9 expression. *Int. J. Mol. Sci.* **2019**, *20*, 1595. [CrossRef] [PubMed]
23. Kwan, M.Y.; Choo, A.; Hanania, T.; Ghavami, A.; Beltran, J.; Shea, J.; Barboza, A.; Hu, A.; Fowler, M.; Neelagiri, V.R.; et al. Biomarker analysis of orally dosed, dual active, matrix metalloproteinase (MMP)-2 and MMP-9 inhibitor, AQU-118, in the spinal nerve ligation (SNL) rat model of neuropathic pain. *Int. J. Mol. Sci.* **2019**, *20*, 811. [CrossRef] [PubMed]
24. Mirastschijski, U.; Lupše, B.; Maedler, K.; Sarma, B.; Radtke, A.; Belge, G.; Dorsch, M.; Wedekind, D.; McCawley, L.J.; Boehm, G.; et al. Matrix metalloproteinase-3 is key effector of TNF-alpha-induced collagen degradation in skin. *Int. J. Mol. Sci.* **2019**, *20*, 5234. [CrossRef] [PubMed]
25. Oku, T.; Shimada, K.; Kenmotsu, H.; Ando, Y.; Kurisaka, C.; Sano, R.; Tsuiji, M.; Hasegawa, S.; Fukui, T.; Tsuji, T. Stimulation of peritoneal mesothelial cells to secrete matrix metalloproteinase-9 (MMP-9) by TNF-alpha: A role in the invasion of gastric carcinoma cells. *Int. J. Mol. Sci.* **2018**, *19*, 3961. [CrossRef]
26. Rogerson, F.M.; Last, K.; Golub, S.B.; Gauci, S.J.; Stanton, H.; Bell, K.M.; Fosang, A.J. ADAMTS-9 in mouse cartilage has aggrecanase activity that is distinct from ADAMTS-4 and ADAMTS-5. *Int. J. Mol. Sci.* **2019**, *20*, 573. [CrossRef]
27. Sammel, M.; Peters, F.; Lokau, J.; Scharfenberg, F.; Werny, L.; Linder, S.; Garbers, C.; Rose-John, S.; Becker-Pauly, C. Differences in shedding of the interleukin-11 receptor by the proteases ADAM9, ADAM10, ADAM17, meprin alpha, meprin beta and MT1-MMP. *Int. J. Mol. Sci.* **2019**, *20*, 3677. [CrossRef]
28. Mehana, E.E.; Khafaga, A.F.; El-Blehi, S.S. The role of matrix metalloproteinases in osteoarthritis pathogenesis: An updated review. *Life Sci.* **2019**, *234*, 116786. [CrossRef]

29. Manka, S.W.; Bihan, D.; Farndale, R.W. Structural studies of the MMP-3 interaction with triple-helical collagen introduce new roles for the enzyme in tissue remodelling. *Sci. Rep.* **2019**, *9*, 18785. [CrossRef]
30. Ågren, M.S.; Schnabel, R.; Christensen, L.H.; Mirastschijski, U. Tumor necrosis factor-alpha-accelerated degradation of type I collagen in human skin is associated with elevated matrix metalloproteinase (MMP)-1 and MMP-3 ex vivo. *Eur. J. Cell Biol.* **2015**, *94*, 12–21. [CrossRef]
31. Gendron, R.; Grenier, D.; Sorsa, T.; Mayrand, D. Inhibition of the activities of matrix metalloproteinases 2, 8, and 9 by chlorhexidine. *Clin. Diagn. Lab. Immunol.* **1999**, *6*, 437–439. [CrossRef]
32. Markland, F.S., Jr.; Swenson, S. Snake venom metalloproteinases. *Toxicon* **2013**, *62*, 3–18. [CrossRef] [PubMed]
33. Sabino, F.; auf dem Keller, U. Matrix metalloproteinases in impaired wound healing. *Met. Med.* **2015**, *2*, 1–8.
34. Syk, I.; Ågren, M.S.; Adawi, D.; Jeppsson, B. Inhibition of matrix metalloproteinases enhances breaking strength of colonic anastomoses in an experimental model. *Br. J. Surg.* **2001**, *88*, 228–234. [CrossRef]
35. Mirastschijski, U.; Haaksma, C.J.; Tomasek, J.J.; Ågren, M.S. Matrix metalloproteinase inhibitor GM 6001 attenuates keratinocyte migration, contraction and myofibroblast formation in skin wounds. *Exp. Cell Res.* **2004**, *299*, 465–475. [CrossRef] [PubMed]
36. Mirastschijski, U.; Johannesson, K.; Jeppsson, B.; Ågren, M.S. Effect of a matrix metalloproteinase activity and TNF-alpha converting enzyme inhibitor on intra-abdominal adhesions. *Eur. Surg. Res.* **2005**, *37*, 68–75. [CrossRef] [PubMed]
37. Krarup, P.M.; Eld, M.; Jorgensen, L.N.; Hansen, M.B.; Ågren, M.S. Selective matrix metalloproteinase inhibition increases breaking strength and reduces anastomotic leakage in experimentally obstructed colon. *Int. J. Colorectal. Dis.* **2017**, *32*, 1277–1284. [CrossRef]
38. Mirastschijski, U.; Schnabel, R.; Claes, J.; Schneider, W.; Ågren, M.S.; Haaksma, C.; Tomasek, J.J. Matrix metalloproteinase inhibition delays wound healing and blocks the latent transforming growth factor-beta1-promoted myofibroblast formation and function. *Wound Repair. Regen.* **2010**, *18*, 223–234. [CrossRef]
39. Ågren, M.S.; Mirastschijski, U.; Karlsmark, T.; Saarialho-Kere, U.K. Topical synthetic inhibitor of matrix metalloproteinases delays epidermal regeneration of human wounds. *Exp. Dermatol.* **2001**, *10*, 337–348. [CrossRef]
40. Trøstrup, H.; Holstein, P.; Karlsmark, T.; Moser, C.; Ågren, M.S. Uncontrolled gelatin degradation in non-healing chronic wounds. *J. Wound Care* **2018**, *27*, 724–734. [CrossRef]
41. Krishnaswamy, V.R.; Mintz, D.; Sagi, I. Matrix metalloproteinases: The sculptors of chronic cutaneous wounds. *Biochim. Biophys. Acta Mol. Cell Res.* **2017**, *1864*, 2220–2227. [CrossRef]
42. Kyriakides, T.R.; Wulsin, D.; Skokos, E.A.; Fleckman, P.; Pirrone, A.; Shipley, J.M.; Senior, R.M.; Bornstein, P. Mice that lack matrix metalloproteinase-9 display delayed wound healing associated with delayed reepithelization and disordered collagen fibrillogenesis. *Matrix Biol.* **2009**, *28*, 65–73. [CrossRef] [PubMed]
43. Pilcher, B.K.; Dumin, J.A.; Sudbeck, B.D.; Krane, S.M.; Welgus, H.G.; Parks, W.C. The activity of collagenase-1 is required for keratinocyte migration on a type I collagen matrix. *J. Cell Biol.* **1997**, *137*, 1445–1457. [CrossRef] [PubMed]
44. Hayden, D.M.; Forsyth, C.; Keshavarzian, A. The role of matrix metalloproteinases in intestinal epithelial wound healing during normal and inflammatory states. *J. Surg. Res.* **2011**, *168*, 315–324. [CrossRef] [PubMed]
45. Krampert, M.; Bloch, W.; Sasaki, T.; Bugnon, P.; Rulicke, T.; Wolf, E.; Aumailley, M.; Parks, W.C.; Werner, S. Activities of the matrix metalloproteinase stromelysin-2 (MMP-10) in matrix degradation and keratinocyte organization in wounded skin. *Mol. Biol. Cell* **2004**, *15*, 5242–5254. [CrossRef]
46. Schlage, P.; Egli, F.E.; Nanni, P.; Wang, L.W.; Kizhakkedathu, J.N.; Apte, S.S.; auf dem Keller, U. Time-resolved analysis of the matrix metalloproteinase 10 substrate degradome. *Mol. Cell Proteom.* **2014**, *13*, 580–593. [CrossRef]
47. Rohani, M.G.; McMahan, R.S.; Razumova, M.V.; Hertz, A.L.; Cieslewicz, M.; Pun, S.H.; Regnier, M.; Wang, Y.; Birkland, T.P.; Parks, W.C. MMP-10 regulates collagenolytic activity of alternatively activated resident macrophages. *J. Invest. Dermatol.* **2015**, *135*, 2377–2384. [CrossRef]
48. Ågren, M.S.; Andersen, T.L.; Andersen, L.; Schiødt, C.B.; Surve, V.; Andreassen, T.T.; Risteli, J.; Franzén, L.E.; Delaissé, J.M.; Heegaard, A.M.; et al. Nonselective matrix metalloproteinase but not tumor necrosis factor-alpha inhibition effectively preserves the early critical colon anastomotic integrity. *Int. J. Colorectal. Dis.* **2011**, *26*, 329–337. [CrossRef]

49. Rehn, M.; Krarup, P.M.; Christensen, L.H.; Seidelin, J.B.; Ågren, M.S.; Syk, I. GM6001 increases anastomotic leakage following colonic obstruction possibly by impeding epithelialization. *Surg. Infect. (Larchmt)* **2015**, *16*, 702–708. [CrossRef]
50. Houghton, A.M.; Hartzell, W.O.; Robbins, C.S.; Gomis-Ruth, F.X.; Shapiro, S.D. Macrophage elastase kills bacteria within murine macrophages. *Nature* **2009**, *460*, 637–641. [CrossRef]
51. Pasternak, B.; Rehn, M.; Andersen, L.; Ågren, M.S.; Heegaard, A.M.; Tengvall, P.; Aspenberg, P. Doxycycline-coated sutures improve mechanical strength of intestinal anastomoses. *Int. J. Colorectal Dis.* **2008**, *23*, 271–276. [CrossRef]
52. Schlage, P.; auf dem Keller, U. Proteomic approaches to uncover MMP function. *Matrix Biol.* **2015**, *44–46*, 232–238. [CrossRef] [PubMed]
53. Dufour, A.; Overall, C.M. Missing the target: Matrix metalloproteinase antitargets in inflammation and cancer. *Trends Pharmacol. Sci.* **2013**, *34*, 233–242. [CrossRef] [PubMed]

International Journal of
Molecular Sciences

MDPI

Review

Post-Translational Modification-Dependent Activity of Matrix Metalloproteinases

Elizabeta Madzharova, Philipp Kastl, Fabio Sabino and Ulrich auf dem Keller *

Department of Biotechnology and Biomedicine, Technical University of Denmark, DK-2800 Kongens Lyngby, Denmark; elimad@dtu.dk (E.M.); phikas@dtu.dk (P.K.); famisa@dtu.dk (F.S.)
* Correspondence: uadk@dtu.dk; Tel.: +45-45-25-2791

Received: 27 May 2019; Accepted: 18 June 2019; Published: 24 June 2019

Abstract: Due to their capacity to process different proteins of the extracellular matrix (ECM), matrix metalloproteinases (MMPs) were initially described as a family of secreted proteases, functioning as main ECM regulators. However, through proteolytic processing of various biomolecules, MMPs also modulate intra- and extracellular pathways and networks. Thereby, they are functionally implicated in the regulation of multiple physiological and pathological processes. Consequently, MMP activity is tightly regulated through a combination of epigenetic, transcriptional, and post-transcriptional control of gene expression, proteolytic activation, post-translational modifications (PTMs), and extracellular inhibition. In addition, MMPs, their substrates and ECM binding partners are frequently modified by PTMs, which suggests an important role of PTMs in modulating the pleiotropic activities of these proteases. This review summarizes the recent progress towards understanding the role of PTMs (glycosylation, phosphorylation, glycosaminoglycans) on the activity of several members of the MMP family.

Keywords: MMPs; PTMs; glycosylation; phosphorylation; glycosaminoglycans

---

1. MMP Domain Structure and Classification

Matrix metalloproteinases (MMPs) comprise a family of 23 distinct secreted or membrane-anchored endopeptidases in humans that belong to the metzincin superfamily of metalloproteases. MMPs were initially described as regulators of the extracellular matrix due to their capacity to degrade ECM proteins like collagen, gelatin, laminin, aggrecan, fibronectin, elastin, and proteoglycans. Based on structure and substrate specificity, the MMPs are divided into collagenases (MMP1, MMP8, MMP13), gelatinases (MMP2, MMP9), stromelysins (MMP3, MMP10, MMP11), matrilysins (MMP7, MMP26), membrane-type MMPs (MMP14, MMP15, MMP16, MMP17, MMP24, MMP25), and other non-classified family members (MMP12, MMP19, MMP20, MMP21, MMP23, MMP27, MMP28). MMPs share a multidomain structure consisting of signal peptide, zymogenic pro-peptide domain, a catalytic domain, a variable linker 'hinge' region and a hemopexin domain. Some MMPs show variation in the domain arrangements, e.g., MMP7 and MMP26 are lacking the linker and hemopexin domain, MMP23 has a unique cysteine-rich and immunoglobulin-like instead of a hemopexin domain, and the two gelatinases MMP2 and MMP9 have additional fibronectin type-II-related domains. Membrane-type MMPs are anchored to the membrane either via glycosylphosphatidylinositol (GPI) or with help of a transmembrane domain (Figure 1). Additionally, MMPs are distinguished by the highly conserved HExGHxxGxxH motif in the catalytic domain, which contains three histidines that coordinate the zinc molecule in the active site, and the PRCGxPD motif identified in the pro-domain whose cysteine residue coordinates with the active zinc molecule to inhibit proteolysis. Each of these domains is associated with a specific function and very frequently post-translationally modified. This suggests an important role of PTMs in modulating the pleiotropic activities of MMPs [1,2].

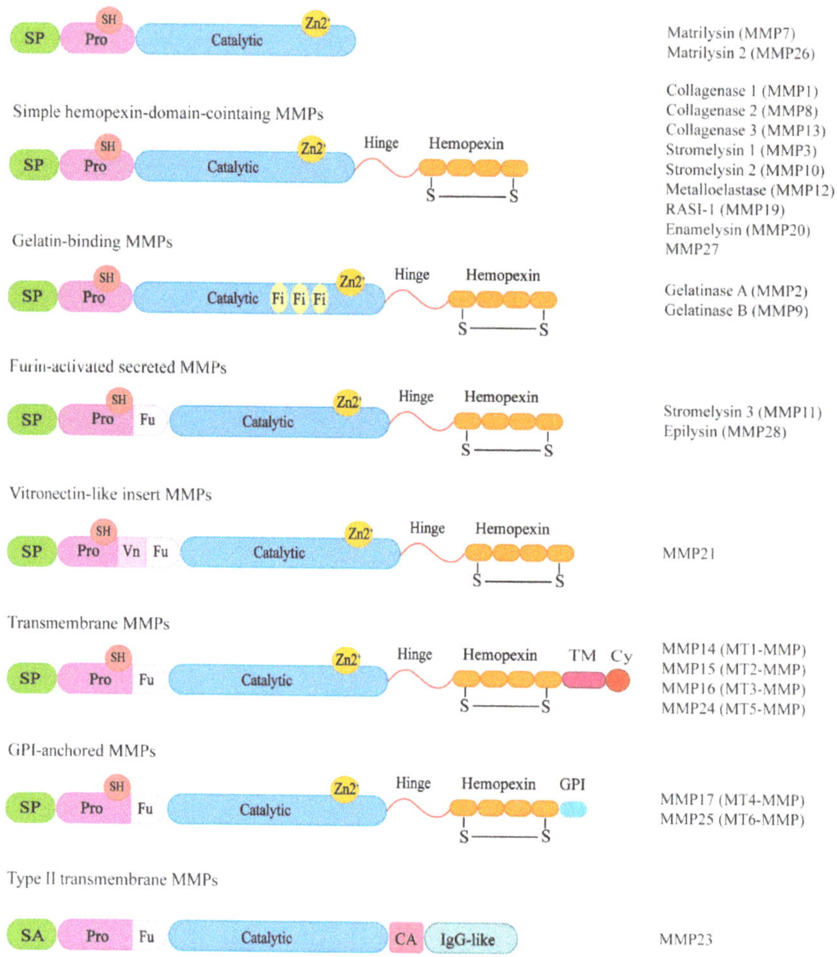

Figure 1. MMP multidomain organization. Based on their domain structure, MMPs can be subdivided into eight different groups. All MMPs share a common structure comprising a signal peptide (SP), a pro-domain (Pro), containing a thiol group (SH), a catalytic domain (Catalytic) with a zinc (Zn) binding site, a linker region (Hinge) and a hemopexin domain (Hemopexin), which has a disulfide bond (S-S). Exceptions to this are the two gelatinases, which contain three fibronectin repeats (Fi) within their catalytic domain and furin-activated MMPs which have a furin-recognition site (Fu) within their pro-domain and MMP21 with an additional vitronectin-like insert (Vn). Some membrane-type MMPs are anchored to the membrane via glycosylphosphatidylinositol (GPI), whereas some MT-MMPs have transmembrane (TM) and cytosolic domains (Cy). In type II MT-MMPs, an N-terminal signal anchor (SA), a cysteine array (CA) domain, and an immunoglobulin-like (Ig-like) domain are present. Adapted by permission from "Springer Nature: New functions for the matrix metalloproteinases in cancer progression. *Nat. Rev. Cancer* **2002**, *2*, 161–174. Egeblad & Werb, Copyright (2002)".

2. MMP Substrates and Function

MMPs are primarily extracellular proteases, supporting the initial concept that they are generally associated with degradation and regulation of the ECM, thereby influencing many fundamental cellular events involving ECM remodeling [2–5]. More recently, it was observed that they also can irreversibly process growth factors, cell-surface receptors, cytokines, and chemokines, as well as other MMPs, other proteases and protease inhibitors and even act inside the cell [6]. The activation or inactivation of bioactive molecules by MMP proteolytic processing unraveled unexpected roles of these proteases in the regulation of extra- and intracellular signaling pathways [4,7–9]. Hence, the field of MMP research progressed from perceiving these enzymes solely as regulators of ECM to the conception that MMPs are functionally implicated in the regulation of multiple physiological and pathological signaling processes [8,10]. Accordingly, the physiological relevance of their function is emphasized by a direct association between altered expression and/or dysregulation of MMPs and development of pathological conditions, such as chronic inflammatory diseases, vascular diseases, neurological disorders, and cancer [10,11].

3. Multilayered Regulation of MMP Activity

Since MMPs regulate major physiological processes, a strict spatiotemporal control of their activity is essential to avoid possible detrimental activities of these proteases. Primarily, expression of MMPs is regulated at the transcriptional level, keeping these enzymes at very low levels in normal tissue homeostasis. MMPs share cis-regulatory elements in their promoter sequences, which allow induction of their expression by stimuli—e.g., in the form of growth factors, cytokines, or hormones. The cooperation between these cis-regulatory elements, coupled with the integration of multiple signaling pathways, provides a wide range of potential interactions between transcriptional regulators, ensuring tissue-specific expression of diverse MMP family members and facilitating a strict control of MMP transcriptional activity. In addition, transcription of MMP encoding genes is regulated by epigenetic mechanisms, such as DNA methylation or histone acetylation [4,5,11,12]. At the post-transcriptional level, regulation of MMP expression is mediated by modulation of mRNA stability and miRNA-based mechanisms that interfere with MMP expression by either transcriptional inhibition or mRNA degradation [13–16]. An important level of MMP regulation is achieved at the post-translational level, since most MMPs are secreted as inactive pro-enzymes. The 'cysteine-switch' region in the pro-peptide domain shields the active site, thereby preventing substrate access and maintaining the enzyme in an inactive state. The activation of proMMPs occurs upon direct proteolytic cleavage of the pro-domain (within the secretory pathway or extracellularly), or induction of conformational changes that disrupt the chelating cysteine residue and enable auto-proteolysis for removal of the pro-domain region [9,12]. Moreover, allosteric activation of proMMPs can be triggered by interaction with ECM components and cell surface molecules and by low-density lipoprotein receptor-related protein (LRP1)-mediated endocytosis [12,17–20]. Upon activation, mature MMPs are exclusively controlled by endogenous inhibitors, such as tissue inhibitors of metalloproteases (TIMPs) and α_2-macroglobulin [21,22].

4. PTMs—An Additional Level of Protein Regulation

The enormous functional complexity of the cellular proteome is regulated by diverse mechanisms, including transcription, alternative splicing, translation, and PTMs [23]. Among these regulatory mechanisms, PTMs provide a significant genome-independent expansion and diversification of the proteome, thereby creating a continuously fine-tuned regulatory network implicated in many cellular processes [24–26]. Based on the type of modification, PTMs can be assigned to several categories: chemical modifications, including methylation, phosphorylation, acetylation, and oxidation; polypeptide modifications, including ubiquitination, SUMOylation and other ubiquitin-like protein conjugation; modifications by complex molecules, including glycosylation, lipids (e.g., acylation, prenylation) and extended structures (e.g., glycosaminoglycans (GAGs)); and modifications of the amino acids or of the polypeptide backbone, including deamidation, eliminylation, and protein cleavage through proteolysis [27–30]. By reversible or irreversible addition of these functional groups PTMs can modify protein function by altering protein structure, subcellular localization, protein–protein interactions, and degradation, thereby influencing many cellular processes in health and disease [31–36].

Like many enzymes, MMPs as well as their substrates are modified by a variety of PTMs. Different domains of the MMP structure are associated with specific functions and undergo different, highly specific PTM modifications, which suggests an important role of PTMs in modulating the pleiotropic activities of MMPs (Figure 2) [1]. Understanding the influence of specific PTMs on the activity of MMPs is necessary to fully understand MMP regulation. In this review, we provide an overview of different PTMs with a focus on glycosylation, phosphorylation, and interaction with extracellular GAGs, and describe their effects on activity of various MMP family members.

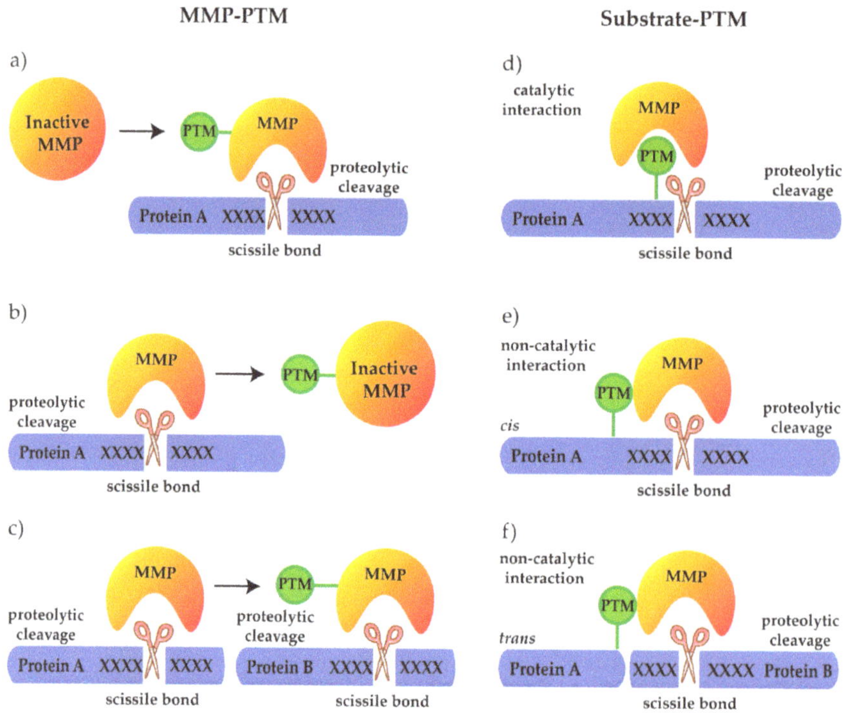

Figure 2. Possibilities of PTM crosstalk with MMPs. PTMs can regulate the activity of MMPs on the protease level by (**a**) activating the protease to allow cleavage of the target protein, (**b**) inactivating the protease, or (**c**) modulating its substrate specificity. Additionally, PTMs may regulate substrate cleavage by MMPs, through (**d**) directly modifying the cleavage motif (catalytic interaction), or substrate domains outside the cleavage site (non-catalytic interaction), guiding MMPs to cleave the same (*cis*) (**e**) or another (*trans*) protein (**f**).

4.1. Glycosylation of MMP

Glycosylation describes the enzymatic reaction that links saccharides to lipids, other saccharides or proteins and represents the most abundant and diverse PTM. The most common forms of protein glycosylation are the *N*- and *O*-linked glycosylation [37,38]. *N*-linked glycosylation is a glycosidic linkage of glycans to the side chains of asparagine (Asn) residues. It is initiated by a single oligosaccharyltransferase complex with a transfer of *N*-acetylglucosamine (GlcNAc) phosphate at the consensus Asn-Xaa-Ser/Thr sequence (X, indicating any amino acid excluding proline) on the cytosolic side of the endoplasmic reticulum (ER) and extensively modified further in the lumen of the ER and Golgi prior to sorting to secretory vesicles [37,39–41]. *O*-linked glycosylation is a glycosidic linkage of glycans to the side chains of serine/threonine (Ser/Thr) or tyrosine (Tyr) residues. *O*-linked glycosylation shows a higher diversity than *N*-glycans, resulting from the complex synthesis of *N*-acetylgalactosamine (GalNAc)-type *O*-linked glycans in the Golgi, which can be initiated by up to 20 different polypeptide GalNAc-transferase isoforms that may compete for the same glycosylation site and the same glycan. Upon transfer of a GalNAc moiety from a donor substrate uridine diphosphate (UDP)-GalNAc to Ser/Thr residues, the *O*-linked glycan is branched out by different core-forming enzymes and further extended by fucosylation and sialylation to the final *O*-linked glycan structure [41–45]. Protein glycosylation regulates essential biological processes, such as protein folding, secretion, cell adhesion, and inter- and intra-cellular trafficking [46–50]. Consequently, alterations in glycosylation patterns are often associated with different pathological conditions like neurodegenerative diseases, diabetes, inflammatory conditions, and cancer [51,52]. The majority of MMPs are glycosylated, and *N*- and *O*-linked glycosylation are present across the MMP family [39]. The following section summarizes the current structural and functional information about glycosylation of different MMPs.

4.1.1. MMP9

MMP9 is the most extensively glycosylated MMP. This protease contains two *N*-linked glycosylation sites, Asn^{38} and Asn^{120} (asparagine residues 38 and 120), in the pro-domain and the catalytic domain, respectively [53]. These *N*-linked glycans have been described as core-fucosylated biantennary structures, partially sialylated with variable fucosylation branches [54]. Despite their similar composition, the individual functions of these *N*-linked glycosylations are distinct, primarily due to their location within the protease. Since *N*-linked glycosylation is generally required for protein secretion, initially a potential role of these two glycans was related to the secretion of MMP9 [55]. Indeed, abrogation of MMP9 glycosylation at Asn^{120} reduced the efficiency of its secretion by increasing the interaction between MMP9 and calreticulin (a protein that prevents misfolded proteins from entering the secretory pathway), thereby inducing MMP9 retention in the ER. The glycosylation at Asn^{38} does not influence the secretion of MMP9, but Asn^{38}-glycosylation-deficient MMP9 shows strong amino-acid dependency towards interaction with calreticulin, likely affecting MMP9 secretion in an *N*-glycosylation-independent manner [56]. Additionally, *N*-glycosylation at Asn^{38} was commonly associated with the activation of proMMP9, yet the process of MMP9 activation occurs independently of the glycosylation at Asn^{38}. Recent molecular dynamics simulations suggest that the glycosylation at Asn^{38} is indirectly involved in the activation of proMMP9 by inducing conformational changes within the pro-domain, enabling MMP3 to access the two cleavage sites for proteolytic activation [57]. Moreover, interactions between the glycosylation at Asn^{38} and galectin 3 decrease the proteolytic activation of MMP9, whereas interactions with galectin 8 enhance the MMP3-mediated processing, suggesting that the presence of *N*-glycosylation is important for a fine-tuned regulation of MMP9 activity [58,59]. Furthermore, MMP9 has a proline-rich linker sequence between the active site and the hemopexin domain, which contains 14 *O*-linked glycans (denominated OG domain) [60]. Based on their glycan composition, these *O*-linked glycans comprise a heterogenic mixture, varying from core-1 (Galβ1-3GalNAc) to core-2 (Gal-(GlcNAc-)GalNAc) structures and further elongated to larger glycans [61]. The presence of these *O*-linked glycans increases the domain flexibility of MMP9, allowing the protease to adopt multiple enzyme conformations and facilitating individual movements of the catalytic and the hemopexin domain. This influences the recognition, binding, and processing of substrates, cell receptors, and endogenous inhibitors [62,63]. A study by den Steen et al. described that the OG domain is indispensable for correct orientation of the hemopexin domain for MMP9 internalization and degradation by LRP-1 and LRP-2, as well as inhibition by endogenous TIMP1 [60]. Consequently, deletion of this linker region significantly reduced the affinity towards TIMP1 and disrupted the interactions with LRP-1 and LRP-2 [60]. However, since deletion of the OG domain did not affect the activity of MMP9, the authors reported that the OG domain functions as a regulator of extracellular bioavailability of the protease, rather than as a regulator of its activity [60]. An independent analysis of MMP9 lacking the OG domain performed by Vandooren et al. showed a reduced gelatinolytic activity of the truncated protease [64]. Furthermore, a study by Dufour et al.

reported reduced MMP9-induced cell migration in endothelial cells expressing MMP9 without the OG domain, which was independent of the proteolytic activity of the enzyme [26].

4.1.2. MMP14

Similar to MMP9, the transmembrane collagenase MMP14 (MT1-MMP) contains *O*-linked glycosylation sites (Thr^{291}, Thr^{299}, Thr^{300}, and Ser^{301}) identified in the linker region. The presence of these *O*-linked glycans does not affect zymogen activation, collagenase activity or the autolytic processing of MMP14. However, the *O*-linked glycosylation is important for the formation of a stable complex between MMP14, TIMP2, and proMMP2, which is required for subsequent cell-surface activation of MMP2. Being unable to recruit TIMP2, which connects the catalytic domain of MMP14 and the hemopexin domain of proMMP2, the glycan-deficient MMP14 cannot present a stable trimeric complex on the cell surface and thereby activate MMP2 [65–67]. Interestingly, although it has not appeared that glycosylation could directly affect MMP14 activity, recent reports describe increase of MMP14 activity upon alterations in the glycosylation pattern. Indeed, hyper-glycosylation of MMP14 led to higher proteolytic activity and promoted tumor growth [68].

4.1.3. MMP1

MMP1 has two potential glycosylation sites in the active site, but only *N*-glycosylation at Asn^{120} has been experimentally confirmed. Comparison between the glycosylated and non-glycosylated MMP1 showed no significant differences in activity, substrate specificity or inhibitory profiles of the two proteoforms. However, the presence of specific glycan motifs (e.g., α1,3-fucosylated LacdiNAc) can initiate MMP1 selectin-mediated binding to the surface of activated cells through a selectin/glycan interface and therefore may have a profound effect on cell migration [39,69].

4.1.4. MMP2

MMP2 has two potential *N*-linked glycosylation sites, Asn^{573} and Asn^{642}, in the hemopexin domain [70,71]. The function of these *N*-linked oligosaccharides remains unclear. However, there are strong implications regarding their involvement in the regulation of MMP2, since the hemopexin domain is involved in MMP2 activation/inhibition, localization of its catalytic activity, and induction of cell signaling upon interaction of MMP2 with cell-surface receptors [71,72].

4.1.5. MMP3

Based on the consensus sequence for *N*-linked glycosylation, MMP3 also has two potential *N*-linked glycosylation sites, Asn^{120} in the catalytic domain and Asn^{398} in the hemopexin domain. However, only a small portion of MMP3 (~20%) is glycosylated, and no correlation has been established between these glycans and the function of the protease [39,73].

4.1.6. MMP13

MMP13 was shown to be potentially glycosylated at two asparagine residues (Asn^{117} and Asn^{152}) in the catalytic domain. The *N*-linked glycosylation site at Asn^{117} has been experimentally verified, but the function of this *N*-linked glycan has not been determined since no differences were observed between glycosylated and non-glycosylated recombinant MMP13 [74,75].

4.1.7. MMP17

MMP17 (MT4-MMP), a glycosylphosphatidylinositol-anchored matrix metalloproteinase, has two *N*-linked glycans at Asn^{137} and Asn^{318} in the catalytic site and the linker region, respectively. The presence of these *N*-linked oligosaccharides stabilizes the dimeric form of MMP17 by promoting non-covalent interactions or facilitating folding and formation of disulfides [76–78].

4.2. Phosphorylation of MMPs

Protein phosphorylation is one of the most important PTMs for regulation of biological processes [79]. This dynamic PTM is characterized by the reversible enzymatic addition of a phosphate group to amino acid side chains of serine (Ser), threonine (Thr), or tyrosine (Tyr), resulting in alterations of protein structure, stability, and dynamics [79]. Hence, phosphorylation can initiate different conformational changes associated with differential activity and binding specificity, leading to an activation or deactivation of proteins [80,81]. Furthermore, the dynamic nature of phosphorylation resulting from the opposing activities of kinases (which add phosphate groups) and phosphatases (which remove phosphate groups) is functionally compatible with the modulation of intramolecular interactions, implicated in essential cellular processes, such as cell division, cellular proliferation and differentiation, apoptosis, and signal transduction [79]. Consequently, dysregulation of phosphorylation dynamics is often related with pathologies, e.g., tumor formation, chronic inflammatory diseases, autoimmune diseases, and neurodegenerative disorders [81,82].

As many secreted and extracellular proteins, the MMPs are phosphorylated. Ser-, Thr-, and Tyr phosphorylation sites are identified across the different structural domains of the MMPs. Yet, the biological significance of these phosphate groups is overall poorly characterized. Among the MMPs, MMP2 and membrane-type MMP14 metalloproteinase are the ones with most extensively studied phosphorylation-dependent effects.

4.2.1. MMP2

MMP2 contains 29 potential phosphorylation sites distributed across the pro-peptide domain, collagen-binding domain, collagenase-like domain-1 and -2, and hemopexin domain. However, only five of the predicted phosphorylation sites (S32, S160, Y271, T250, and S365) have been confirmed by mass spectrometry (MS). The phosphorylation of MMP2 noticeably diminishes its proteolytic activity, while dephosphorylation increases MMP2 activity. This is possibly due to the conformational changes observed in the secondary structure of dephosphorylated MMP2, with α-helices 50% longer and β-strands 17% shorter than phosphorylated MMP2. However, it is still unclear if protease activity is directly influenced by these conformational changes [83,84].

4.2.2. MMP14

MMP-14 phosphorylation is known to play an important role in fine-tuning the activity of MMP14 at the cell surface and in the activation of intracellular signaling, and it is critical for regulating the pro-metastatic function of this metalloproteinase. This membrane-type metalloprotease has nine potential phosphorylation sites in its multidomain structure (PhosphoSitePlus [85], UniProt [86]). The cytoplasmic domain of MMP14 regulates its internalization and trafficking, thereby modulating enzymatic activity at the cell surface. A study by García-Pardo et al. reported that this domain was essential for MMP14 mediated cellular invasion and migration [72]. The cytoplasmic domain of MMP14 contains two phosphorylation sites, at Thr^{567} and Tyr^{573}, which have a significant influence on MMP14-induced cellular invasion and migration. A study by Williams et al. in fibrosarcoma cells showed that substitution of the residue Thr^{567} by alanine increased retention at the cell surface and reduced internalization of the protease significantly compromising invasion and migration. Contrarily, mimicking protease phosphorylation by substitution of Thr^{567} by glutamic acid reduced retention of MMP14 at the cell surface, increased efficiency of internalization, and was correlated with an increase in migration and invasion [87]. A study by Moss et al. showed that phospho-mimetic Thr^{567} mutants exhibit higher collagenolytic activity and three-dimensional growth within a collagen matrix, thereby promoting enhanced matrix invasion in ovarian cancer cells [88]. In addition, phosphorylation of Thr^{567} impacted the integrity of cell monolayer, cell motility and multicellular aggregate dynamics in ovarian cancer cells, promoting metastasis-associated behaviors [89]. Furthermore, a study by Nyalendo et al. reported that phosphorylation at Tyr^{573} influenced cell migration, suggested by the ability of a phospho-defective mutant to inhibit migration of cells endogenously expressing MMP14 [90]. Additionally, epidermal growth factor (EGF)-induced phosphorylation of Tyr^{573} prompted internalization of MMP14 together with pericellular collagen, establishing an environment for expansive growth in three–dimensional collagen matrix, whereas lack of responding to EGFR signaling triggered invasive growth. Therefore, the phosphorylation of Tyr^{573} modulates cell surface dynamics of MMP14, thereby regulating the transition between invasive and expansive growth [91].

4.2.3. Extracellular Phosphorylation of MMPs

A subset of MMPs (MMP1, 12, 13, 14, 16, 24, and 27) are tyrosine phosphorylated by the extracellular vertebrate lonesome kinase (VLK). These phosphotyrosines are found exclusively within the hemopexin domains of the MMPs. Interestingly, proteins distantly related to MMPs, which contain hemopexin-like domains, were found to be tyrosine phosphorylated at identical positions by VLK. The structural conservation of these phosphorylation sites suggests that they play a role in regulating MMP activity, a potential function, which still remains largely undiscovered [92,93]. Additionally, a study by Bordoli et al. showed that co-expression of VLK with MMPs supported their tyrosine phosphorylations that have been extensively observed in vivo. Bordoli also showed that by introducing a mutation in the ATP binding site of VLK or deletion of the proline glycine-rich domain (a conserved domain close to the kinase domain), the phosphorylation of the MMPs and other co-expressed substrates in the extracellular environment was eliminated. Similarly, shRNA-mediated downregulation of VLK expression and substitution of Tyr^{360} with phenylalanine in MMP1 reduced MMP13 and MMP1 tyrosine phosphorylation, respectively [93]. While VLK generates the majority of extracellular phosphotyrosines, additional secreted kinases have been identified, which may also be responsible for the extracellular protein phosphorylation events observed in vivo. This indicates a major impact of the secreted kinome on tissue homeostasis and disease pathogenesis. Discovery of novel extracellular kinases, identification of their substrates, and dissection of the regulatory mechanisms involved will provide us with a better understanding of their functions [94–96].

4.3. *Glycosaminoglycans*

An additional layer of regulation of extracellular proteolysis is mediated by interactions of MMPs with glycans in the extracellular space rather than direct glycosylation or phosphorylation of the protease or substrate. Glycosaminoglycans (GAGs) are extracellular glycans, which are composed of linear, unbranched repeats of disaccharide units. Heparan sulfate (HS) is a GAG-family member expressed by virtually every cell of a multicellular organism composed of repeats of glucoronic acid (GlcA) and *N*-acetylglucosamine (GlcNAc) dimers, which in most cases are attached to a cell-membrane associated core protein. During GAG-chain synthesis HS undergoes several sulfotransferase-mediated *N*- and *O*-sulfations. These modifications do not undergo completion, which results in strongly negatively charged sugar chains with variable length and degree of sulfation interrupted by stretches of unmodified, neutral regions. The sulfated regions bind basic peptide motifs of many signaling molecules, including morphogens (Hedgehogs, Wnts, TGFs), cytokines, and most chemokines [97–99]. This electrostatic interaction leads to partial neutralization of the protein surface and can have different effects on the biological functions of the interacting protein by facilitating, enhancing, or inhibiting the interaction with other proteins. It is important to note that these interactions, even though they are mainly based on electrostatics, are in most cases very specific and do not occur randomly. This is shown by the fact that many proteins contain specific GAG-binding motifs, with which they exclusively bind HS, even though they contain many lysine and arginine residues (for an in-depth review of the molecular basis of HS-protein interactions see Xu & Esko, 2014 [100]). Examples of HS-binding proteins are members of the fibroblast growth factor (FGF) family. Here, HS is part of the trimeric FGF/FGF-receptor/HS co-receptor complexes and activates FGF signaling in a sulfation pattern-dependent manner [101–104]. HS-induced oligomerization has been shown in many other processes including dimerization of amyloid precursor protein (APP) complexes or nearly all chemokines [105–108]. Interestingly, HS does not only mediate protein complex formation, but many proteins binding to HS, including the aforementioned candidates, are also known to undergo proteolytic processing on or near the cell surface. This indicates a crucial role of HS in regulating extracellular cleavage processes.

4.3.1. GAG-regulated Substrate Proteolysis

Signaling processes in the extracellular space usually involve at least one form of substrate proteolysis. Therefore, co-factors have to act as decision makers to ensure regulated processing of signaling proteins. Multiple cases have been reported, where this is mediated by substrate binding to GAGs. For example, similar to *N*-glycans, GAGs can modulate the stability of substrates. FGF2 binding to heparin, an extremely sulfated HS variant expressed by connective-tissue type macrophages, stabilizes FGF2 and protects it from degradation by proteolysis [109,110]. The same stabilizing effect has been shown for the degradation of stromal cell-derived factor 1 (SDF1)/CXCL12 and C-terminal processing of interferon-γ [111,112]. Despite this stabilizing effect GAGs have been demonstrated to play additional important roles in extracellular proteolysis. In the case of shedding of morphogens from the Hedgehog (Hh) family, HS acts as an assembly platform for protease-release complexes [113]. Here, HS recruits Hh co-factors that co-localize proteases and allow for cleavage of the substrate (Figure 3) [114]. Furthermore, direct HS binding of the N-terminal cleavage site of Hh also has a stabilizing effect by inhibiting N-terminal Hh processing and reduces its release in vitro and in *Drosophila* in vivo [114,115]. An example of how GAG binding can positively regulate substrate cleavage (Figure 3) was shown for the cleavage of viral capsid proteins. Binding of the human papillomavirus viral capsid protein L1 to highly sulfated HS is essential to induce a conformational change and leads to cleavage of L1 by the human trypsin-like serine proteinase kallikrein 8, a process which is required for virus internalization [116].

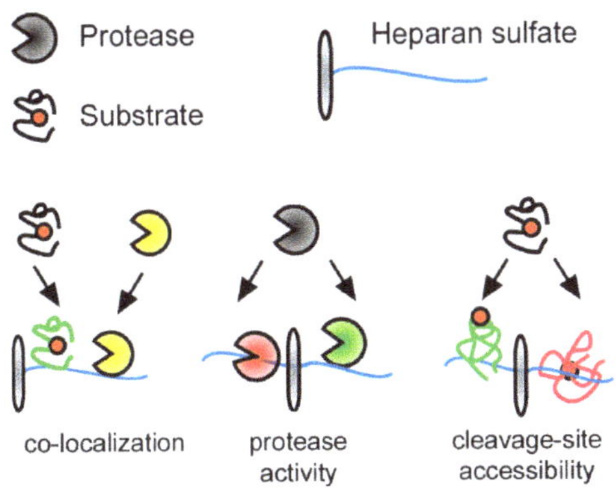

Figure 3. Heparan sulfate (HS)-protein interaction as a post translational regulator of extracellular proteolysis. Electrostatic interactions of proteins with negatively charged HS affect extracellular proteolysis at three different levels: (1) by mediating co-localization of reactants; (2) on the protease level by affecting protease activity/accessibility; or (3) on the substrate level by modulating accessibility to cleavage sites. The substrate cleavage site is indicated by a red circle, green substrate/protease color denotes activity; red substrate/protease color denotes inactivity/inaccessibility.

4.3.2. GAG-regulated MMP Activity

GAGs also regulate extracellular proteolysis at the protease level by modulating activation, localization, and protease–substrate interactions (Figure 3 middle). Many secreted vertebrate MMPs associate to the cell surface via binding to HS, including MMP9, which is specific, since excess soluble heparin extracts and solubilizes MMP2, MMP7, MMP9, and MMP13 [117,118]. Moreover, heparin, a highly sulfated form of HS, affects protease activity by increasing TIMP3 affinity to MMP2, MMP7, and MMP9, and HS affects MMP9 expression and plasma levels, depending on its sulfation and expressing cell type [119–123]. Finally, in vivo studies showed that MMPs in secretory granules of mast cells interact with heparin and that loss of heparin sulfation reduces MMP stability [124,125]. Like all MMPs, the wound repair and immune reaction associated MMP7 has to be converted into its proteolytic active form by removal of its inhibitory pro-sequence. MMP7 is able to activate itself by intermolecular autolytic processing. This is amplified by binding of MMP7 to highly sulfated GAGs (e.g., heparin or chondroitin sulfate), which significantly increases MMP7 auto-processing and also increases cleavage of specific physiological substrates [19].

5. Conclusions

MMP biology has been revolutionized with the recognition of extracellular proteolysis not as a simple mechanism of ECM degradation but as a regulatory mechanism for precise cellular control of biological processes. The paradigm shift for MMP functions from ECM degradative proteases to important regulators of essential cellular processes has highlighted the physiological relevance of these proteases, directly implicated by the relationship between MMP expression and disease development [4,8]. Hence, the multilayered regulation of MMPs emphasizes the tremendous importance of the balance between synthesis of active enzymes and their inhibition, which is pivotal to avoid the destructive activity of these proteases. Characterization of these regulatory mechanisms will aid the development of new therapeutics for various pathologies. MMP activity is regulated at the transcriptional level, post-translationally by pro-enzyme activation, by PTMs, and through extracellular inhibition by TIMPs and by non-specific proteinase inhibitors. Among these regulatory mechanisms, PTMs have recently obtained wide attention in the MMP community. MMPs are modified by PTMs at multiple sites (Table 1), which affects the activity of individual MMPs to a different extent, but the biological relevance of many of them is still unknown. However, many of the studies referenced in this review have been performed prior to the rapid progression of proteomics method development within the last decade. Many of the challenges that still limit our understanding of PTM function in MMP biology can now be approached by modern technologies of state-of-the-art proteomics. Advances in MS-based methods including multiplexed chemical labeling, novel label-free quantification strategies, improvement in PTM enrichment, more robust PTM analysis workflows and streamlined bioinformatics strategies will contribute to reliable identification and quantification of high numbers of PTMs [126,127]. The remaining challenge lies in defining the functional role of the physiologically relevant PTMs to understand PTM-dependent activity of MMPs in complex biological systems [128].

Table 1. List of identified PTMs in MMPs.

MMP	Modification	Biological Effect	Reference
MMP1	*N*-linked glycosylation at Asn^{120}	Tumor cell invasion and angiogenesis	[39,69,129]
	Phosphorylation at Tyr^{360}	Not reported	[93]
MMP2	*O*-linked glycosylation at Ser^{32}, $Thr^{96, 262, 458, 460}$	Upregulation of MMP2	[39,71,72]
	N-linked glycosylation at Asn^{573} and Asn^{642}	Not reported	[130]
	Phosphorylation at Ser^{32}, Ser^{160}, Tyr^{271}, Thr^{250} and Ser^{365}	Phosphorylation decreases, while dephosphorylation increases protease activity	[83,84]
	Heparan sulfate	Cell surface localization; affects protease activity by increasing TIMP3 affinity	[119–122]
MMP3	*N*-linked glycosylation at Asn^{120} and Asn^{398}	Not reported	[39,73]
	Three potential *O*-linked glycosylation at Ser^{56}, Ser^{269} and Thr^{277}		

Table 1. *Cont.*

MMP	Modification	Biological Effect	Reference
MMP7	Heparan sulfate, Chondroitin sulfate	Cell surface localization; affects protease activity by increasing TIMP3 affinity; increases MMP7 auto-processing and activity	[19,120,121]
MMP9	*N*-linked glycosylation at Asn^{38} and Asn^{120}	MMP9 secretion and activation	[53–59]
	O-linked glycosylation in the linker region	Increases the domain flexibility; necessary for internalization and degradation; protects against proteolytic degradation; reduces gelatinolytic activity	[60,62–64]
	Heparan sulfate	Cell surface localization; affects protease activity by increasing TIMP3 affinity; affects MMP9 expression and plasma levels	[119–123]
MMP12	Phosphorylation at Tyr^{414}	Not reported	[93]
MMP13	*N*-linked glycosylation at Asn^{117} and Asn^{152}	Not reported	[1,74,75]
	O-linked glycosylation at Ser^{24} and Ser^{62}	Not reported	[1,74,75]
	Phosphorylation at Tyr^{366}	Not reported	[93]
MMP14	*N*-linked glycosylation at Asn^{229} and Asn^{311}	Not reported	[39]
	O-linked glycosylation at Thr^{291}, Thr^{299}, Thr^{300}, and Ser^{301}	Required for formation of a stable complex with proMMP2 and TIMP2; increases activity upon glycosylation perturbation	[65–68]
	Phosphorylation at Thr^{567}, Tyr^{573} and Tyr^{353}	Regulates MMP14 induced cellular invasion and migration; cell surface dynamics and internalization; mimetic mutants exhibit higher collagenolytic activity and three-dimensional growth; promotes metastasis-associated behaviors	[72,87–91,93]
MMP16	Phosphorylation at Tyr^{377} and Tyr^{521}	Not reported	[93]
MMP17	*N*-linked glycosylation at Asn^{137} and Asn^{318}	Stabilizes the dimeric form of MMP17	[76–78]
MMP24	Phosphorylation at Tyr^{534}	Not reported	[93]
MMP27	Phosphorylation at Tyr^{360}	Not reported	[93]

Funding: Ulrich auf dem Keller acknowledges support by a Novo Nordisk Foundation Young Investigator Award (NNF16OC0020670). Philipp Kastl is supported by a research fellowship from the German Research Foundation (project no. 415888450), and Fabio Sabino is recipient of a H.C. Ørsted COFUND Postdoc Fellowship (COFUNDfellowsDTU, grant agreement no. 713683).

Conflicts of Interest: The authors declare no conflicts of interest.

Abbreviations

Asn	Asparagine amino acid
CA	Cysteine array
Cy	Cytosolic domain
ECM	Extracellular matrix
EGFR	Epidermal growth factor receptor
ER	Endoplasmic reticulum
Fu	Furin-recognition site
GalNAc	*N*-acetyl galactosamine
GlcNAc	*N*-acetyl glucosamine
GPI	Glycosylphosphatidylinositol
LRP	Low-density lipoprotein receptor-related protein
MMP	Matrix metalloproteinase
Pro	Pro-domain
PTMs	Post-translational modifications
SA	Signal anchor
Ser	Serine amino acid

SH	Thiol group
SP	Signal peptide
Thr	Threonine amino acid
TIMPs	Tissue inhibitors of metalloproteinases
TM	Transmembrane domain
Tyr	Tyrosine amino acid
UDP	Uridine diphosphate

References

1. Nagase, H.; Visse, R.; Murphy, G. Structure and function of matrix metalloproteinases and TIMPs. *Cardiovasc. Res.* **2006**, *69*, 562–573. [CrossRef] [PubMed]
2. Jabłońska-Trypuć, A.; Matejczyk, M.; Rosochacki, S. Matrix metalloproteinases (MMPs), the main extracellular matrix (ECM) enzymes in collagen degradation, as a target for anticancer drugs. *J. Enzyme Inhib. Med. Chem.* **2016**, *31*, 177–183. [CrossRef] [PubMed]
3. Bonnans, C.; Chou, J.; Werb, Z. Remodelling the extracellular matrix in development and disease. *Nat. Rev. Mol. Cell Biol.* **2014**, *15*, 786–801. [CrossRef] [PubMed]
4. Loffek, S.; Schilling, O.; Franzke, C.-W.; Claustrat, B.; Brun, J.; Chazot, G. Biological role of matrix metalloproteinases: a critical balance. *Eur. Respir. J.* **2011**, *38*, 191–208. [CrossRef] [PubMed]
5. Stamenkovic, I. Extracellular matrix remodelling: The role of matrix metalloproteinases. *J. Pathol.* **2003**, *200*, 448–464. [CrossRef] [PubMed]
6. Marchant, D.J.; Bellac, C.L.; Moraes, T.J.; Wadsworth, S.J.; Dufour, A.; Butler, G.S.; Bilawchuk, L.M.; Hendry, R.G.; Robertson, A.G.; Cheung, C.T.; et al. A new transcriptional role for matrix metalloproteinase-12 in antiviral immunity. *Nat. Med.* **2014**, *20*, 493–502. [CrossRef] [PubMed]
7. Page-McCaw, A.; Ewald, A.J.; Werb, Z. Matrix metalloproteinases and the regulation of tissue remodelling. *Nat. Rev. Mol. Cell Biol.* **2007**, *8*, 221–233. [CrossRef] [PubMed]
8. Cauwe, B.; Opdenakker, G. Intracellular substrate cleavage: A novel dimension in the biochemistry, biology and pathology of matrix metalloproteinases. *Crit. Rev. Biochem. Mol. Biol.* **2010**, *45*, 351–423. [CrossRef]
9. Tallant, C.; Marrero, A.; Gomis-Rüth, F.X. Matrix metalloproteinases: Fold and function of their catalytic domains. *Biochim. Biophys. Acta Mol. Cell Res.* **2010**, *1803*, 20–28. [CrossRef]
10. Rodríguez, D.; Morrison, C.J.; Overall, C.M. Matrix metalloproteinases: What do they not do? New substrates and biological roles identified by murine models and proteomics. *Biochim. Biophys. Acta Mol. Cell Res.* **2010**, *1803*, 39–54. [CrossRef]
11. Fanjul-Fernández, M.; Folgueras, A.R.; Cabrera, S.; López-Otín, C. Matrix metalloproteinases: Evolution, gene regulation and functional analysis in mouse models. *Biochim. Biophys. Acta Mol. Cell Res.* **2010**, *1803*, 3–19. [CrossRef] [PubMed]
12. Yamamoto, K.; Murphy, G.; Troeberg, L. Extracellular regulation of metalloproteinases. *Matrix Biol.* **2015**, *44–46*, 255–263. [CrossRef] [PubMed]
13. Huwiler, A.; Akool, E.S.; Aschrafi, A.; Hamada, F.M.A.; Pfeilschifter, J.; Eberhardt, W. ATP Potentiates Interleukin-1??-induced MMP-9 Expression in Mesangial Cells via Recruitment of the ELAV Protein HuR. *J. Biol. Chem.* **2003**, *278*, 51758–51769. [CrossRef] [PubMed]
14. Fähling, M.; Steege, A.; Perlewitz, A.; Nafz, B.; Mrowka, R.; Persson, P.B.; Thiele, B.J. Role of nucleolin in posttranscriptional control of MMP-9 expression. *Biochim. Biophys. Acta Gene Struct. Expr.* **2005**, *1731*, 32–40. [CrossRef]
15. Clark, I.M.; Swingler, T.E.; Sampieri, C.L.; Edwards, D.R. The regulation of matrix metalloproteinases and their inhibitors. *Int. J. Biochem. Cell Biol.* **2008**, *40*, 1362–1378. [CrossRef]
16. Meng, F.; Zhang, Z.; Chen, W.; Huang, G.; He, A.; Hou, C.; Long, Y.; Yang, Z.; Zhang, Z.; Liao, W. MicroRNA-320 regulates matrix metalloproteinase-13 expression in chondrogenesis and interleukin-1β-induced chondrocyte responses. *Osteoarthr. Cartil.* **2016**, *24*, 932–941. [CrossRef]
17. Bannikov, G.A.; Karelina, T.V.; Collier, I.E.; Marmer, B.L.; Goldberg, G.I. Substrate binding of gelatinase B induces its enzymatic activity in the presence of intact propeptide. *J. Biol. Chem.* **2002**, *277*, 16022–16027. [CrossRef]

18. Emonard, H.; Hornebeck, W. Binding of 92 kDa and 72 kDa progelatinases to insoluble elastin modulates their proteolytic activation. *Biol. Chem.* **1997**, *378*, 265–271. [CrossRef]
19. Ra, H.J.; Harju-Baker, S.; Zhang, F.; Linhardt, R.J.; Wilson, C.L.; Parks, W.C. Control of promatrilysin (MMP7) activation and substrate-specific activity by sulfated glycosaminoglycans. *J. Biol. Chem.* **2009**, *284*, 27924–27932. [CrossRef]
20. Geurts, N.; Martens, E.; Van Aelst, I.; Proost, P.; Opdenakker, G.; Van Den Steen, P.E. β-hematin interaction with the hemopexin domain of gelatinase B/MMP-9 provokes autocatalytic processing of the propeptide, thereby priming activation by MMP-3. *Biochemistry* **2008**, *47*, 2689–2699. [CrossRef]
21. Murphy, G.; Nagase, H. Progress in matrix metalloproteinase research. *Mol. Aspects Med.* **2009**, *29*, 290–308. [CrossRef] [PubMed]
22. Baker, A.H. Metalloproteinase inhibitors: biological actions and therapeutic opportunities. *J. Cell Sci.* **2002**, *115*, 3719–3727. [CrossRef] [PubMed]
23. Harper, J.W.; Bennett, E.J. Proteome complexity and the forces that drive proteome imbalance. *Nature* **2016**, *537*, 328–338. [CrossRef] [PubMed]
24. Kim, K.L.; Park, K.M.; Murray, J.; Kim, K.; Ryu, S.H. Direct Profiling the Post-Translational Modification Codes of a Single Protein Immobilized on a Surface Using Cu-free Click Chemistry. *ACS Cent. Sci.* **2018**, *4*, 614–623. [CrossRef] [PubMed]
25. Deribe, Y.L.; Pawson, T.; Dikic, I. Post-translational modifications in signal integration. *Nat. Struct. Mol. Biol.* **2010**, *17*, 666–672. [CrossRef] [PubMed]
26. Wang, Y.C.; Peterson, S.E.; Loring, J.F. Protein post-translational modifications and regulation of pluripotency in human stem cells. *Cell Res.* **2014**, *24*, 143–160. [CrossRef]
27. Ricard-Blum, S. Protein–Glycosaminoglycan interaction networks: Focus on heparan sulfate. *Perspect. Sci.* **2017**. [CrossRef]
28. Audagnotto, M.; Dal Peraro, M. Protein post-translational modifications: In silico prediction tools and molecular modeling. *Comput. Struct. Biotechnol. J.* **2017**, *15*, 307–319. [CrossRef]
29. Spoel, S.H. Orchestrating the proteome with post-translational modifications. *J. Exp. Bot.* **2018**, *69*, 4499–4503. [CrossRef]
30. Rogers, L.D.; Overall, C.M. Proteolytic Post-translational Modification of Proteins: Proteomic Tools and Methodology. *Mol. Cell. Proteomics* **2013**, *12*, 3532–3542. [CrossRef]
31. Beltrao, P.; Bork, P.; Krogan, N.J.; van Noort, V.; Abu-Qarn, M.; Eichler, J.; Sharon, N.; Alexander, J.; Lim, D.; Joughin, B.; et al. Evolution and functional cross-talk of protein post-translational modifications. *Mol. Syst. Biol.* **2013**, *9*, 714. [CrossRef] [PubMed]
32. Seo, J.; Lee, K.J. Post-translational modifications and their biological functions: proteomic analysis and systematic approaches. *J. Biochem. Mol. Biol.* **2004**, *37*, 35–44. [CrossRef] [PubMed]
33. Vanheule, V.; Metzemaekers, M.; Janssens, R.; Struyf, S.; Proost, P. How post-translational modifications influence the biological activity of chemokines. *Cytokine* **2018**, *109*, 29–51. [CrossRef] [PubMed]
34. Yang, F. Post-translational modification control of HBV biological processes. *Front. Microbiol.* **2018**, *9*, 2661. [CrossRef] [PubMed]
35. Arbez, N.; Ratovitski, T.; Roby, E.; Chighladze, E.; Stewart, J.C.; Ren, M.; Wang, X.; Lavery, D.J.; Ross, C.A. Post-translational modifications clustering within proteolytic domains decrease mutant huntingtin toxicity. *J. Biol. Chem.* **2017**, *292*, 19238–19249. [CrossRef] [PubMed]
36. Ryan, B.J.; Nissim, A.; Winyard, P.G. Oxidative post-translational modifications and their involvement in the pathogenesis of autoimmune diseases. *Redox Biol.* **2014**, *2*, 715–724. [CrossRef] [PubMed]
37. Jaeken, J. Glycosylation and its Disorders: General Overview. *Ref. Modul. Biomed. Sci.* **2016**, 1–8. [CrossRef]
38. Moremen, K.W.; Tiemeyer, M.; Nairn, A.V. Vertebrate protein glycosylation: Diversity, synthesis and function. *Nat. Rev. Mol. Cell Biol.* **2012**, *13*, 448–462. [CrossRef]
39. Boon, L.; Ugarte-Berzal, E.; Vandooren, J.; Opdenakker, G. Glycosylation of matrix metalloproteases and tissue inhibitors: present state, challenges and opportunities. *Biochem. J.* **2016**, *473*, 1471–1482. [CrossRef]
40. Schwarz, F.; Aebi, M. Mechanisms and principles of N-linked protein glycosylation. *Curr. Opin. Struct. Biol.* **2011**, *21*, 576–582. [CrossRef]

41. Steentoft, C.; Vakhrushev, S.Y.; Joshi, H.J.; Kong, Y.; Vester-Christensen, M.B.; Schjoldager, K.T.B.G.; Lavrsen, K.; Dabelsteen, S.; Pedersen, N.B.; Marcos-Silva, L.; et al. Precision mapping of the human O-GalNAc glycoproteome through SimpleCell technology. *EMBO J.* **2013**, *32*, 1478–1488. [CrossRef] [PubMed]
42. Spiro RG Protein glycosylation: nature, distribution, enzymatic formation, and disease implications of glycopeptide bonds. *Glycobiology* **2002**, *12*, 43R–56R. [CrossRef] [PubMed]
43. Tian, E.; Ten Hagen, K.G. Recent insights into the biological roles of mucin-type O-glycosylation. *Glycoconj. J.* **2009**, *26*, 325–334. [CrossRef] [PubMed]
44. Gill, D.J.; Clausen, H.; Bard, F. Location, location, location: New insights into O-GalNAc protein glycosylation. *Trends Cell Biol.* **2011**, *21*, 149–158. [CrossRef] [PubMed]
45. Hurtado-Guerrero, R. Recent structural and mechanistic insights into protein O-GalNAc glycosylation. *Biochem. Soc. Trans.* **2016**, *44*, 61–67. [CrossRef] [PubMed]
46. Helenius, A.; Aebi, M. Intracellular functions of N-linked glycans. *Science* **2001**, *291*, 2364–2369. [CrossRef]
47. Lee, H.S.; Qi, Y.; Im, W. Effects of N-glycosylation on protein conformation and dynamics: Protein Data Bank analysis and molecular dynamics simulation study. *Sci. Rep.* **2015**, *5*, 8926. [CrossRef] [PubMed]
48. Goettig, P. Effects of glycosylation on the enzymatic activity and mechanisms of proteases. *Int. J. Mol. Sci.* **2016**, *17*, 1969. [CrossRef] [PubMed]
49. Bergstrom, K.; Fu, J.; Xia, L. Biological Functions of C1GalT1 and Mucin-Type *O*-Glycans. In *Glycoscience: Biology and Medicine*; Springer: Tokyo, Japan, 2015; pp. 1073–1080. ISBN 9784431548416.
50. Tran, D.T.; Ten Hagen, K.G. Mucin-type o-glycosylation during development. *J. Biol. Chem.* **2013**, *288*, 6921–6929. [CrossRef] [PubMed]
51. Pinho, S.S.; Reis, C.A. Glycosylation in cancer: Mechanisms and clinical implications. *Nat. Rev. Cancer* **2015**, *15*, 540–555. [CrossRef] [PubMed]
52. Lee, L.Y.; Hincapie, M.; Packer, N.; Baker, M.S.; Hancock, W.S.; Fanayan, S. An optimized approach for enrichment of glycoproteins from cell culture lysates using native multi-lectin affinity chromatography. *J. Sep. Sci.* **2012**, *35*, 2445–2452. [CrossRef] [PubMed]
53. Vandooren, J.; Van Den Steen, P.E.; Opdenakker, G. Biochemistry and molecular biology of gelatinase B or matrix metalloproteinase-9 (MMP-9): The next decade. *Crit. Rev. Biochem. Mol. Biol.* **2013**, *48*, 222–272. [CrossRef] [PubMed]
54. Kotra, L.P.; Zhang, L.; Fridman, R.; Orlando, R.; Mobashery, S. N-glycosylation pattern of the zymogenic form of human matrix metalloproteinase-9. *Bioorg. Chem.* **2002**, *30*, 356–370. [CrossRef]
55. Roth, J. Protein N-glycosylation along the Secretory Pathway: Relationship to organelle topography and function, protein quality control, and cell interactions. *Chem. Rev.* **2002**, *102*, 285–303. [CrossRef] [PubMed]
56. Duellman, T.; Burnett, J.; Yang, J. Functional Roles of N-Linked Glycosylation of Human Matrix Metalloproteinase 9. *Traffic* **2015**, *16*, 1108–1126. [CrossRef]
57. Kumar, S.; Cieplak, P. Role of N-glycosylation in activation of proMMP-9. A molecular dynamics simulations study. *PLoS ONE* **2018**, *13*, e0191157. [CrossRef] [PubMed]
58. Nishi, N.; Shoji, H.; Seki, M.; Itoh, A.; Miyanaka, H.; Yuube, K.; Hirashima, M.; Nakamura, T. Galectin-8 modulates neutrophil function via interaction with integrin αM. *Glycobiology* **2003**, *13*, 755–763. [CrossRef]
59. Boon, L.; Ugarte-Berzal, E.; Martens, E.; Vandooren, J.; Rybakin, V.; Colau, D.; Gordon-Alonso, M.; van der Bruggen, P.; Stöcker, W.; Becker-Pauly, C.; et al. Propeptide glycosylation and galectin-3 binding decrease proteolytic activation of human proMMP-9/progelatinase B. *FEBS J.* **2019**, *285*, 930–945. [CrossRef]
60. Van Den Steen, P.E.; Van Aelst, I.; Hvidberg, V.; Piccard, H.; Fiten, P.; Jacobsen, C.; Moestrup, S.K.; Fry, S.; Royle, L.; Wormald, M.R.; et al. The hemopexin and O-glycosylated domains tune gelatinase B/MMP-9 bioavailability via inhibition and binding to cargo receptors. *J. Biol. Chem.* **2006**, *281*, 18626–18637. [CrossRef]
61. Mattu, T.S.; Royle, L.; Langridge, J.; Wormald, M.R.; Van den Steen, P.E.; Van Damme, J.; Opdenakker, G.; Harvey, D.J.; Dwek, R.A.; Rudd, P.M. O-glycan analysis of natural human neutrophil gelatinase B using a combination of normal phase- HPLC and online tandem mass spectrometry: Implications for the domain organization of the enzyme. *Biochemistry* **2000**, *39*, 15695–15704. [CrossRef]
62. Rosenblum, G.; Van den Steen, P.E.; Cohen, S.R.; Grossmann, J.G.; Frenkel, J.; Sertchook, R.; Slack, N.; Strange, R.W.; Opdenakker, G.; Sagi, I. Insights into the Structure and Domain Flexibility of Full-Length Pro-Matrix Metalloproteinase-9/Gelatinase B. *Structure* **2007**, *15*, 1227–1236. [CrossRef] [PubMed]

63. Vandooren, J.; Knoops, S.; Buzzo, J.L.A.; Boon, L.; Martens, E.; Opdenakker, G.; Kolaczkowska, E. Differential inhibition of activity, activation and gene expression of MMP-9 in THP-1 cells by azithromycin and minocycline versus bortezomib: A comparative study. *PLoS ONE* **2017**, *12*, e0174853. [CrossRef] [PubMed]
64. Vandooren, J. Gelatin degradation assay reveals MMP-9 inhibitors and function of O-glycosylated domain. *World J. Biol. Chem.* **2011**, *2*, 14. [CrossRef] [PubMed]
65. Zucker, S.; Drews, M.; Conner, C.; Foda, H.D.; DeClerck, Y.A.; Langley, K.E.; Bahou, W.F.; Docherty, A.J.P.; Cao, J. Tissue inhibitor of metalloproteinase-2 (TIMP-2) binds to the catalytic domain of the cell surface receptor, membrane type 1-matrix metalloproteinase 1 (MT1-MMP). *J. Biol. Chem.* **1998**, *273*, 1216–1222. [CrossRef] [PubMed]
66. Kinoshita, T.; Sato, H.; Okada, A.; Ohuchi, E.; Imai, K.; Okada, Y.; Seiki, M. TIMP-2 promotes activation of progelatinase A by membrane-type 1 matrix metalloproteinase immobilized on agarose beads. *J. Biol. Chem.* **1998**, *273*, 16098–16103. [CrossRef] [PubMed]
67. Wu, Y.I.; Munshi, H.G.; Sen, R.; Snipas, S.J.; Salvesen, G.S.; Fridman, R.; Stack, M.S. Glycosylation Broadens the Substrate Profile of Membrane Type 1 Matrix Metalloproteinase. *J. Biol. Chem.* **2004**, *279*, 8278–8289. [CrossRef] [PubMed]
68. Growth, T.; Nguyen, A.T.; Chia, J.; Ros, M.; Hui, K.M.; Saltel, F.; Bard, F. Organelle Specific O-Glycosylation Drives MMP14 Activation, Tumor Growth, and Metastasis. *Cancer Cell* **2017**, *32*, 639–653.e6. [CrossRef]
69. Saarinen, J.; Welgus, H.G.; Flizar, C.A.; Kalkkinen, N.; Helin, J. N-Glycan structures of matrix metalloproteinase-1 derived from human fibroblasts and from HT-1080 fibrosarcoma cells. *Eur. J. Biochem.* **1999**, *259*, 829–840. [CrossRef] [PubMed]
70. Piccard, H.; Van den Steen, P.E.; Opdenakker, G. Hemopexin domains as multifunctional liganding modules in matrix metalloproteinases and other proteins. *J. Leukoc. Biol.* **2007**, *81*, 870–892. [CrossRef]
71. Dufour, A.; Sampson, N.S.; Zucker, S.; Cao, J. Role of the hemopexin domain of matrix metalloproteinases in cell migration. *J. Cell. Physiol.* **2008**, *217*, 643–651. [CrossRef]
72. García-Pardo, A.; Opdenakker, G. Nonproteolytic functions of matrix metalloproteinases in pathology and insights for the development of novel therapeutic inhibitors. *Met. Med.* **2015**, *2*, 19–28. [CrossRef]
73. Nagase, H. Matrix Metalloproteinase 3/Stromelysin 1. In *Handbook of Proteolytic Enzymes*, 3rd ed.; Elsevier: Amsterdam, The Netherlands, 2013; Volume 1, pp. 763–774. [CrossRef]
74. Henriet, P.; Eeckhout, Y. Eeckhout, Y. Matrix Metallopeptidase-13/Collagenase 3. In *Handbook of Proteolytic Enzymes*, 3rd ed.; Elsevier: Amsterdam, The Netherlands, 2013; Volume 1, pp. 734–744. [CrossRef]
75. Knäuper, V.; López-Otin, C.; Smith, B.; Knight, G.; Murphy, G. Biochemical characterization of human collagenase-3. *J. Biol. Chem.* **1996**, *271*, 1544–1550. [CrossRef] [PubMed]
76. Hieronimus, B.; Pfohl, J.; Busch, C.; Graeve, L. Expression and characterization of membrane-type 4 matrix metalloproteinase (MT4-MMP) and its different forms in melanoma. *Cell. Physiol. Biochem.* **2017**, *42*, 198–210. [CrossRef] [PubMed]
77. Itoh, Y.; Kajita, M.; Kinoh, H.; Mori, H.; Okada, A.; Seiki, M. Membrane type 4 matrix metalloproteinase (MT4-MMP, MMP-17) is a glycosylphosphatidylinositol-anchored proteinase. *J. Biol. Chem.* **1999**, *274*, 34260–34266. [CrossRef] [PubMed]
78. Sohail, A.; Marco, M.; Zhao, H.; Shi, Q.; Merriman, S.; Mobashery, S.; Fridman, R. Characterization of the dimerization interface of membrane type 4 (MT4)-matrix metalloproteinase. *J. Biol. Chem.* **2011**, *286*, 33178–33189. [CrossRef] [PubMed]
79. Ardito, F.; Giuliani, M.; Perrone, D.; Troiano, G.; Muzio, L. Lo The crucial role of protein phosphorylation in cell signalingand its use as targeted therapy (Review). *Int. J. Mol. Med.* **2017**, *40*, 271–280. [CrossRef]
80. Nishi, H.; Shaytan, A.; Panchenko, A.R. Physicochemical mechanisms of protein regulation by phosphorylation. *Front. Genet.* **2014**, *5*, 270. [CrossRef]
81. Cutillas, P.R. Targeted In-Depth Quantification of Signaling Using Label-Free Mass Spectrometry. In *Methods in Enzymology*; Elsevier: Amsterdam, The Netherlands, 2017; Volume 585, pp. 245–268. ISBN 9780128097427.
82. Cohen, P. The role of protein phosphorylation in human health and disease. *Eur. J. Biochem.* **2001**, *268*, 5001–5010. [CrossRef]
83. Sariahmetoglu, M.; Crawford, B.D.; Leon, H.; Sawicka, J.; Li, L.; Ballermann, B.J.; Holmes, C.; Berthiaume, L.G.; Holt, A.; Sawicki, G.; et al. Regulation of matrix metalloproteinase-2 (MMP-2) activity by phosphorylation. *FASEB J.* **2007**, *21*, 2486–2495. [CrossRef]

84. Jacob-Ferreira, A.L.; Kondo, M.Y.; Baral, P.K.; James, M.N.G.; Holt, A.; Fan, X.; Schulz, R. Phosphorylation Status of 72 kDa MMP-2 Determines Its Structure and Activity in Response to Peroxynitrite. *PLoS ONE* **2013**, *8*, e71794. [CrossRef]
85. 3rd Millenium, PhosphoSitePlus. Cell Signalling Technology Inc. 2011. Available online: http://www.phosphosite.org/proteinAction.do?id=662&showAllSites=true (accessed on 22 May 2019).
86. UniProt Consortium, T. UniProt: the universal protein knowledgebase. *Nucleic Acids Res.* **2018**, *46*, 2699. [CrossRef] [PubMed]
87. Williams, K.C.; Coppolino, M.G. Phosphorylation of membrane type 1-matrix metalloproteinase (MT1-MMP) and its vesicle-associated membrane protein 7 (VAMP7)-dependent trafficking facilitate cell invasion and migration. *J. Biol. Chem.* **2011**, *286*, 43405–43416. [CrossRef] [PubMed]
88. Moss, N.M.; Wu, Y.I.; Liu, Y.; Munshi, H.G.; Stack, M.S. Modulation of the membrane type 1 matrix metalloproteinase cytoplasmic tail enhances tumor cell invasion and proliferation in three-dimensional collagen matrices. *J. Biol. Chem.* **2009**, *284*, 19791–19799. [CrossRef] [PubMed]
89. Yang, J.; Kasberg, W.C.; Celo, A.; Liang, Z.; Quispe, K.; Sharon Stack, M. Post-translational modification of the membrane type 1 matrix metalloproteinase (MT1-MMP) cytoplasmic tail impacts ovarian cancer multicellular aggregate dynamics. *J. Biol. Chem.* **2017**, *292*, 13111–13121. [CrossRef] [PubMed]
90. Nyalendo, C.; Michaud, M.; Beaulieu, E.; Roghi, C.; Murphy, G.; Gingras, D.; Béliveau, R. Src-dependent Phosphorylation of Membrane Type I Matrix Metalloproteinase on Cytoplasmic Tyrosine 573. *J. Biol. Chem.* **2007**, *282*, 15690–15699. [CrossRef]
91. Moss, N.M.; Liu, Y.; Johnson, J.J.; Debiase, P.; Jones, J.; Hudson, L.G.; Munshi, H.G.; Stack, M.S. Epidermal Growth Factor Receptor-Mediated Membrane Type 1 Matrix Metalloproteinase Endocytosis Regulates the Transition between Invasive versus Expansive Growth of Ovarian Carcinoma Cells in Three-Dimensional Collagen. *Mol. Cancer Res.* **2009**, *7*, 809–820. [CrossRef] [PubMed]
92. Rinschen, M.M.; Yu, M.-J.; Wang, G.; Boja, E.S.; Hoffert, J.D.; Pisitkun, T.; Knepper, M.A. Quantitative phosphoproteomic analysis reveals vasopressin V2-receptor–dependent signaling pathways in renal collecting duct cells. *Proc. Natl. Acad. Sci. USA* **2010**, *107*, 3882–3887. [CrossRef]
93. Bordoli, M.R.; Yum, J.; Breitkopf, S.B.; Thon, J.N.; Italiano, J.E.; Xiao, J.; Worby, C.; Wong, S.K.; Lin, G.; Edenius, M.; et al. A secreted tyrosine kinase acts in the extracellular environment. *Cell* **2014**, *158*, 1033–1044. [CrossRef]
94. Tagliabracci, V.S.; Wiley, S.E.; Guo, X.; Kinch, L.N.; Durrant, E.; Wen, J.; Xiao, J.; Cui, J.; Nguyen, K.B.; Engel, J.L.; et al. A Single Kinase Generates the Majority of the Secreted Phosphoproteome. *Cell* **2015**, *161*, 1619–1632. [CrossRef]
95. Sreelatha, A.; Kinch, L.N.; Tagliabracci, V.S. The secretory pathway kinases. *Biochim. Biophys. Acta Proteins Proteomics* **2015**, *1854*, 1687–1693. [CrossRef]
96. Ishikawa, H.O.; Takeuchi, H.; Haltiwanger, R.S.; Irvine, K.D. Four-jointed is a Golgi kinase that phosphorylates a subset of cadherin domains. *Science* **2008**, *321*, 401–404. [CrossRef] [PubMed]
97. Lin, X. Functions of heparan sulfate proteoglycans in cell signaling during development. *Development* **2004**, *131*, 6009–6021. [CrossRef] [PubMed]
98. Proudfoot, E.A.; Johnson, Z.; Bonvin, P.; Handel, M.T. Glycosaminoglycan Interactions with Chemokines Add Complexity to a Complex System. *Pharmaceuticals* **2017**, *10*, 70. [CrossRef]
99. Raman, R.; Sasisekharan, V.; Sasisekharan, R. Structural Insights into Biological Roles of Protein-Glycosaminoglycan Interactions. *Chem. Biol.* **2005**, *12*, 267–277. [CrossRef]
100. Xu, D.; Esko, J.D. Demystifying Heparan Sulfate–Protein Interactions. *Annu. Rev. Biochem.* **2014**, *83*, 129–157. [CrossRef]
101. Lin, X.; Buff, E.M.; Perrimon, N.; Michelson, A.M. Heparan sulfate proteoglycans are essential for FGF receptor signaling during Drosophila embryonic development. *Development* **1999**, *126*, 3715–3723. [PubMed]
102. Guimond, S.; Maccarana, M.; Olwin, B.B.; Lindahl, U.; Rapraeger, A.C. Activating and inhibitory heparin sequences for FGF-2 (basic FGF). Distinct requirements for FGF-1, FGF-2, and FGF-4. *J. Biol. Chem.* **1993**, *268*, 23906–23914.
103. Pye, D.A.; Vives, R.R.; Turnbull, J.E.; Hyde, P.; Gallagher, J.T. Heparan Sulfate Oligosaccharides Require 6-O-Sulfation for Promotion of Basic Fibroblast Growth Factor Mitogenic Activity. *J. Biol. Chem.* **1998**, *273*, 22936–22942. [CrossRef] [PubMed]

104. Rapraeger, A.C. In the clutches of proteoglycans: how does heparan sulfate regulate FGF binding? *Chem. Biol.* **1995**, *2*, 645–649. [CrossRef]
105. Xue, Y.; Lee, S.; Wang, Y.; Ha, Y. Crystal Structure of the E2 Domain of Amyloid Precursor Protein-like Protein 1 in Complex with Sucrose Octasulfate. *J. Biol. Chem.* **2011**, *286*, 29748–29757. [CrossRef]
106. Gralle, M.; Botelho, M.G.; Wouters, F.S. Neuroprotective Secreted Amyloid Precursor Protein Acts by Disrupting Amyloid Precursor Protein Dimers. *J. Biol. Chem.* **2009**, *284*, 15016–15025. [CrossRef] [PubMed]
107. Hoogewerf, A.J.; Kuschert, G.S.V.; Proudfoot, A.E.I.; Borlat, F.; Clark-Lewis, I.; Power, C.A.; Wells, T.N.C. Glycosaminoglycans Mediate Cell Surface Oligomerization of Chemokines. *Biochemistry* **1997**, *36*, 13570–13578. [CrossRef] [PubMed]
108. Salanga, C.L.; Handel, T.M. Chemokine oligomerization and interactions with receptors and glycosaminoglycans: The role of structural dynamics in function. *Exp. Cell Res.* **2011**, *317*, 590–601. [CrossRef]
109. Gospodarowicz, D.; Cheng, J. Heparin protects basic and acidic FGF from inactivation. *J. Cell. Physiol.* **1986**, *128*, 475–484. [CrossRef] [PubMed]
110. Saksela, O.; Moscatelli, D.; Sommer, A.; Rifkin, D.B. Endothelial cell-derived heparan sulfate binds basic fibroblast growth factor and protects it from proteolytic degradation. *J. Cell Biol.* **1988**, *107*, 743–751. [CrossRef] [PubMed]
111. Sadir, R.; Imberty, A.; Baleux, F.; Lortat-Jacob, H. Heparan Sulfate/Heparin Oligosaccharides Protect Stromal Cell-derived Factor-1 (SDF-1)/CXCL12 against Proteolysis Induced by CD26/Dipeptidyl Peptidase IV. *J. Biol. Chem.* **2004**, *279*, 43854–43860. [CrossRef] [PubMed]
112. Lortat-Jacob, H.; Baltzer, F.; Grimaud, J.-A. Heparin Decreases the Blood Clearance of Interferon-γ and Increases Its Activity by Limiting the Processing of Its Carboxyl-terminal Sequence. *J. Biol. Chem.* **1996**, *271*, 16139–16143. [CrossRef]
113. Jakobs, P.; Schulz, P.; Ortmann, C.; Schürmann, S.; Exner, S.; Rebollido-Rios, R.; Dreier, R.; Seidler, D.G.; Grobe, K. Bridging the gap: heparan sulfate and Scube2 assemble Sonic hedgehog release complexes at the surface of producing cells. *Sci. Rep.* **2016**, *6*, 26435. [CrossRef]
114. Jakobs, P.; Schulz, P.; Schürmann, S.; Niland, S.; Exner, S.; Rebollido-Rios, R.; Manikowski, D.; Hoffmann, D.; Seidler, D.G.; Grobe, K. Ca^{2+} coordination controls sonic hedgehog structure and its Scube2-regulated release. *J. Cell Sci.* **2017**, *130*, 3261–3271. [CrossRef]
115. Kastl, P.; Manikowski, D.; Steffes, G.; Schürmann, S.; Bandari, S.; Klämbt, C.; Grobe, K. Disrupting Hedgehog Cardin-Weintraub sequence and positioning changes cellular differentiation and compartmentalization in vivo. *Development* **2018**, *145*, dev167221. [CrossRef]
116. Cerqueira, C.; Samperio Ventayol, P.; Vogeley, C.; Schelhaas, M. Kallikrein-8 Proteolytically Processes Human Papillomaviruses in the Extracellular Space To Facilitate Entry into Host Cells. *J. Virol.* **2015**, *89*, 7038–7052. [CrossRef] [PubMed]
117. Koyama, Y.; Naruo, H.; Yoshitomi, Y.; Munesue, S.; Kiyono, S.; Kusano, Y.; Hashimoto, K.; Yokoi, T.; Nakanishi, H.; Shimizu, S.; et al. Matrix Metalloproteinase-9 Associated with Heparan Sulphate Chains of GPI-Anchored Cell Surface Proteoglycans Mediates Motility of Murine Colon Adenocarcinoma Cells. *J. Biochem.* **2008**, *143*, 581–592. [CrossRef] [PubMed]
118. Yu, W.-H.; Woessner, J.F. Heparan Sulfate Proteoglycans as Extracellular Docking Molecules for Matrilysin (Matrix Metalloproteinase 7). *J. Biol. Chem.* **2000**, *275*, 4183–4191. [CrossRef] [PubMed]
119. Butler, G.S.; Apte, S.S.; Willenbrock, F.; Murphy, G. Human Tissue Inhibitor of Metalloproteinases 3 Interacts with Both the N- and C-terminal Domains of Gelatinases A and B: REGULATION BY POLYANIONS. *J. Biol. Chem.* **1999**, *274*, 10846–10851. [CrossRef] [PubMed]
120. Mannello, F.; Jung, K.; Tonti, G.A.; Canestrari, F. Heparin affects matrix metalloproteinases and tissue inhibitors of metalloproteinases circulating in peripheral blood. *Clin. Biochem.* **2008**, *41*, 1466–1473. [CrossRef] [PubMed]
121. Di Simone, N.; Di Nicuolo, F.; Sanguinetti, M.; Ferrazzani, S.; D'Alessio, M.C.; Castellani, R.; Bompiani, A.; Caruso, A. Low-molecular Weight Heparin Induces In Vitro Trophoblast Invasiveness: Role of Matrix Metalloproteinases and Tissue Inhibitors. *Placenta* **2007**, *28*, 298–304. [CrossRef] [PubMed]
122. Isnard, N.; Robert, L.; Renard, G. Effect of Sulfated GAGs on the Expression and Activation of MMP-2 and MMP-9 in Corneal and Dermal Explant Cultures. *Cell Biol. Int.* **2003**, *27*, 779–784. [CrossRef]

123. Rababah, M.; Worthmann, H.; Deb, M.; Tryc, A.B.; Ma, Y.T.; El Bendary, O.M.; Hecker, H.; Goldbecker, A.; Heeren, M.; Brand, K.; et al. Anticoagulants affect matrix metalloproteinase 9 levels in blood samples of stroke patients and healthy controls. *Clin. Biochem.* **2012**, *45*, 483–489. [CrossRef] [PubMed]
124. Forsberg, E.; Pejler, G.; Ringvall, M.; Lunderius, C.; Tomasini-Johansson, B.; Kusche-Gullberg, M.; Eriksson, I.; Ledin, J.; Hellman, L.; Kjellén, L. Abnormal mast cells in mice deficient in a heparin-synthesizing enzyme. *Nature* **1999**, *400*, 773. [CrossRef] [PubMed]
125. Humphries, D.E.; Wong, G.W.; Friend, D.S.; Gurish, M.F.; Qiu, W.-T.; Huang, C.; Sharpe, A.H.; Stevens, R.L. Heparin is essential for the storage of specific granule proteases in mast cells. *Nature* **1999**, *400*, 769. [CrossRef] [PubMed]
126. Olsen, J.V.; Mann, M. Status of Large-scale Analysis of Post-translational Modifications by Mass Spectrometry. *Mol. Cell. Proteomics* **2013**, *12*, 3444–3452. [CrossRef] [PubMed]
127. Ryšlavá, H.; Doubnerová, V.; Kavan, D.; Vaněk, O. Effect of posttranslational modifications on enzyme function and assembly. *J. Proteomics* **2013**, *92*, 80–109. [CrossRef] [PubMed]
128. Pascovici, D.; Wu, J.X.; McKay, M.J.; Joseph, C.; Noor, Z.; Kamath, K.; Wu, Y.; Ranganathan, S.; Gupta, V.; Mirzaei, M. Clinically Relevant Post-Translational Modification Analyses-Maturing Workflows and Bioinformatics Tools. *Int. J. Mol. Sci.* **2018**, *20*, 16. [CrossRef] [PubMed]
129. Wilhelm, S.M.; Eisen, A.Z.; Teter, M.; Clark, S.D.; Kronberger, A.; Goldberg, G. Human fibroblast collagenase: glycosylation and tissue-specific levels of enzyme synthesis. *Proc. Natl. Acad. Sci. USA* **2006**, *83*, 3756–3760. [CrossRef] [PubMed]
130. Huanna, T.; Tao, Z.; Xiangfei, W.; Longfei, A.; Yuanyuan, X.; Jianhua, W.; Cuifang, Z.; Manjing, J.; Wenjing, C.; Shaochuan, Q.; et al. GALNT14 mediates tumor invasion and migration in breast cancer cell MCF-7. *Mol. Carcinog.* **2015**, *54*, 1159–1171. [CrossRef] [PubMed]

International Journal of *Molecular Sciences*

MDPI

Review

Matrix Metalloproteinases in Pulmonary and Central Nervous System Tuberculosis—A Review

Ursula K. Rohlwink [1], Naomi F. Walker [2], Alvaro A. Ordonez [3,4], Yifan J. Li [5], Elizabeth W. Tucker [4,6,7], Paul T. Elkington [8], Robert J. Wilkinson [9,10,11] and Katalin A. Wilkinson [9,10,*]

1 Neuroscience Institute, University of Cape Town, Faculty of Health Sciences, Anzio Road, Observatory 7925, South Africa; uk.rohlwink@uct.ac.za
2 TB Centre and Department of Clinical Research, London School of Hygiene and Tropical Medicine, Keppel St, London WC1E 7HT, UK; Naomi.Walker@lshtm.ac.uk
3 Department of Pediatrics, Johns Hopkins University School of Medicine, Baltimore, MD 21287, USA; aordone2@jhmi.edu
4 Center for Tuberculosis Research, Johns Hopkins University School of Medicine, Baltimore, MD 21287, USA; etucker9@jhmi.edu
5 Faculty of Health Sciences, University of Cape Town, Anzio Road, Observatory 7925, South Africa; joshualyf17@gmail.com
6 Department of Anesthesiology and Critical Care Medicine, Johns Hopkins University School of Medicine, Baltimore, MD 21287, USA
7 Division of Pediatric Critical Care, Johns Hopkins All Children's Hospital, St. Petersburg, FL 33701, USA
8 NIHR Biomedical Research Centre, School of Clinical and Experimental Sciences, Faculty of Medicine, University of Southampton, Southampton, SO16 6YD, UK; P.Elkington@soton.ac.uk
9 Wellcome Centre for Infectious Diseases Research in Africa, Institute of Infectious Diseases and Molecular Medicine, University of Cape Town, Cape Town 7925, South Africa
10 The Francis Crick Institute, 1 Midland Road, London NW1 1AT, UK
11 Department of Medicine, Imperial College London, London W2 1PG, UK; r.j.wilkinson@imperial.ac.uk
* Correspondence: Katalin.Wilkinson@crick.ac.uk

Received: 30 January 2019; Accepted: 3 March 2019; Published: 18 March 2019

Abstract: Tuberculosis (TB) remains the single biggest infectious cause of death globally, claiming almost two million lives and causing disease in over 10 million individuals annually. Matrix metalloproteinases (MMPs) are a family of proteolytic enzymes with various physiological roles implicated as key factors contributing to the spread of TB. They are involved in the breakdown of lung extracellular matrix and the consequent release of *Mycobacterium tuberculosis* bacilli into the airways. Evidence demonstrates that MMPs also play a role in central nervous system (CNS) tuberculosis, as they contribute to the breakdown of the blood brain barrier and are associated with poor outcome in adults with tuberculous meningitis (TBM). However, in pediatric TBM, data indicate that MMPs may play a role in both pathology and recovery of the developing brain. MMPs also have a significant role in HIV-TB-associated immune reconstitution inflammatory syndrome in the lungs and the brain, and their modulation offers potential novel therapeutic avenues. This is a review of recent research on MMPs in pulmonary and CNS TB in adults and children and in the context of co-infection with HIV. We summarize different methods of MMP investigation and discuss the translational implications of MMP inhibition to reduce immunopathology.

Keywords: tuberculosis; matrix metalloproteinases; tuberculous meningitis; HIV-TB-associated IRIS; extracellular matrix breakdown; adult; pediatric; lung; central nervous system

1. Introduction

Tuberculosis (TB) remains one of the top 10 causes of death globally, with 1.6 million deaths attributed to TB in 2017 [1]. Although the overall incidence of the disease is decreasing, there were over 10 million new cases of TB reported in 2017 [1], suggesting that more work needs to be done on disease prevention and cure. Tuberculous meningitis (TBM) is the most severe form of this disease and leaves many surviving adults and children with severe neurological impairment [2]. The inflammatory processes occurring during TB infection in the lungs and the neuro-inflammatory response as a result of dissemination to the brain play key roles in disease outcome. In pulmonary TB (pTB), matrix metalloproteinases (MMPs) are considered important in cavitary lung disease and the subsequent spread of *Mycobacterium tuberculosis* (*Mtb*) from the lung parenchyma into the airways from whence they can be expectorated to perpetuate infection [3]. They also contribute to the breakdown of the protective blood brain barrier (BBB) and brain tissue destruction consequent to TBM [3]. In this review, we discuss the current status of MMP research in pulmonary and central nervous system (CNS) TB, and their potential dual role in pathology as well as neurodevelopment and recovery. We also summarize potential novel treatment strategies that may target MMPs.

MMPs

MMPs are a superfamily of zinc- and calcium-dependent proteolytic enzymes that is well conserved across species with 25 vertebrate MMPs (24 of which are found in humans) characterized to date [4]. Several MMPs, including MMP-1, -3, -7, and -8, are located within a single gene cluster, highlighting their essential role in fundamental physiologic processes [4]. Their primary function in humans is degradation of the extracellular matrix (ECM) [5], which not only contributes to health by providing tissue homeostasis but also has a role in many pathologic conditions [4]. MMPs fall under the general category of metalloproteinases (MPs) of the Metzincin family, which function both on cell surfaces (as sheddases, which release growth factors, death receptors, and death-inducing ligands), and within the ECM [6]. They are divided into five sub-families, namely collagenases (MMP-1, -8, and -13), gelatinases (MMP-2 and MMP-9), stromelysins (MMP-3 and MMP-10), elastases (MMP-7 and MMP-12) and membrane-type MMPs (MT-MMP-1 to -5) [7,8].

MMPs are synthesized and released as proenzymes, known as zymogens, that are subsequently activated when the zinc-thiol interaction between the catalytic and the pro-peptide domain is disrupted by cleavage of the pro-peptide domain [8–10]. Due to the destructive nature of excessive activity, MMPs are controlled by gene expression (transcriptional and post-transcriptional regulation), proenzyme activation, and innate inhibitors. They are inhibited by α2-macroglobulin in the plasma or tissue inhibitors of metalloproteinase (TIMPs) in the tissue [4]. TIMPs are endogenous protein regulators and show tissue-specific, constitutive, or inducible expression [11]. There are four members of the TIMP family (TIMPs 1-4), which bind MMPs non-covalently to inhibit activity. TIMP-3 acts on both MMPs and tumor necrosis factor (TNF)-α converting enzyme (TACE) [9,11]. However, despite preference for specific MMPs, they are able to inhibit all MMPs, binding them in a 1:1 molar stoichiometry [11].

Physiologically, MMPs have an important role in development (blastocyst implantation, embryonic development, nerve growth), reproduction (ovulation, endometrial cycling, cervical dilatation), and maintenance of homoeostasis (wound-healing, bone remodeling, angiogenesis, and nerve regeneration) [4]. However, in various pathologies, MMPs have demonstrated a role in both immuno-protection and tissue destruction. They contribute to activation of immune mediators and leukocyte migration to sites of inflammation and infection by modulating cytokine and chemokine activity [12]. They are also implicated in numerous pathologies such as cancer invasion and metastasis, hypertension, rheumatoid and osteo-arthritis, CNS diseases, and may cause direct tissue destruction at excessive concentrations [10,12]. In the lung, a number of MMPs contribute to tissue homeostasis, such as MMP-7, -16, -19, -21, -24, -25, and -28 [4]. For example, lung epithelial cells express MMP-7 that is essential in normal wound repair and plays a part in the chemotactic recruitment of neutrophils [13]. In the CNS, MMPs are normally undetectable or present at very low levels (except for MMP-2) [10,14]. They can play a beneficial role in the CNS through their involvement with tissue repair after injury [9,14] and are also found to be elevated postnatally in rat and mouse CNS, suggesting pivotal roles in neuroplasticity and CNS development [15,16] such as synaptogenesis, synaptic plasticity, and long-term potentiation [17].

2. Methods of Investigation

Quantification of MMPs can be an analytical challenge for several reasons. In TBM, a number of methods of MMP investigation have been used, including ELISA, zymography and reverse zymography for serum, and cerebrospinal fluid (CSF) (Table 1). ELISAs for MMPs are, however, often limited by the absence of differentiation between active and inactive or degraded forms of MMP, and multiplexed assays are often complicated by cross-reactivity between analytes due to the protein's common domains [8]. Additionally, the method of blood collection can impact the concentration of MMPs, with higher concentrations reported in serum compared to plasma (both heparin and EDTA) [8] due to release from leukocytes and platelets during clot activation. For MMP-1, -8, and -9, plasma is recommended [18]. Brain tissue or granuloma samples have been examined using immunohistochemistry [19,20] and MMP gene expression using PCR [20,21]. Newer techniques such as analysis of gene polymorphisms, near-infrared optical imaging, and HPLC (high performance liquid chromatography) together with inductively-coupled plasma mass spectrometry (ICPMS) are evolving and promising analytic techniques that could improve the ability to profile all MMPs [8]. In addition to human data, in vitro cell and animal models have also offered valuable insights.

Table 1. Summary of recent matrix metalloproteinase (MMP) studies in pulmonary tuberculosis (TB) and central nervous system (CNS) TB.

Reference	CNS or pTB	Subjects	Samples	Method of Investigation	Analytes	Key Findings
Pulmonary TB (pTB)						
Elkington et al. [22,23]	pTB	Adults (n = 6 proven pTB patients, 6 controls with cancer diagnosis)	Lung tissue from biopsy	Immunohistochemistry	MMP-1 MMP-7	Study examined affected lung in *Mtb* vs unaffected lung in cancer patients • MMP-1 and MMP-7 present in macrophages and Langhans giant cells in granuloma, and MMP-1 in adjacent epithelial cells, in PTB cases only
Kuo et al. [24]	pTB	Adults (n = 101 confirmed pTB cases—38 with endobronchial TB, 68 without). All HIV negative	Blood	Genotyping	MMP-1 DNA (G-1607 GG) sequence single nucleotide polymorphisms	MMP-1 1G genotype was associated with endobronchial TB on bronchoscopy • MMP-1 1G genotypes were associated with a 9.86-fold increased risk of developing tracheobronchial stenosis, in patients with endobronchial TB
Wang et al. [25]	pTB	Adult (n = 98 pTB cases, 49 healthy controls). All HIV negative	Blood	Genotyping	*MMP-1*(G-1067GG) single nucleotide *MMP-12*(Asn357Ser), *MMP-9*(C-1562T) polymorphisms	MMP-1 (-1607G) polymorphism increased the risk of moderate and advanced lung fibrosis at one year in pTB cases—the odds increased by 3.80 and 6.02 fold, respectively for one copy and this remained after adjustment for age, sex, initial disease score on chest radiograph, sputum bacterial load, smoking status and presence of diabetes • MMP-9 and -12 polymorphisms were not associated with increased risk of developing lung fibrosis
Ganachari et al. [26]	pTB	Adults (n = 894 pTB cases, 1039 PPD+ controls collected from 2 sites). All HIV negative	Blood	Genotyping Immunohistochemistry of lymph node samples for MMP-1 and MCP-1	-2518A>G SNP in MCP-1 (rs1024611) -1607_1608insG variant in *MMP-1* (rs1799750), and 42 genomic control SNPs	• MMP-1 allele 2G associated with TB disease • MMP-1 2G/2G genotype associated with increased lymph node MMP-1 in active TB cases compared to other genotypes
Elkington et al. [27]	pTB	Adults (n = 33 HIV uninfected pTB cases, 32 respiratory symptomatic controls	Induced sputum and BAL	Luminex (concentrations normalized to total protein)	MMP-1 MMP-2 MMP-3 MMP-7 MMP-8 MMP-9 MMP-12 TIMP-1 TIMP-4	• MMP-1, -3: pTB > controls • MMP-2, -8, -9, -12, TIMP-1 and-2: pTB $\approx$ controls • MMP-13, TIMP-3 and -4 undetectable

Table 1. *Cont.*

Reference	CNS or pTB	Subjects	Samples	Method of Investigation	Analytes	Key Findings
Walker et al. [28]	pTB	Adults (n = 23 pTB cases, 21 controls—mixed healthy and respiratory symptomatic). Mixed HIV status	Induced sputum	Luminex (concentrations normalized to total protein)	MMP-1 MMP-2 MMP-3 MMP-7 MMP-8 MMP-9 MMP-12 MMP-13	• MMP-1, -2, -3, -8: pTB > controls irrespective of HIV status • MMP-7 and -9: pTB ≈ controls • MMP-12 and -13 undetectable • In advanced HIV (CD4 < 200), pTB patients had relatively lower sputum MMP-1, -2, -8 and -9 • MMP-1 and -2 increased in pTB patients with cavities versus no cavities, and correlated with chest x-ray inflammation score
Ganachari et al. [29]	pTB	Adults (n = 224 pTB cases, 42 controls). HIV negative	Blood	Genotyping	-2518A>G SNP in *MCP-1* (rs1024611) -1607_1608insG variant in *MMP-1* (rs1799750) and 42 genomic control SNPs	• Greater proportion of patients with severe disease carried the two locus genoype -2518 *MCP-1* GG and -1607 *MMP-1* 2G/2G, which was also associated with delayed sputum smear conversion and increased fibrosis
Seddon et al. [30]	pTB	Adults (n = 78). Mixed HIV status	Induced sputum Plasma	Luminex ELISA	MMP-1 MMP-2 MMP-3 MMP-7 MMP-8 MMP-9 MMP-10 Also PINP, PIIINP, PIIICP, CTX-I, CTX-III, EMMPRIN	• Plasma PIIINP correlated with radiographic inflammation score and sputum MMP-1 in pTB • Plasma MMP-8 and PIIINP predictive of pTB diagnosis. AUC of 0.82 (95% confidence interval 0.742–0.922, $p <$ 0.001) by ROC curve analysis
Ugarte-Gil et al. [31]	pTB	Adults (n = 68 HIV negative pTB cases, 69 healthy controls) Longitudinal study, follow up at 2, 8 and 24 weeks	Induced sputum	Luminex (adjusted for total protein) ELISA	MMP-1 MMP-2 MMP-3 MMP-7 MMP-8 MMP-9 TIMP-1 TIMP-2	• MMP-1, -2, -3, -8, and -9: pTB cases > controls • TIMP-1 and -2: pTB cases > controls • Sputum MMP concentrations decreased with TB treatment but TIMP concentrations initially rose • Elevated sputum MMP-2, -8, -9 and TIMP-2 at TB diagnosis, and elevated sputum MMP-3, MMP-8 and TIMP-1 at two weeks were associated with two-week sputum culture positivity

Table 1. *Cont.*

Reference	CNS or pTB	Subjects	Samples	Method of Investigation	Analytes	Key Findings
Kubler et al. [32]	pTB	Adults (n = 97 pTB cases, 14 latent TB and 20 healthy controls without latent TB)	Plasma	ELISA	MMP-1 MMP-3 MMP-7 MMP-8 MMP-9 TIMP-1 TIMP-2 TIMP-3 TIMP-4	• MMP-1/TIMP1-4 ratios: pTB > latent TB and healthy controls
Singh et al. [33]	pTB	Adults (n = 17 confirmed pTB cases, 18 respiratory symptomatic controls. All HIV uninfected)	BAL Fluid	Not specified	MMP-1 MMP-2 MMP-3 MMP-7 MMP-8 MMP-9 MMP-12 MMP-13	• median MMP-1, -2, -3, -7, -8, and -9: pTB cases > controls
Chen et al. [34]	Pleural TB	Adults (n = 18 TB pleuritis cases, 18 controls with congestive heart failure and pleural effusion)	Pleural fluid	ELISA	MMP-1 MMP-7 MMP-9 Also TNF-α	• Elevated TNF-α, MMP-1 and -9, which correlated with the size of the effusion in cases • MMP-7: cases ≈ controls • TNF-α and MMP-1 positively correlated with degree of pleural fibrosis at 6 months in cases
Sathyamoorthy et al. * [35]	pTB	Adults (n = 151 pTB cases, 109 symptomatic controls and 120 healthy controls)	Plasma	Luminex	MMP-1 MMP-3 MMP-7 MMP-8 MMP-9 MMP-10 MMP-12 MMP-13	• Plasma MMP-8: pTB cases > respiratory symptomatic and healthy controls • MMP-1: pTB cases > healthy controls only • Gender specific differences in MMPs: MMP-8 in men > women
Lee et al. [36]	pTB	Adults (n = 167, HIV negative, culture-confirmed, drug sensitive pTB)	Blood	Luminex	MMP-1 MMP-3 MMP-8 MMP-9 MMP-12 Also cytokines and cytotoxic mediators	• 26 cases were smear positive at 2 months (15.6%) • RANTES concentration at diagnosis and MMP-8 concentration at 2 months predicted 2-month culture status, AUC: 0.725 (0.624–0.827), and 0.632 (0.512–0.753) respectively

Table 1. *Cont.*

Reference	CNS or pTB	Subjects	Samples	Method of Investigation	Analytes	Key Findings
Andrade et al. * [37]	pTB	Brazilian adults (n = 63 active PTB, 15 individuals with LTBI, 10 healthy controls) Indian adults (n = 97 active PTB, 39 with LTBI, 40 uninfected healthy controls) North American adults (n = 18 culture-confirmed TB, 11 non-tuberculous mycobacteria infection [NTM], 48 pulmonary sarcoidosis) All HIV negative	Plasma	Luminex and ELISA	MMP-1 MMP-8 MMP-9 TIMP-1 TIMP-2 TIMP-3 TIMP-4 HO-1 and others	• MMP-1 (Brazilian cohort) and HO-1 (Indian cohort) were elevated but inversely related (both cohorts) in PTB patients compared to healthy controls (including those with latent TB infection) • MMP-1: pTB > sarcoid but pTB ≈ NTM (in American cohort)
Ong et al. [38]	pTB	Adults (n = 5 pTB cases)	Lung biopsies	Immunohistochemistry	H&E and anti-neutrophil elastase	• MMP-8 co-localized with neutrophils at the inner surface of TB cavities • Neutrophils were also immunoreactive for MMP-9
		Adults (n = 51 pTB cases, 57 healthy controls or a subset of 11 patients in each group for collagenase experiments). All HIV negative	Induced sputum	Luminex	MMP-8 MMP-9 (adjusted for total protein) Also myeloperoxidase (MPO) and neutrophil gelatinase associated lipocalin (NGAL)	• Neutrophil activation markers MPO and NGAL: pTB cases > controls and strongly correlated with sputum MMP-8 • Sputum MMP-8: cavities > no cavities
				DQ collagen degradation assay		• Induced sputum collagenase activity: pTB cases > controls • Collagenase activity correlates with MMP-8 concentration and is reduced by MMP-8 neutralization

Table 1. *Cont.*

Reference	CNS or pTB	Subjects	Samples	Method of Investigation	Analytes	Key Findings
Sathyamoorthy et al. [39]	pTB	Adults (n = 15 pTB cases, 10 controls	Induced sputum	RT-PCR	MT-MMP-1 (MMP-14)	• MT-MMP-1 RNA: pTB cases > controls
		Adults (n = 5 pTB cases, 5 controls)	Lung biopsy	Immunohistochemistry		• Granuloma cell MT-MMP-1 immunoreactivity: pTB cases > controls—only alveolar macrophages were positive
Brilha et al. [40]	pTB	Adult PTB vs control (respiratory symptomatic and healthy) South African cohort; Induced sputum, mixed serostatus as described in Walker et al. [28] Indian cohort: BAL as described in Singh et al. [33] Second South African adult cohort: Induced sputum for RNA (11 pTB patients and 17 healthy controls—all HIV negative)	Induced sputum and BAL	Luminex for MMP concentrations, RT-PCR for RNA	MMP-10	• Induced sputum secreted MMP-10 and MMP-10 RNA: pTB cases > controls • BAL MMP-10: pTB cases > controls
Fox et al. [41]	pTB	Peruvian cohort: Plasma from adults (n = 50 pTB patients 50 and matched asymptomatic PPD negative controls) Indian cohort: BAL fluid from adults (n = 15 pTB patients and 15 matched respiratory symptomatic controls)	Plasma and BAL Fluid	Luminex	MMP-9 and platelet-derived growth factor (PDGF)-BB, RANTES, P-selectin, platelet factor-4 (PF4), Pentraxin-3 (PTX3)	• Plasma MMP-9 correlated with platelet factors, PF4, PDGF-BB and PTX3 • In BAL fluid, P-selectin concentrations correlated with IL-1β, MMP-1, -3, -7, -8, and -9; PDGF-BB concentrations correlated with MMP-1, -3, -8, and -9; RANTES concentrations correlated with MMP-1, -8, and -9 as well as IL-1β

Table 1. *Cont.*

Reference	CNS or pTB	Subjects	Samples	Method of Investigation	Analytes	Key Findings
Singh et al. [42]	pTB	Adults (n = 5 pTB cases, 5 non-TB controls	Lung tissue	Immunohistochemistry	MMP-3 IL-17	• IL-17 and MMP-3: pTB > control tissue
Tuberculous meningitis (TBM)						
Matsuura et al. [43]	CNS	Adults (n = 21 meningitis cases [7 TBM], 30 controls)	CSF	Gelatin zymography Immunohistochemistry (demonstrated immunoreactivity for MMP-9 and -2 for infiltrating mononuclear cells)	MMP-9 MMP-2 TIMP-1 TIMP-2	• MMP-9, -2 and TIMP-1: cases > controls • TIMP-2: cases ~ controls • MMPs correlated positively with respective TIMPs • No correlations between analytes and proteins/cell counts • MMP-9 and TIMP-1 concentrations positively associated with neurological complications
Price et al. [44]	CNS	Human monocytic (THP-1) cells (in-vitro study) Adults (n = 23 TBM, 12 bacterial meningitis, 20 viral meningitis)	CSF	Northern Blot Western Blot Gelatin zymography ELISA (for TIMP-1)	MMP-9 TIMP-1 (MMP-2)	*In-vitro study:* • MMP-9 secretion increased in TB-infected cells at 24 h (compared to controls) • No difference in TIMP-1 secretion between TB-infected cells and controls at 24 h (suggesting net proteolytic activity). Moderate increase (5× compared to controls) at 48 h • MMP-9 mRNA—undetectable in controls, detected at 24 h in TB-infected cells and increased at 48 h • TIMP-1 mRNA—detected in controls. Only moderate increase at 48 h in infected cells *Human data* (Represented as activity on zymogram and as MMP/CSF-leukocyte ratio): • MMP-9 activity in TBM > other meningitides • MMP/CSF leukocyte ratio in TBM > other meningitides • MMP-9/CSF leukocyte ratio positively associated with neurological complications • MMP-2 was constitutionally expressed in the CSF, not affected by infection • TIMP-1 was not significantly elevated compared to other meningitides or controls

Table 1. *Cont.*

Reference	CNS or pTB	Subjects	Samples	Method of Investigation	Analytes	Key Findings
Thwaites et al. [45]	CNS	Adults (n = 21 TBM)	CSF Serum	ELISA	MMP-9 TIMP-1 Also several cytokines	Measured pre- and post-treatment analyte concentrations: All patients received streptomycin (20 mg/kg intramuscularly daily; maximum, 1 g) and an oral regimen of 5 mg/kg isoniazid, 10 mg/kg rifampicin, and 30 mg/ kg pyrazinamide for 3 months, followed by 3 drugs (isoniazid, rifampicin, and pyrazinamide) for 6 months • Pre-treatment: MMP-9 = 146 ng/mL, TIMP-1 = 463 ng/mL • Post-treatment: MMP-9 = 70 ng/mL ($p < 0.05$), TIMP-1 = 269 ng/mL ($p > 0.05$) • MMP-9 was not associated with outcome • post-treatment was not significantly different to pre-treatment concentrations
Lee et al. [46]	CNS	Adults (n = 24 TBM, 23 acute aseptic meningitis, 10 controls [4 pTB and 6 non-inflammatory neurological disorders])	CSF	ELISA Gelatin zymography	MMP-9 MMP-2	Measured MMP concentrations early (<7 days after treatment) and late (after 7 days of treatment—range 10–106 days) • MMP-9: early = 74 ng/mL, late = 123 ng/mL Both early and late TBM concentrations > aseptic meningitis and controls ($p < 0.001$) • MMP-2: early = 75 ng/mL, late = 120 ng/mL Early TBM > controls ($p < 0.01$) and late TBM > aseptic meningitis ($p = 0.01$) and controls ($p < 0.001$) • Both MMP-9 and -2 appear to increase temporally (after treatment), but not evaluated statistically • MMP-9 and -2 significantly higher in patients with delayed neurological complications ($p < 0.001$ and $p < 0.01$ respectively) • MMP-9 correlated with CSF protein and white cell count

Table 1. *Cont.*

Reference	CNS or pTB	Subjects	Samples	Method of Investigation	Analytes	Key Findings
Green et al. [47]	CNS	Adults (n = 37 TBM)	CSF	ELISA	MMP-1 MMP-2 MMP-3 MMP-7 MMP-8 MMP-9 MMP-10 TIMP-1 TIMP-2 TIMP-4	Study compared the effect of dexamethasone on analyte concentrations relative to a placebo group. Concentrations were measured pre-treatment, on day 5 (3–8), day 30, 60, and 270 • Significant decrease in MMP-9 early in dexamethasone treatment (day 5, $p = 0.01$)—suggested this as potential mechanism in which steroids improve outcome in TBM • No relationship found between early decrease in MMP-9 and outcome • Did not find any relationship between pre-treatment MMP or TIMP concentrations and outcome, except: lower MMP-2 associated with hemiparesis ($p = 0.02$) • MMP-9 correlated with CSF neutrophil count ($R^2 = 0.52$, $p < 0.001$)
Rai et al. [48]	CNS	Adults (n = 36 HIV negative, 28 HIV positive)	CSF	ELISA	MMP-2 MMP-9	• TBM case MMP > Controls for HIV− and HIV+ • MMP-2 and -9 concentrations not associated with patient outcome • MMP-2 associated with visual impairment MMP-2 (ng/mL) • Control: 60.5 (47.1–69.5) • HIV− TBM: 77.3 (38.8–221.9) • HIV+ TBM: 83.2 (48.8–286.5) • HIV− versus HIV+ $p > 0.05$ MMP-9 (pg/mL) • Control: 466.8 (432.1–986) • HIV− TBM: 4962.2 (200–5966) • HIV+ TBM: 3476.2 (491–5999.4) • HIV− versus HIV+ $p > 0.05$

Table 1. *Cont.*

Reference	CNS or pTB	Subjects	Samples	Method of Investigation	Analytes	Key Findings
Marais et al. [49]	CNS	Adults (*n* = 34 HIV-associated TBM) Sampled longitudinally and stratified into TBM-IRIS vs TBM-non-IRIS	CSF serum	Luminex ELISA	MMP-1 MMP-2 MMP-3 MMP-7 MMP-9 MMP-10 MMP-12 MMP-13 TIMP-1 TIMP-2 Plus a large number of cytokines and chemokines.	CSF was analyzed 3–5 time points in HIV-TBM (*n* = 34) at TBM diagnosis, initiation of ART (day 14), 14 days after ART initiation, at presentation of TBM-IRIS, and 14 days thereafter. 40 mediators in CSF were compared to blood and between patients who developed TBM-IRIS (*n* = 16) vs those who did not (TBM-non-IRIS; *n* = 18) • MMP-9 and TIMP-1 were elevated in CSF at TBM diagnosis, while MMP-1, -3, -7, and -10 were significantly higher in blood • HIV-TBM patients who subsequently developed TBM-IRIS showed significantly elevated MMP-1, MMP-7, MMP-10, TIMP-1 and TIMP-2 in the baseline CSF, compared to those who did not
Majeed et al. [50]	CNS	C6 glioma cells (in-vitro study) Adults (*n* = 91 TBM cases, 16 controls)	CSF Serum	Zymography Reverse zymography	MMP-9 TIMP-1	*In-vitro study:* Infected C6 glioma cells with *Mtb*, and treated them with MMP9-inhibitor (SB-3CT) and dexamethasone • MMP-9: infected cells > uninfected cells • Both drugs decreased MMP-9 concentrations *Human data*: measured MMP-9 and TIMP-1 levels at presentation (grouped according to TBM staging) • CSF and serum MMP-9: ○ cases > controls for all 3 stages ○ higher MMP-9 concentrations at more severe TBM stages • CSF TIMP-1 detectable in controls and stage I TBM only. Serum TIMP undetectable • MMP-9 (ng/mL) *CSF* concentrations: Controls: 0.62 ± 0.4, Stage I TBM: 9 ± 0.87, Stage II TBM: 12 ± 1.34, Stage III TBM: 16.9 ± 2.7 • MMP-9 (ng/mL) *serum* concentrations: Controls: 5.67 ± 2.45, Stage I TBM: 830.66 ± 83.07, Stage II TBM: 1202.55 ± 136.81, Stage III TBM: 1679 ± 277.4

Table 1. *Cont.*

Reference	CNS or pTB	Subjects	Samples	Method of Investigation	Analytes	Key Findings
Marais et al. [51]	CNS	Adults (n = 34 as described above in Marais et al. [49])	RNA from blood, Proteins from CSF and plasma	Microarray analysis (RNA) ELISA (CSF and plasma)	>47,000 probes	• Whole blood transcriptional signature confirmed elevated MMP-8 and MMP-9, with MMP-9 elevated at protein level in baseline CSF in TBM-IRIS patients • MMP-9 transcript remained elevated despite anti-TB and anti-inflammatory (prednisone) treatment in TBM-IRIS patients, possibly leading to degradation of the extracellular matrix as a mechanism of tissue damage in the central nervous system that is systemically reflected in the blood
Mailankody et al. [52]	CNS	Adults (n = 40 TBM)	CSF	ELISA	MMP-9 TIMP-1	• MMP-9 positively correlated with GCS (r = 0.3, p = 0.04) and CSF albumin levels (r = 0.354, p = 0.025) • MMP-9 negatively correlated with CSF glucose (r = −0.4, p = 0.007) • MMP-9 did not correlate with Q-alb or CSF neutrophil count • Higher median MMP-9 in patients with good outcome (254 ng/mL) compared to bad outcome (192 ng/mL), but not significant
Li et al. [53]	CNS	Children (n = 40 TBM, 8 controls)	CSF (lumbar & ventricular) Serum	Luminex	MMP-9 MMP-2 TIMP-1 TIMP-2	First study to measure MMP (and TIMP) concentrations in children with TBM • Persistently elevated gelatinase (and inhibitor) concentrations in paediatric TBM • MMP-9 levels decreased early in response to TB treatment (including steroid usage), but increased later during the treatment phase • No correlation between admission gelatinase levels and Q-alb, clinical characteristics or poor outcomes • Better outcomes (risk ratio 2.1, 95% CI 1.23–3.53, p = 0.01) at 6 months for patients whose MMP-9 levels increased during sampling period

Data presented as mean ± standard deviation or median (range) depending on information available from the study. * Provide diagnostic accuracy analysis. Matrix metalloproteinase (MMP), tissue inhibitor of matrix metalloproteinase (TIMP), Enzyme-linked immunosorbent assay (ELISA), broncho-alveolar lavage (BAL), interleukin (IL), cerebrospinal fluid (CSF). Anti-retroviral therapy (ART). Area under the curve (AUC). Cross-linked C-telopeptide of type I (CTX-I) and type III (CTX-III) collagen. Extracellular matrix metalloproteinase inducer (EMMPRIN). Procollagen I N-terminal propeptide (PINP). Procollagen III N-terminal propeptide (PIIINP). Procollagen III C-terminal propeptide (PIIICP). Platelet-derived growth factor-BB (PDGF-BB). Platelet factor-4 (PF4). Pentraxin-3 (PTX3). Receiver operating characteristics (ROC). Tumour necrosis factor (TNF). Regulated upon Activation, Normal T cell Expressed, and Secreted (RANTES). Membrane type (MT). Monocyte chemoattractant protein 1 (MCP-1). Haematoxylin and Eosin (H&E). Single nucleotide polymorphism (SNP). Albumin quotient (Q-alb).

3. MMPs in pTB

Destructive pulmonary pathology is the hallmark of human tuberculosis, and *Mtb* is relatively unique amongst pathogens for its ability to drive progressive destructive pulmonary pathology, including cavitation, in immunocompetent adults. Cavities are critical in *Mtb* pathogenesis, as they drive transmission (the most highly infectious patients are those with cavities), are immune-privileged sites with high bacterial burden, and are poorly penetrated by anti-mycobacterial drugs, leading to higher risk of treatment failure and relapse [54]. MMPs have been implicated in the pathology of pulmonary tissue damage and cavitation, as collectively, they can degrade all fibrillary components of the ECM (Figure 1). Previous reviews of MMPs in TB and mechanisms of cavitation have presented substantial evidence of MMP activity in pTB disease [3,55], summarized below:

1. Elevated MMP concentrations (including MMP-1, -2, -8, -9) are consistently reported in respiratory fluids [sputum, broncho-alveolar lavage (BAL), and pleural fluid] from TB patients compared to patients with respiratory symptoms and/or healthy controls. Increased MMPs are associated with various markers of pTB disease severity, most significantly MMP-1 with sputum smear status, radiographic disease extent, and cavitation number in human pTB [28].
2. Significantly increased MMP gene expression is found in human respiratory cells (alveolar macrophages, bronchial epithelial cells, and fibroblasts) and macrophages in response to *Mtb* infection and/or stimulation by conditioned media from *Mtb*-infected monocytes (CoMtb), resulting in increased MMP secretion [19,22,56]. Specifically, intracellular signaling involving p38 and extracellular signal-regulated kinase (ERK) mitogen-activated protein kinase pathway (MAPK) are important for MMP upregulation in macrophages in response to *Mtb* infection.
3. Genetic associations implicate MMPs in TB disease risk (2G/2G MMP-1 genotype), endobronchial TB and tracheobronchial stenosis (1G MMP-1 allele), TB dissemination (MMP-9 1562C/C genotype), and post-TB chronic lung fibrosis (MMP-1 G-1607GG polymorphism).
4. Whilst some animal models of tuberculosis have failed to fully replicate the spectrum of human pTB disease, following *Mtb* infection in human MMP-1-expressing transgenic mice, pathology was found to be more similar to pTB in humans with increased alveolar tissue damage and collagen destruction compared to wild type mice [27].

As MMPs have different cellular sources and substrate specificity, it is unsurprising that specific MMPs have been associated with different roles within the same disease. For example, MMP-1 has been most strongly implicated in the development of pulmonary cavities, whereas MMP-9 has been implicated in *Mtb* dissemination.

Here, we review in detail more recent (since 2013) studies of pTB that have further delineated cellular sources of MMPs in TB, mechanisms by which MMP activity is regulated in TB, and the importance of the cellular and extracellular environment. We also describe studies that have evaluated MMPs and matrix degradation products as TB biomarkers (Table 1, Section A).

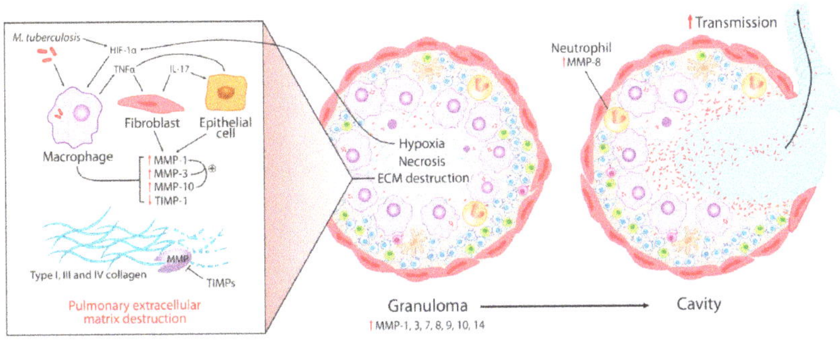

Figure 1. MMPs contribute to extracellular matrix (ECM) destruction in pulmonary TB (pTB). Lung TB granulomas comprise mycobacterium tuberculosis (*Mtb*) infected macrophages (purple), dendritic cells (beige), T (blue) and B (green) lymphocytes, and neutrophils (orange). These cells secrete numerous MMPs including MMP-1, -3, -7, -9, and -10, which may act in a proteolytic cascade. Evidence demonstrates their association with lung ECM breakdown, necrosis, and pTB disease severity. MMP-1 drives degradation of fibrillary type I, III, and IV collagen and is the key pulmonary collagenase. Neutrophil derived MMP-8 also contributes to collagen degradation, and both MMP-1 and -8 are involved in cavity formation, which facilitates the transmission of *Mtb* by releasing the bacilli into the airways. Hypoxia augments monocyte and neutrophil MMP secretion acting through the hypoxia inducible factor (HIF)-1α transcription factor, an important regulator of the host response to oxygen deprivation. *Mtb* also promotes MMP activity via cellular networks involving immune and stromal cells. For example, lung fibroblasts have demonstrated upregulated MMP gene expression in response to *Mtb* infection or in the presence of monocytes infected with *Mtb* and MMP-1 secretion in response to *Mtb*-induced TNF-α.

3.1. Collagen Degradation is an Early Pathological Event Promoted by Cell-Matrix Adhesion in TB

Fibrillar type I, III, and IV collagen are major structural components of the human lung. In order to develop cavities, these collagen fibers must be cleaved. MMP-1 (also known as interstitial collagenase, collagenase-1) has emerged as the key MMP causing collagen degradation in pTB. In a recent study, of all the MMPs evaluated, MMP-1 was the most significantly increased in induced sputum of TB patients compared to respiratory symptomatic patients and healthy controls, along with MMP-3, which activates MMP-1 [57]. Further, MMP-1 positively correlated with chest radiograph inflammation score in HIV-uninfected patients and with sputum acid fast bacilli score and cavity frequency in HIV-infected and uninfected patients [57]. Procollagen III N-terminal propeptide (PIIINP), a matrix degradation product released during Type III collagen turnover, was increased in sputum of TB patients compared to controls and correlated with sputum MMP-1 but not with other sputum MMPs or cytokines measured, suggesting that MMP-1 is the key collagenase degrading Type III collagen in the lung [30].

Kubler et al. [32] investigated cavitary formation in a novel rabbit model of pTB, demonstrating increased MMP -1, -3, -7, -12, and -13 transcription in abnormal lung tissue. MMP-1, followed by MMP-3, was the most highly upregulated MMP with increased expression in the cavity walls compared to granulomatous tissue, and MMP-1 was found to be highly abundant in cavitary areas, supporting human data implicating MMP-1 in cavity formation. In contrast, TIMP-3 expression was lower in the cavitary compared to granulomatous areas of lung.

Al Shammari et al. [58] examined lung biopsy specimens from pTB patients and demonstrated that collagen and elastin absence mapped to areas of caseous necrosis. In transgenic mice expressing human MMP-1 on the scavenger receptor A promoter/enhancer, infection with a pathological strain of *Mtb* (recently isolated from a patient) resulted in MMP-1 expression in the lung and development of granuloma with central areas of tissue destruction containing amorphous cellular debris, consistent with caseous necrosis. Wild-type or MMP-9-expressing mice infected with the same strain of *Mtb* did not develop this pathology despite a similar extent of pulmonary inflammation, mycobacterial burden, and cytokine profiles. The primary pathological difference was loss of collagen within the MMP-1-expressing mouse granulomas. In a 3D extracellular matrix *Mtb* infection model, the addition of collagen, but not gelatin, improved monocyte survival after *Mtb* infection, suggesting that loss of collagen may promote cell death in TB. Together, this work supports the hypothesis that MMP-1 activity leads to collagen degradation as a prelude to caseous necrosis in TB rather than as a consequence, and that the integrity of the ECM may be an important determinant of the host immune response to *Mtb* [58].

Following up on this, Brilha et al. [59] found that the ECM regulated MMP activity in TB. Culture filtrate from *Mtb*-infected monocytes (CoMtb) increased monocyte adhesion to the extracellular matrix. Adhesion to Type I collagen increased MMP-1 secretion from *Mtb*-infected monocytes by 60%, MMP-7 by 57%, and MMP-10 secretion by 90%. Similarly, adhesion to fibronectin (but not Type IV collagen) increased MMP-1 secretion by 63% and MMP-10 secretion by 55%. Type I collagen adhesion increased *Mtb*-driven TIMP-1, but TIMP-2 decreased. Monocyte migration, adhesion, and type I collagen degradation were dependent on $\alpha V\beta 3$-integrin activation, and integrin $\alpha V\beta 3$ cell surface expression increased in *Mtb*-infected monocytes following adhesion to type I collagen and fibronectin. This suggests that *Mtb* infection promotes immune cellular interaction with the ECM propagating ECM destruction. Whilst plasma concentrations of TIMP-1 are elevated in pTB in adults and children [60,61], the concentration of antiproteases at the site of disease is evidently insufficient to prevent matrix destruction, and TIMP-1 concentrations in pulmonary secretions are reduced in TB [27]. In cell culture experiments, the increase in MMP secretion is not countered by an increase in TIMP-1 secretion [23], consistent with *Mtb* skewing the protease-antiprotease balance towards matrix breakdown.

In addition to surface-bound collagenases, MT-MMPs may drive collagen degradation in TB. MT-MMP-1 (also known as MMP-14) is expressed in TB granulomas [39]. Increased MT-MMP-1 transcript abundance was found in the sputum of pTB patients compared to controls. *Mtb* infection increased monocyte cell surface MT-MMP-1 expression and resulted in collagen degradation, which was largely MT-MMP-1-dependent. CoMtb similarly upregulated MT-MMP-1, which was p38 MAPK dependent. MT-MMP-1 expression increased cell migration in a cellular model of TB infection, therefore MT-MMP-1 may contribute to both collagenase activity and cell migration in TB [39].

3.2. Neutrophil-Derived MMP-8 Drives TB Immunopathology

Neutrophils are abundant in the lungs in pTB disease and can be both helpful and harmful in TB [62–64]. Neutrophil-derived MMP-8 (neutrophil collagenase) has now emerged as the second main collagenase contributing to TB immunopathology. A systematic investigation of the role of MMP-8 in TB [38] demonstrated that neutrophils secrete MMP-8 in response to *Mtb* infection via NF-κβ and in response to CoMtb, resulting in collagen degradation. Furthermore, elevated MMP-8 found in sputum from pTB patients correlated well with neutrophil activation markers and TB severity score and was associated with cavitation and involved with collagen breakdown compared to control sputum. Neutrophils containing MMP-8 were found at the inner wall of pulmonary cavities in TB patients and at the center of cavities in areas of caseous necrosis. Pro-catabolic AMP-activated protein kinase (AMPK) phosphorylation was identified as a key regulator of neutrophil-derived MMP-8, but there was no evidence of involvement of Akt/phosphoinositide-3-kinase (PI3K) pathway and mTOR/p70S6 kinase [38]. A further study found elevated sputum and plasma MMP-8 in patients with and without HIV infection, and two studies have implicated MMP-8 in pulmonary immunopathology associated with TB-associated immune reconstitution inflammatory syndrome (TB-IRIS—see below) [57,65]. In HIV-uninfected patients, elevated plasma MMP-8 was found in TB patients compared to respiratory symptomatic patients and healthy controls and was found to differ by gender in TB patients, with elevated MMP-8 in men compared to women [35]. Although men had a longer duration of cough than women, the association with gender was independent of cough duration. Taken with the MMP-1 data, these findings suggest that multiple collagenases contribute to lung matrix destruction in TB, potentially with macrophage-derived MMP-1 causing initial matrix breakdown and then neutrophil-derived MMP-8 augmenting lung tissue damage once early cavitation occurs.

3.3. Hypoxia Drives Collagenase Activity in pTB

TB granulomas in several animal models have been shown to be hypoxic [66], but the interplay with matrix turnover in human TB has only recently been investigated. The MMP-1 promoter sequence has numerous putative hypoxia response elements, which are hypoxia-inducible factor (HIF)-1-binding sites as well as NF-κB and activated protein-1 (AP-1)-binding sites [67]. HIF-1α is a heterodimeric transcription factor, which is a key regulator of the host response to oxygen deprivation. Positron emission tomography–computed tomography (PET-CT) study of human pTB lesions with the hypoxia-specific tracer [^{18}F] fluoromisonidazole ([^{18}F]FMISO) has demonstrated severe hypoxia in radiologically abnormal areas [67]. Experimental hypoxia or HIF-1α stabilization using dimethyloxalyl glycine (DMOG) increased expression and secretion of MMP-1 in *Mtb*-infected macrophages without affecting *Mtb* growth. A similar effect of experimental hypoxia on MMP-1 secretion in response to *Mtb*-stimulation (via CoMtb) on normal human bronchial epithelial cells was observed. *Mtb* increases HIF-1α accumulation, even in normoxia, and HIF-1α is found in TB granuloma co-localizing with epithelioid macrophages and multinucleate giant cells [67], demonstrating a complex interplay between *Mtb* infection, hypoxia, and matrix breakdown, which together can be predicted to cause lung cavitation.

Ong et al. [68] recently demonstrated that hypoxia (or HIF-1α stabilization) exacerbates neutrophil-derived *Mtb*-driven MMP-8 secretion. The formation of neutrophil extracellular traps (NETs) in response to direct infection of neutrophils by *Mtb* was decreased in the presence of hypoxia, but neutrophil phagocytosis of *Mtb* was not affected. Furthermore, hypoxia increased neutrophil survival following *Mtb* infection compared to normoxia and increased neutrophil-driven matrix destruction, suggesting that hypoxia in pTB lesions may facilitate *Mtb* survival and promote tissue destruction.

3.4. MMP and Cytokine Networks Enhance Tissue Damage in TB

Cellular networks lead to a tissue degrading phenotype in pTB, upregulating MMPs by a number of pathways without a compensatory increase in TIMPs. Cell-cell networks involving respiratory epithelial cells and fibroblasts in addition to immune cells result in synergistic MMP upregulation in TB [3].

MMP-10 (stromelysin-2) was not formerly well studied in TB. However, as a key activator of MMP-1, recent work demonstrated that MMP-10 drives collagenase activity in cell culture, and its upregulation in macrophages was induced by virulent *Mtb* in an ESAT-6-dependent manner via p38 and ERK MAPK [40]. Elevated MMP-10 was found in respiratory fluids (induced sputum and BAL fluid) in TB patients compared to respiratory symptomatic controls, suggesting it may play an important role in human disease operating within a proteolytic cascade.

A recent study by Fox et al. [41] identified a novel role for platelets in amplifying *Mtb*-driven MMP-1, -3, -7, and MMP-10, and interleukin (IL)-1β secretion from monocytes without increasing TIMP-1 and -2 levels. Similar to MMP-10, MMP-3 can activate MMP-1, and thus a predicted outcome was increased collagenase activity. When *Mtb* was co-cultured with monocytes and platelets in a fluorescence model using dye-quenched (DQ) collagen, significantly increased type 1 collagen degradation was found compared to incubation with monocytes or platelets alone. Platelet-derived mediators (P-selectin, RANTES, platelet-derived growth factor (PDGF)-BB) were present in BAL fluid from TB patients at increased concentrations compared to BAL from non-TB respiratory symptomatic patients and correlated with multiple MMPs (including MMP-1, -8, -9) and IL-1β.

IL-17 is expressed in human TB granulomas [42]. Supplementation with IL-17 enhanced CoMtb-induced MMP-3 secretion from human distal small airway epithelial cells without compensatory TIMP-1/2 upregulation. MMP-9 secretion was inhibited by IL-17 while MMP-1, -7, and -8 were unchanged. IL-17 enhanced CoMtb-driven MMP-3 upregulation in normal human bronchial epithelial cells via p38 MAPK-dependent and the PI3K pathway. In fibroblasts, CoMtb-driven MMP-3 secretion was augmented by both IL-17 and IL-22 [42].

In pleural TB, MMP-1, MMP-9, and TNF-α concentrations were elevated in pleural fluid compared to pleural fluid from heart failure controls [34]. MMP-1 and -9 were secreted by pleural mesothelial cells in response to *Mtb*-stimulation. TNF-α strongly correlated with MMP-1 secretion and was found to be a potent stimulus of MMP-1 upregulation by pleural mesothelial cells in response to *Mtb*. In pericardial TB, concentrations of MMP-1, MMP-2, and TIMP-1 were significantly elevated in pericardial fluid, while MMP-7 and MMP-9 were higher in blood of patients with and without HIV-infection [69]. Differential abundance was also detected at mRNA level for MMP-2, MMP-9, TIMP-1, and TIMP-2.

In summary, *Mtb* promotes MMP activity and collagen degradation via cellular networks, likely involving monocytes, macrophages, neutrophils, platelets, and epithelial cells in the airways and mesothelial cells in the pleural space. In addition to previously described roles for monocyte-derived Oncostatin M and TNF-α enhancing MMP-1 secretion from fibroblasts [70], *Mtb*-driven (ESAT-6-dependent) MMP-10 and IL-17-augmented MMP-3 may propagate collagenase activity in cellular networks via MMP-1 activation. An outstanding question is the relative contribution of direct cellular infection versus intercellular networks in causing lung cavitation in vivo.

Furthermore, whilst macrophages and stromal cells are established sources of MMPs in TB, the role of T cells in MMP-related pathology is unclear. T cells are critical to an efficacious host immune response to *Mtb*, yet they also contribute to immune-mediated tissue damage. To exemplify, a greater purified protein derivative (PPD) response associates with subsequent development of pulmonary TB [71] and immune checkpoint inhibition for cancer associates with development of active TB [72,73]. In contrast, in advanced HIV-induced immunocompromise, pulmonary cavitation rarely occurs—despite high mycobacterial load within the lung—unless immune reconstitution takes place with therapy. The knowledge gap defining what comprises a protective T cell response versus what contributes to MMP-mediated tissue damage is a challenge to the development of rational strategies that may limit pathology after treatment initiation or alternatively prevent cavitation to break the cycle of transmission.

3.5. Intracellular Regulation of MMP Activity in pTB

Regulation of MMP-1 along with MMP-3 and MMP-9 by the PI3K pathway in TB has been investigated in normal human bronchial epithelial cells stimulated by intercellular networks [74]. Inhibition of the PI3K p110α subunit led to increased MMP-1, whereas downstream inhibition of Akt, mTOR, and p70S6 kinase led to reduced MMP-1. MMP-3 was inhibited by blockade at all three points in the pathway, whereas MMP-9 was increased by blockade at PI3K p110α subunit, decreased by blockade at Akt, and increased by blockade at mTOR or p70S6 kinase, indicating that regulation was complex and possibly involved epigenetic mechanisms [74].

Subsequent work has demonstrated epigenetic control of *Mtb*-driven MMP-1 and -3 expression by changes in histone acetylation in macrophages and normal human bronchial epithelial cells in a cell-type specific manner [75]. *Mtb* also inhibits negative regulatory pathways in directly infected human macrophages that function to limit MMP-1 secretion by inducing microRNAs that target PI3K, mTOR-containing complex 1 (mTORC1), and MAPK-interacting kinase (MNK) pathways, which together inhibit MMP-1 gene expression [76]. Therefore, *Mtb* not only increases collagenase activity by inducing macrophage MMP-1 and -10 secretion via p38 and ERK MAPK, it also effectively switches off inhibition of MMP-1 gene expression by interfering with PI3K and MNK pathway- regulation of MMP-1 inhibition [76].

Andrade et al. [37] observed elevated plasma MMP-1 and heme oxygenase (HO)-1 in TB patients compared to latently TB-infected asymptomatic controls. There was an inverse relationship between plasma MMP-1 and HO-1, and active TB patients were found to be either plasma MMP-1-high, plasma HO-1 low, or plasma MMP-1-low, HO-1-high. HO-1 is an antioxidant that catalyses the first step of the oxidative reaction that degrades the heme group contained in several proteins (including haemoglobin and myoglobin) and is immunomodulatory [77]. In vitro, both MMP-1 and HO-1 were induced in a dose-dependent fashion by *Mtb*. HO-1 was determined to be ESAT-6-dependent and required replicating bacilli, whereas MMP-1 secretion did not. In murine macrophages, HO-1-induced carbon monoxide (CO) production inhibited MMP-1 at the transcriptional level by inhibiting activation of c-JUN/AP-1, suggesting that *Mtb*-driven HO-1 may serve to limit MMP-1 [37]. Together, these findings demonstrate the complexity of MMP regulation with a balance between stimulatory and inhibitory pathways.

3.6. MMPs as Biomarkers of TB Disease

MMPs have been evaluated as diagnostic biomarkers of active TB. The previously described study by Sathyamoorthy et al. [35] reported that MMP-8 was the most discriminatory MMP for active TB in plasma, reporting an area under the curve (AUC) of 0.77 (0.80 for men and 0.72 for women) in comparison to non-TB respiratory symptomatic patients by receiver operating characteristics (ROC) curve analysis.

Following the demonstration that plasma PIIINP correlated with radiographic inflammation score and sputum MMP-1 in pTB, PIIINP was explored alongside other matrix degradation products as a peripheral biomarker of pTB disease [30]. Plasma PIIINP was combined with procollagen III C-terminal propeptide (PIIICP), MMP-2, MMP-7, MMP-8, and body mass index (BMI) in a TB prediction model in comparison to a mixed group of non-TB respiratory symptomatics and healthy controls. The final model, which included only those variables predictive after adjustment for all the others, included only MMP-8 and PIIINP, giving an AUC of 0.82 (95% confidence interval CI 0.742–0.922, $p < 0.001$) by ROC curve analysis.

In the aforementioned study by Andrade et al. [37], elevated plasma MMP-1 and HO-1 were highly discriminatory for active TB compared to latent TB. However, the comparator group of latent TB patients was asymptomatic. In an analysis including patients with other granulomatous pulmonary conditions, MMP-1 was significantly elevated in active TB ($n = 18$) compared to patients with sarcoidosis ($n = 48$) but not compared with patients with non-TB mycobacterial infection ($n = 11$). TIMP-1 has been evaluated in active TB compared to patients with pneumonia. However, the sensitivity was only 62.3% for a specificity of 45.95% [60]. It is possible that a panel of biomarkers that combine elements of tissue remodeling (MMPs, TIMPs, and matrix degradation products) may be more helpful than focus on a single component given the complexity of their interactions.

A limited number of studies to date have evaluated MMPs as biomarkers of treatment response in pTB. Ugarte-Gil et al. examined sputum MMPs longitudinally in relation to culture conversion and TBScore (which is associated with mortality risk) in a study of Peruvian patients [31]. Elevated sputum MMP-2, -8, -9, and TIMP-2 at diagnosis (but not TBScore) and elevated sputum MMP-3, MMP-8, and TIMP-1 at two weeks were associated with two-week sputum culture positivity. A study of Taiwanese patients found a modest association between elevated plasma MMP-8 and sputum culture positivity, both tested at two months [36]. An association between elevated serum MMP-1 and -8 over the first six weeks of treatment and delayed sputum culture conversion has been reported [78]. Further studies to evaluate MMPs and matrix degradation products as biomarkers of TB disease are in progress.

3.7. MMPs in TB-IRIS

TB is the leading cause of death among HIV-1-infected persons, and the World Health Organization (WHO) African region accounted for 72% of the global burden of HIV-associated TB in 2017 [1]. There were 464,633 cases of HIV-TB notified in 2017 globally, and 84% were on antiretroviral therapy (ART), which is an essential, life-saving intervention for HIV infection. However, ART-mediated immune system recovery in the setting of *Mtb* infection may lead to a pro-inflammatory state in the form of IRIS, presenting as TB-associated IRIS (TB-IRIS), which frequently complicates the otherwise beneficial dual therapy for HIV-1 and TB [79]. Two forms of TB-IRIS are recognized: paradoxical, which occurs in patients established on antituberculosis therapy before ART, but who develop acute, recurrent, or new TB symptoms and pathology after ART initiation; and unmasking TB-IRIS in patients not receiving treatment for TB when ART is started but who present with active TB within three months of starting ART [80]. The most common and most frequently studied form of HIV-associated IRIS is the paradoxical TB-IRIS with incidence in adults ranging between (19–57%) and lower incidence in children [81]. A recent randomized controlled trial demonstrated that prednisone therapy during the first four weeks after the initiation of ART for HIV infection resulted in a lower incidence of TB-IRIS than placebo without evidence of an increased risk of severe infections or cancers [82]. The pathophysiology of paradoxical TB-IRIS remains incompletely defined. The majority of studies evaluating MMPs have been conducted in adults, and this section of the review focuses on studies related to adult TB-IRIS.

The hypothesis that dysregulated MMP activity may play a role in immunopathology and tissue destruction in TB-IRIS patients was investigated for the first time by Tadokera et al. [83] through a comprehensive analysis of MMP-1, -2, -3, -7, -8, -9, -10, -11, -12, and -13 as well as TIMP-1 and -2 gene expression in *Mtb* stimulated peripheral blood mononuclear cells (PBMC) from 22 patients who developed paradoxical TB-IRIS compared to 22 similar HIV-TB co-infected, non-IRIS control patients. Protein secretion in the form of MMP pro-enzyme content was also investigated using multiplex assays in peripheral blood mononuclear cell (PBMC) culture supernatants and serum samples using the Luminex platform and ELISA for TIMP-1 and -2. Transcript levels were increased in TB-IRIS patients for MMP-3, -7, and -10, while protein secretion was higher for MMP-1, -7, -8, and -10 in stimulated cultures compared to the control patients. Serum MMP-7 concentration was elevated in TB-IRIS, and two weeks of clinically beneficial corticosteroid therapy [84] decreased this level, although not significantly.

The association between the development of paradoxical TB-IRIS and increasing MMP-1, -2, -3, -8, and -9 concentrations during ART was investigated in plasma from a cohort of 148 HIV-infected adults with advanced HIV and TB before ART initiation and four weeks later [65]. This study found that median concentrations of MMP-1, -2, and -3 decreased, while MMP-8 and -9 increased significantly during ART, and a greater early MMP-8 increase was associated with both TB-IRIS development and decreased long term lung function following TB cure. The results were broadly consistent with the Tadokera study [83], although in that study, MMP-8 transcript abundance was not increased in IRIS in *Mtb*-stimulated PBMC. This is possibly because MMP-8 is predominantly secreted by neutrophils and present in pre-synthesized granules, thus these changes would not be identified by RNA analysis of PBMC [38].

A large systematic study of HIV-associated TB that comprised of (i) a cross-sectional study of HIV-1-infected and uninfected TB patients and controls, and (ii) a prospective cohort study of HIV-1-infected TB patients at risk of TB-IRIS evaluated plasma and sputum MMP concentrations using Luminex multiplex assays [57]. The results were analyzed in conjunction with plasma PIIINP (determined by ELISA), a degradation product of collagen type III known to be released by matrix destruction in pTB [30]. MMP activity differed between HIV-1-infected and -uninfected TB patients. In HIV-uninfected TB patients, MMP activity was prominent in the pulmonary compartment and MMP-1 was dominant, whereas HIV-1-infected TB patients had reduced pulmonary MMP-1, -2, and -9 concentrations and reduced cavitation but increased plasma PIIINP compared to HIV-1-uninfected TB patients. TB-IRIS was associated with elevated extrapulmonary extracellular MMP concentrations (measured in plasma) both before and during TB-IRIS onset. Plasma MMP-8 was the most significantly increased and correlated with plasma PIIINP concentration as well as neutrophil count. This was likely driven by *Mtb* antigen, as TB-IRIS patients with a positive urinary LAM (lipoarabinomannan) result had higher plasma MMP-3, -7, and -8 concentrations compared to urine LAM negative TB-IRIS patients. Consequently, the compartmentalization of MMP activity differs between immunocompetent adults, where it is primarily within the lung, whereas in advanced HIV infection, MMP dysregulation is more systemic. In an in vitro culture model, *Mtb*-induced matrix degradation was suppressed by the MMP inhibitor doxycycline, suggesting MMP inhibition as a possible novel host-directed therapeutic strategy for preventing and/or treating TB-IRIS [57].

Overall, these studies indicate that TB-IRIS is associated with a distinct pattern of MMP gene and protein activation. In addition to the beneficial effect of corticosteroid therapy [82], modulation of dysregulated MMP activity may represent a novel therapeutic approach to alleviate TB-IRIS in HIV-TB patients undergoing treatment.

4. MMPs in TBM

4.1. The Blood Brain Barrier (BBB)

The BBB is a highly-selective permeability barrier comprised of specialized capillary endothelial cells with tight junctions, astrocytic foot processes, and pericytes [85]. It is integral in maintaining homeostasis in the CNS by the stringent regulation of molecular access to prevent entry of toxins and organisms into the brain. Increased permeability of the BBB is considered a hallmark of neuroinflammation and CNS infection [85], and research has implicated MMPs in BBB compromise [86,87]. Astrocytes, which form an integral component of the BBB, are known to release MMPs in response to disease and may therefore contribute to local damage [12]. Research into BBB disruption in TBM is sparse, but evidence indicates that MMP levels are increased in TBM [19,43,44,46,47], suggesting that they play a key role in BBB-breakdown, thereby contributing to the pathogenesis of TBM. The level and presence of specific MMPs and their inhibitors in the cerebrospinal fluid (CSF) and blood of patients may therefore be markers of BBB compromise. However, human studies have not shown association between MMP concentrations and the CSF-serum albumin index, an accepted index of BBB breakdown [47,52,53].

4.2. MMPs in Neuro-Inflammation

As summarized in Figure 2, MMPs contribute to ECM destruction in CNSTB. The two MMPs that have been the focus of human TBM research to date include MMP-2 and MMP-9 and their respective inhibitors, TIMP-2 and TIMP-1. MMP-2, also known as gelatinase-A, is a constitutively-expressed molecule that is present in large quantities in the normal brain and is found in astrocytes and CSF [9]. MMP-9, or gelatinase-B, is normally absent or present at low levels in the brain. It is found to be up-regulated during neuroinflammation [6], specifically in conditions such as viral and bacterial meningitis [88,89]. Together, these gelatinases digest tenascins, proteoglycans, fibronectin, type IV collagen, and laminin found in basement membranes of the BBB and ECM of the CNS, and thus are likely to contribute to the degradation of the BBB and tissue destruction [6,19,46]. This BBB degradation drives cerebral vasogenic oedema and facilitates further influx of blood-derived inflammatory cells into the CNS. The resultant chronic fibrosis and adhesions (due to exudate formation) contribute to the complications and neurological sequelae of TBM discussed below. In contrast to MMPs, TIMPs are widely expressed in many mammalian tissues. In the CNS, TIMP-1 expression is restricted to regions of persistent neuronal plasticity, such as the hippocampus, cerebellum, and olfactory bulb [11].

Cell culture models suggest that astrocyte and microglial secretion of MMPs does not occur with direct *Mtb* infection of these cells but rather in response to *Mtb*-infected monocytes as part of a leukocyte-astrocyte and leukocyte-microglia interaction, leading to the release of MMP-9, -1, and -3, respectively [12,20,21]. This network-dependent MMP expression is driven by pro-inflammatory mediators including TNF-α and IL-1β and IL-1β's synergistic interaction with IFN-γ in the case of MMP-9. The transcription of MMP-9 is regulated by MAPK and NF-$\kappa\beta$ [12], while MMP-1 and -3 transcriptional regulation occurs through p38 MAPK pathway, NF-$\kappa\beta$ and AP-1 [20,90]. In contrast to other MMPs, MMP-2 secretion from microglia is inhibited by cytokines, such as TNF-α, through p38, caspase 8, and NF-$\kappa\beta$ signaling [91]. Additionally, TIMP expression is not influenced by these cellular networks, suggesting a possible shift to matrix degradation in response to TB. MMP-3 may also be released by apoptotic neurons, resulting in the activation of microglia [12] and contributing to BBB breakdown by digesting the basal lamina and tight-junction proteins, specifically claudin-5, occludin, and laminin-α1 [92]. The breakdown of the BBB facilitates an influx of neutrophils that exacerbate BBB damage by secreting MMP-9 (also activated by MMP-3) through a monocyte-neutrophil network regulated by MAPK and Akt-PI3K [21,92]. MMP-9 is also reported to specifically target myelin basic protein in the white matter, which can negatively affect neuronal myelination [93]. The source of MMP secretion in the CNS is therefore likely a combination of infiltrating monocytes and neutrophils as well as resident astrocytes and microglia.

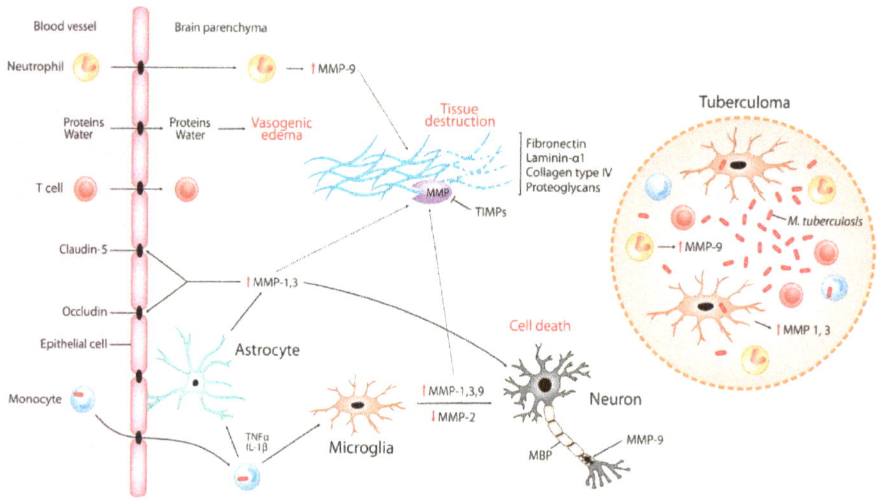

Figure 2. MMPs contribute to extracellular matrix (ECM) destruction in CNS TB. *Mtb* bacilli cross the protective blood brain barrier (BBB) through various mechanisms, including in infected monocytes. In the presence of these infected monocytes and driven by TNF-α and interleukin (IL)-1β, brain microglia and astrocytes respond as part of a leukocyte-microglia and leukocyte-astrocyte interaction by secreting MMP-1, -2, and -9, and MMP-1 and -3, respectively. These MMPs contribute to the breakdown of ECM proteins of the brain parenchyma, including fibronectin, laminin-α1, collagen type IV, and proteoglycans, and are pro-apoptotic to neurons (indicated in red text as "cell death") with the resultant tissue destruction adding to the cerebral inflammatory response. MMP-9 is also known to attack myelin basic protein (MBP), an integral component of the myelin sheath insulating neurons. Further, MMPs degrade claudin-5 and occludin, thereby compromising the tight BBB and contributing to the influx of proteins and water resulting in vasogenic edema, as well as further blood-derived inflammatory cells, including neutrophils. Neutrophils secrete MMP-9 in the parenchyma as well as in TB granulomas (tuberculomas), in which *Mtb*-infected microglia have also been found to secrete MMP-1 and -3. Contrary to the other MMPs, MMP-2 expression is suppressed through cytokines like TNF-α. Up-arrow indicates increased concentrations, down-arrow indicates decreased concentrations.

4.3. MMPs in TBM

The role of MMPs in TBM is summarized in Table 1, section B. Elevated MMP-9 concentrations have been found in the CSF of TBM patients with the degree of elevation correlated with the severity of disease [50], neurological compromise [19,44], and brain tissue injury [43,44]. Matsuura et al. demonstrated that concentrations of MMP-2 and -9 as well as TIMP-1 (but not TIMP-2) in the CSF were increased in patients with subacute (involving tubercle bacilli or mycotic organisms) meningitis compared with controls, and MMP-9 levels associated with CNS complications (such as disturbances of consciousness and psychiatric symptoms) [43]. Similarly, a study by Lee et al. demonstrated that increased MMP-2 and -9 levels in the CSF persisted during the late course of TBM and were associated with the development of complications [46]. Price et al. demonstrated that the imbalance of MMP-9 versus TIMP-1 in the CSF of TBM patients correlates with morbidity and mortality [94]. Although Green et al. found that adjunctive dexamethasone was associated with a significant decrease in CSF MMP-9 concentrations, they did not find an association between a decline in MMP-9 concentrations and improved outcome [47].

However, the association of elevated MMP-2 and -9 concentrations with poor outcome is not consistent across TBM studies, and numerous studies have demonstrated no association between MMP concentrations and outcome [45,48,52,53]. In fact, Mailankody et al. demonstrated an association between elevated MMPs and a favorable outcome, and the authors postulated that the MMP contribution to a leaky BBB may be favorable to anti-TB drug penetration and could therefore offer an advantage to recovery [52]. In a pediatric TBM study, the beneficial effect of MMPs was corroborated by a 2.2 times greater likelihood of a better outcome associated with an increase in MMP-9 concentrations in the early weeks after hospitalization [53]. It is known that MMP-9 can mediate both protective and pathological immunity at different concentrations and in different age groups [94], and this raises questions about the role of MMP-9 in pathology—and potentially recovery—at different time points during the disease course. Similar challenges have been encountered in therapeutically targeting MMPs in oncology, reflecting the complexity of protease networks in human disease [95].

TBM is often associated with localized TB granulomas (tuberculomas) [96], which demonstrate elevated MMP-1, -3, and -9 expression [19–21]. In brain biopsies from patients with CNS TB, MMP-1 secretion was highest in the granuloma, and elevated concentrations of MMP-1 and -3 were associated with microglia in the peri-granuloma region and surrounding brain parenchyma and p38 positive microglia infiltrating necrotizing TB granulomas [20,91]. Using similar samples, Ong et al. also demonstrated MMP-9 secreting neutrophil presence in these CNS TB granulomas [21].

4.4. MMPs in TBM-IRIS

TBM-IRIS is a frequent, severe complication of ART in HIV-associated TBM patients. A prospective study of HIV-infected ART-naive patients with TBM showed that 47% (16/34) of TBM patients developed TBM-IRIS [97]. Patients who went on to develop TBM-IRIS had higher CSF neutrophil counts compared with non-TBM-IRIS patients and were more likely to have *Mtb* cultured from CSF at TBM presentation. The relative risk of developing TBM-IRIS was 9.3 (CI: 1.4–62.2) if the CSF was *Mtb* culture positive, indicating that the condition is driven by *Mtb* antigen load. This study also showed that a combination of IFN-γ and TNF-α concentrations measured in the CSF may predict TBM-IRIS, potentially paving the way for CSF immunodiagnostics. Subsequently, the concentration of 40 immune mediators in CSF (33 paired with blood) from HIV-infected patients with TBM was determined at TBM diagnosis, at initiation of antiretroviral therapy (ART, day 14), 14 days after ART initiation, at presentation of TBM-IRIS, and 14 days thereafter [49]. Among others, MMP-1, -2, -3, -7, -9, -10, -12, and -13, and TIMP-1 and -2 were analyzed using Luminex multiplex assays in CSF and plasma of patients who developed TBM-IRIS ($n = 16$) compared to those who did not (TBM-non-IRIS; $n = 18$).

Patients who developed TBM-IRIS had higher concentrations of MMP-1, -3, -7, and -10 in their blood, while MMP-9 and TIMP-1 were higher in CSF relative to blood at the time of TBM diagnosis and at the time of starting ART, as well as at the time of developing TBM-IRIS symptoms, when their concentrations increased further. However, CSF MMP-9 concentrations did not decrease following anti-TB and corticosteroid therapy and continued to rise following ART initiation, suggesting that more potent and specific therapy may be needed for the management of TBM-IRIS [98]. Overall, both at TBM diagnosis and at two weeks after ART initiation, TBM-IRIS was associated with severe compartmentalized inflammation in the CSF, elevated concentrations of cytokines, chemokines, neutrophil-associated mediators, and MMPs compared with TBM-non-IRIS, and driven by a high baseline *Mtb* antigen load [49]. A longitudinal whole-blood microarray analysis of HIV-infected patients with TBM confirmed the more abundant neutrophil-associated transcripts in patients who went on to develop TBM-IRIS from before symptom development through to TBM-IRIS symptom onset. This study also demonstrated a significantly higher abundance of transcripts associated with canonical and noncanonical inflammasomes in patients with TBM-IRIS compared to non-IRIS controls [51]. These findings, reflected at the protein and molecular level, suggest a dominant role for the innate immune system in the pathogenesis of TBM-IRIS, offer mechanistic insights into the disease, and inform potential future directions for host-directed therapy. It is notable that when patients with HIV-associated TBM were compared to those with HIV-associated cryptococcal meningitis, MMP-1 and -3 were more elevated in the former and MMP-10 in the latter, suggesting that MMP profiles may differ across meningitides of varying etiologies [99].

4.5. Adults and Children

Before the era of HIV, the most important risk factor for TBM was age, with young children (aged 0–4) bearing the brunt of the disease [100,101]. This is possibly due to children having an immature immune system less capable of halting TB dissemination after primary infection [100,101]. However, apart from their role in neuro-inflammation, MMPs also play a unique role in normal brain development. Several animal studies have shown that MMPs, specifically MMP-2, -3, and -9, have a crucial role in the developing brain. For example, mice lacking MMP-9 have maladapted neuronal circuitry with simplified dendritic morphology, abnormal synaptic structure and function, and enhanced excitability [102]. MMP-3 and -2 are expressed in the cortical neurons of young mice and are required for appropriate axonal outgrowth and orientation [16,103]. Van Hove et al. also found that MMP-3 deficient mice developed deficits in balance and motor performance related to abnormal cerebellar development [104]. Despite the high prevalence of TBM in children, MMPs during pediatric TBM infection have not been well studied. In our recent pediatric TBM study [53], the positive association between increasing MMP-9 concentrations and good outcome may suggest a role in recovery and ongoing neurodevelopment, possibly including angio- and myelino-genesis, the growth of axons and synaptic plasticity. Similarly, the positive correlation found between MMP-2 and TIMP-2 may also suggest a physiological role in remodeling ECM in the developing CNS [53]. However, further studies with appropriate comparison with adults are required.

5. Treatment

5.1. MMP Inhibition in pTB

Multiple host-directed therapies centered on modulating protease activity have been evaluated in animal models (Table 2) as adjuncts to antibacterial treatment in TB with the prospect of beneficial immune modulation leading to enhanced bacterial clearance and the reduction of tissue damage and inflammation, thereby potentially shortening treatment duration [105]. This is challenging given the complexity of the host immune response. For example, multiple drugs that target upstream inflammatory pathways can have an effect on MMP expression by indirectly inhibiting their activity (e.g., prednisone, doxycycline, vitamin D, rapamycin) [106]. Similarly, non-specific immunomodulation of the host response with corticosteroids has been evaluated in pTB, with most studies showing no added benefit compared to standard treatment alone [107]. Conversely, in cases of extra-pulmonary TB, such as TBM and TB pericarditis, adjuvant use of corticosteroids has been shown to reduce mortality and morbidity, leading to their common use in current clinical practice [108]. Doxycycline, a tetracycline antibiotic, has been shown to non-specifically inhibit multiple MMPs independently of its antimicrobial activity [109]. However, doxycycline also has immunomodulatory effects on other host inflammatory pathways, confounding its potential MMP-specific effects [110]. Walker et al. reported that doxycycline modulated MMP expression in *Mtb*-infected cells and reduced mycobacterial growth in vivo [28]. However, additional studies are necessary to evaluate the potential of doxycycline as MMP-inhibitor therapy in TB. Indirect inhibition of MMPs has also been reported with a phosphodiesterase-4 inhibitor (CC-3052) in a rabbit model of TB [111]. CC-3052 plus isoniazid significantly reduced the extent of immune pathology and the expression of MMPs compared with antibiotics alone.

A more specific approach to evaluating MMP-inhibition in pTB has also been attempted with multiple specific MMP inhibitors (e.g., batimastat, cipemastat, marimastat) using different animal models of pTB. However, these studies have generated conflicting data, and the differences in treatment regimens, bacterial burden, host immune response, and variables measured make direct comparisons difficult (Table 2).

Initial experiments focused on monotherapy with MMP inhibitors in mouse models of pTB. Batimastat (BB-94), a potent broad-spectrum inhibitor, showed promising results in cancer models and was one of the first MMP inhibitors to enter clinical trials [112]. However, its poor solubility and side effects associated with intraperitoneal dosing led to its discontinuation. In pTB models, Hernandez-Pando et al. used *Mtb*-infected Balb/c mice to show a trend towards increased mortality if the treatment with batimastat was started immediately after infection with *Mtb* [113]. However, the increased mortality and histological differences were not evident between treated and control groups if batimastat was initiated a month after infection. Overall, there was a shift towards a type-2 cytokine profile and delayed granuloma formation in treated mice compared to controls. Subsequent studies by Izzo et al. using batimastat in C57BL/6 mice infected with *Mtb* showed reduced bacterial burden and smaller granulomas with more collagen deposition at earlier stages (day 40), but no differences were seen at day 60 post-infection [114]. However, a follow up study by the same group did not observe a reduced bacterial burden in the same mouse model when batimastat was initiated right after *Mtb* infection [115].

Recently, Urbanowski et al. used cipemastat, an orally available MMP inhibitor, in a rabbit model of cavitary pTB with repeated aerosol exposures to *Mtb* [116]. Compared to controls, monotherapy with cipemastat during week five to ten post-infection did not reduce the rates of cavitation evaluated by CT, mitigate disease severity, or visibly affect the collagen structure of the cavitary wall (examined post-mortem). Conversely, the cipemastat-treated group had a trend toward worse cavitary disease. Ordonez et al. used cipemastat monotherapy in the C3HeB/FeJ mouse model that develops cavitary lesions after aerosol infection with *Mtb* [117]. Similar to rabbits, cipemastat-treated mice had worse cavitation and more disease severity compared to controls.

While monotherapy with MMP inhibitors suggested worse outcomes, the use of multiple MMP inhibitors (marimastat, batimastat, or SB-3CT) in combination with anti-TB therapy have had synergistic effects in reducing the *Mtb* burden in a pTB mouse model [118]. While no effect on bacterial burden was observed with marimastat treatment alone, its combination with isoniazid significantly reduced lung colony forming units compared to isoniazid alone. Marimastat also reduced vascular leakage surrounding TB granulomas and increased the concentration of isoniazid in *Mtb*-infected lungs compared to controls. Similarly, targeting MMP-9 with a specific antibody in addition to anti-TB therapy (rifampin, pyrazinamide, and isoniazid) reduced pulmonary bacterial burden and significantly lowered relapse rates in a cavitary mouse model of pTB [117]. Therefore, MMP inhibition in conjunction with antibiotic therapy may be beneficial, whilst MMP inhibition monotherapy seems to be harmful, further demonstrating the complexity of events in vivo and the importance of performing translational preclinical studies that most closely reflect the intended therapeutic use in patients.

Table 2. Evaluation of MMP inhibitors in pulmonary TB animal models.

MMP Inhibitor	Animal Model	Treatment Started *	Combination Therapy	Results in Treated Group Compared to Controls				Reference
				Lung CFU	Lung Pathology	Mortality	Other Findings in Treated Group	
Batimastat (BB-94)	Mouse (Balb/c)	Day 1	No	NR	+	+	Lower TNF-α, IL-2 and IL-1α. Higher IL-4	[113]
		Day 30	No	NR	=	=	No difference in TNF-α, IL-2 and IL-4. Higher IL-1α in pneumonic areas.	[113]
	Mouse (C57BL/6)	Day 18	No	–	–/= [a]	NR	Decreased leukocytes. No differences in IFN-γ, IL-4, IL-12, TNF-α and IL-10. Less CFU in blood.	[114]
		Day 1	No	=	NR	NR	Less CFU in spleen and blood by day 14.	[115]
		Day 7	Isoniazid	–	NR	NR		[118]
Cipemastat (Ro 32-3555, Trocade)	Mouse (C3HeB/FeJ)	Day 1	No	=	+	+	Higher rate of cavitation	[119]
	Rabbit (New Zealand white)	Day 35	No	NR	+	NR	Higher rate of cavitation. No differences in disease severity by gross pathology or histology	[116]
Marimastat (BB-2516)	Mouse (C57BL/6)	Day 7	No	=	=	NR		[118]
		Day 7	Isoniazid	–	–	NR	Improved stability of blood vessels surrounding TB lesions	[118]
Prinomastat	Mouse (C57BL/6)	Day 7	Isoniazid	=	NR	NR		[118]
SB-3CT	Mouse (C57BL/6)	Day 7	Isoniazid	–	NR	NR		[118]
MMP-9 inhibitor I	Mouse (C57BL/6)	Day 7	Isoniazid	–	NR	NR		[118]

Table 2. *Cont.*

MMP Inhibitor	Animal Model	Treatment Started *	Combination Therapy	Results in Treated Group Compared to Controls				Reference
				Lung CFU	Lung Pathology	Mortality	Other Findings in Treated Group	
Anti MMP-9 antibody	Mouse (C3HeB/FeJ)	Day 42	Rifampin, isoniazid and pyrazinamide	–	–	NR	Less relapse rates after 12 weeks of treatment	[117]
CC-3052 (PDE-4 inhibitor) [b]	Rabbit (New Zealand white)	Day 28	No	=	+	NR	Worse disease compared to untreated controls	[111]
		Day 28	Isoniazid	–	–	NR	Less inflammation and fibrosis compared to isoniazid monotherapy controls	[111]
Doxycycline	Guinea pig	Day 14	No	–	–	NR	No MMP-specific effect of doxycycline was identified	[28]

* Days post-infection with *M. tuberculosis*, (−) equals less disease, (+) equals worse disease, (=) no change. NR = not reported, [a] Smaller pulmonary granulomas with more collagen deposition by day 40. No difference in pathology compared to controls by day 60. [b] Phosphodiesterase-4 inhibitor. Tumor necrosis factor (TNF), interleukin (IL), colony forming units (CFU).

5.2. MMP-Inhibition in CNS TB

Given the important role MMPs have in the degradation of the BBB and tissue destruction in CNS TB, MMPs are a possible target for adjunctive treatment. In vitro CNS TB studies have demonstrated that dexamethasone leads to a decrease in MMP-1 and -3 mRNA expression and secretion without impacting TIMP expression. This has been suggested as a mechanism by which steroids improve short-term mortality from TBM [20]. NF-κβ inhibition and anti-TNF-α have been shown to decrease neutrophil expression of MMP-9, which is postulated as one mechanism by which to decrease CNS destruction [21]. Given the pro-inflammatory network involvement in MMP-9 expression, it has been suggested that targeting this network may offer opportunities for MMP regulation without directly interfering with the role of MMPs in the host defense against *Mtb* [12]. In TBM-IRIS, protein and transcriptome studies have highlighted the role of a neutrophil driven and inflammasome-associated inflammatory response in disease pathogenesis [49,51]. ECM destruction due to neutrophil secretion of MMP-9 is suggested to contribute to brain tissue damage, and pyroptosis driven by caspase-1 secretion may cause further injury. Agents specifically targeting the inflammasome could therefore offer another potential avenue for host-directed therapy.

Although MMP secretion may be affected by several medications, there has also been interest in developing drugs to specifically target MMPs. Paul et al. evaluated the effect of the MMP inhibitor Batimastat (BB-94) on BBB breakdown (as measured by Evans Blue leakage) and increased intracranial pressure (ICP—measured via catheter) using a rat model of meningococcal meningitis. The disruption of the BBB (and subsequent pathological changes) was ameliorated in animals treated with BB-94 in a dose-dependent manner [120]. Majeed et al. used an MMP-9 specific inhibitor SB-3CT [50,121], which appeared more effective than dexamethasone in that it suppressed MMP-9 and enhanced the effect of anti-TB drugs on clearing TB bacilli [121].

6. Conclusions

Evidence suggests that MMPs are an important component of the inflammatory response to Mtb infection and the consequent breakdown of the ECM in both the lungs and the brain. Their potential role in brain development suggests that MMPs may be relevant in both normal physiology and pathophysiology. However, they form part of an elaborate network of inflammation and injury and constitute one element of a complex inflammatory process in TB. A better understanding of what contributes to MMP-mediated tissue damage and the role of MMPs in the broader disease context is needed to develop targeted treatment strategies that may limit pathology or prevent cavitation and transmission without the risk of unexpected effects.

Author Contributions: Conceptualization: U.K.R., R.J.W., K.A.W.; Methodology: U.K.R.; N.F.W., Y.J.L.; Software and visualisation: A.A.O.; Resources: P.T.E., R.J.W., K.A.W.; Data curation, writing of the original draft: U.K.R., N.F.W., A.A.O., Y.J.L., E.W.T., K.A.W. All authors contributed to reviewing and editing the manuscript.

Funding: NFW is supported by a National Institute for Health Research Academic Clinical Lecturership, the British Infection Association, and a Starter Grant for Clinical Lecturers (The Academy of Medical Sciences UK, Wellcome, Medical Research Council UK, British Heart Foundation, Arthritis Research UK, Royal College of Physicians and Diabetes UK). AO is supported by NIH grants R01-EB020539 and R01-HL131829. EWT is supported by the Johns Hopkins All Children's Hospital Foundation Institutional Grant Program. PTE is supported by the Medical Research Council Global Challenges Research Fund: Bioengineering to combat the tuberculosis epidemic (MR/P023754/1). RJW and KAW are supported by the Francis Crick Institute, which receives its core funding from Cancer Research UK, the UK Medical Research Council, and the Wellcome Trust (FC00110218). RJW also receives support from the Wellcome (104803 and 203135), National Research Foundation of South Africa (96841) and National Institutes of Health (U01AI115940).

Conflicts of Interest: The authors declare no conflict of interest.

References

1. World Health Organisation. *Global Tuberculosis Report*; WHO: Geneva, Switzerland, 2018.

2. Wilkinson, R.J.; Rohlwink, U.; Misra, U.K.; van Crevel, R.; Thi Hoang Mai, N.; Dooley, K.E.; Caws, M.; Figaji, A.; Savic, R.; Solomons, R.; et al. Tuberculous meningitis. *Nat. Rev. Neurol.* **2017**, *13*, 581. [CrossRef] [PubMed]
3. Elkington, P.T.; Ugarte-Gil, C.A.; Friedland, J.S. Matrix metalloproteinases in tuberculosis. *Eur. Respir. J.* **2011**, *38*, 456–464. [CrossRef] [PubMed]
4. Loffek, S.; Schilling, O.; Franzke, C.W. Series "matrix metalloproteinases in lung health and disease": Biological role of matrix metalloproteinases: A critical balance. *Eur. Respir. J.* **2011**, *38*, 191–208. [CrossRef] [PubMed]
5. Sternlicht, M.D.; Werb, Z. How matrix metalloproteinases regulate cell behavior. *Annu. Rev. Cell Dev. Biol.* **2001**, *17*, 463–516. [CrossRef] [PubMed]
6. Rosenberg GARosenberg, G.A. Matrix metalloproteinases in neuroinflammation. *Glia* **2002**, *40*, 130. [CrossRef] [PubMed]
7. Visse, R.; Nagase, H. Matrix metalloproteinases and tissue inhibitors of metalloproteinases: Structure, function, and biochemistry. *Circul. Res.* **2003**, *92*, 827–839. [CrossRef]
8. Lopez-Avila, V.; Spencer, J.V. Methods for detection of matrix metalloproteinases as biomarkers in cardiovascular disease. *Clin. Med. Cardiol.* **2008**, *2*, S484. [CrossRef]
9. Candelario-Jalil, E.; Yang, Y.; Rosenberg, G. Diverse roles of matrix metalloproteinases and tissue inhibitors of metalloproteinases in neuroinflammation and cerebral ischemia. *Neuroscience* **2009**, *158*, 983–994. [CrossRef]
10. Yong, V.W.; Power, C.; Forsyth, P.; Edwards, D.R. Metalloproteinases in biology and pathology of the nervous system. *Nat. Rev. Neurosci.* **2001**, *2*, 502. [CrossRef]
11. Murphy, G. Tissue inhibitors of metalloproteinases. *Genome Biol.* **2011**, *12*, 233. [CrossRef]
12. Green, J.A.; Friedland, J.S. Astrocyte-leucocyte interactions and the mechanisms regulating matrix degradation in CNS tuberculosis. *Biochem. Soc. Trans.* **2007**, *35 Pt 4*, 686–688. [CrossRef]
13. Chen, P.; Abacherli, L.E.; Nadler, S.T.; Wang, Y.; Li, Q.; Parks, W.C. MMP7 shedding of syndecan-1 facilitates re-epithelialization by affecting $\alpha 2\beta 1$ integrin activation. *PLoS ONE* **2009**, *4*, e6565. [CrossRef]
14. Yong, V.W. Metalloproteinases: Mediators of pathology and regeneration in the CNS. *Nat. Rev. Neurosci.* **2005**, *6*, 931. [CrossRef]
15. Vaillant, C.; Didier-Bazes, M.; Hutter, A.; Belin, M.F.; Thomasset, N. Spatiotemporal expression patterns of metalloproteinases and their inhibitors in the postnatal developing rat cerebellum. *J. Neurosci.* **1999**, *19*, 4994–5004. [CrossRef]
16. Gonthier, B.; Koncina, E.; Satkauskas, S.; Perraut, M.; Roussel, G.; Aunis, D.; Kapfhammer, J.P.; Bagnard, D. A PKC-dependent recruitment of MMP-2 controls semaphorin-3A growth-promoting effect in cortical dendrites. *PLoS ONE* **2009**, *4*, e5099. [CrossRef]
17. Ethell, I.M.; Ethell, D.W. Matrix metalloproteinases in brain development and remodeling: Synaptic functions and targets. *J. Neurosci. Res.* **2007**, *85*, 2813–2823. [CrossRef]
18. Jung, K.; Klotzek, S.; Stephan, C.; Mannello, F.; Lein, M. Impact of blood sampling on the circulating matrix metalloproteinases 1, 2, 3, 7, 8, and 9. *Clin. Chem.* **2008**, *54*, 772–773. [CrossRef]
19. Harris, J.E.; Nuttall, R.K.; Elkington, P.T.; Green, J.A.; Horncastel, D.E.; Graeber, M.B.; Edwards, D.R.; Friedland, J.S. Monocyte-astrocyte networks regulate matrix metalloproteinase gene expression and secretion in central nervous system tuberculosis in vitro and in vivo. *J. Immunol.* **2007**, *178*, 1199–1207. [CrossRef]
20. Green, J.A.; Elkington, P.T.; Pennington, C.J.; Roncaroli, F.; Dholakia, S.; Moores, R.C.; Bullen, A.; Porter, J.C.; Agranoff, D.; Edawrds, D.R.; et al. Mycobacterium tuberculosis upregulates microglial matrix metalloproteinase-1 and -3 expression and secretion via NF-kappaB- and Activator Protein-1-dependent monocyte networks. *J. Immunol.* **2010**, *184*, 6492–6503. [CrossRef]
21. Ong, C.W.; Pabisiak, P.J.; Brilha, S.; Singh, P.; Roncaroli, F.; Elkington, P.; Friedland, J.S. Complex regulation of neutrophil-derived MMP-9 secretion in central nervous system tuberculosis. *J. Neuroinflam.* **2017**, *14*, 31. [CrossRef]
22. Elkington, P.T.; Nuttall, R.K.; Boyle, J.J.; O'Kane, C.M.; Horncastle, D.E.; Edwards, D.R.; Friedland, J.S. Mycobacterium tuberculosis, but not vaccine BCG, specifically upregulates matrix metalloproteinase-1. *Am. J. Respir. Crit. Care Med.* **2005**, *172*, 1596–1604. [CrossRef] [PubMed]
23. Elkington, P.T.; Emerson, J.E.; Lopez-Pascua, L.D.; O'Kane, C.M.; Horncastle, D.E.; Boyle, J.J.; Friedland, J.S. Mycobacterium tuberculosis up-regulates matrix metalloproteinase-1 secretion from human airway epithelial cells via a p38 MAPK switch. *J. Immunol.* **2005**, *175*, 5333–5340. [CrossRef] [PubMed]

24. Kuo, H.P.; Wang, Y.M.; Wang, C.H.; He, C.C.; Lin, S.M.; Lin, H.C.; Liu, C.Y.; Huang, K.H.; Hsieh, L.L.; Huang, C.D. Matrix metalloproteinase-1 polymorphism in Taiwanese patients with endobronchial tuberculosis. *Tuberculosis (Edinb).* **2008**, *88*, 262–267. [CrossRef] [PubMed]
25. Wang, C.; Lin, H.; Lin, S.; Hunag, C.; Liu, C.; Huang, K.; Hsieh, L.; Chung, K.F.; Kuo, H. MMP-1 (-1607G) polymorphism as a risk factor for fibrosis after pulmonary tuberculosis in Taiwan. *Int. J. Tuberc. Lung Dis.* **2010**, *14*, 627–634. [PubMed]
26. Ganachari, M.; Ruiz-Morales, J.A.; Gomez de la Torre Pretell, J.C.; Dinh, J.; Granados, J.; Flores-Villanueva, P.O. Joint effect of MCP-1 genotype GG and MMP-1 genotype 2G/2G increases the likelihood of developing pulmonary tuberculosis in BCG-vaccinated individuals. *PLoS One.* **2010**, *5*, e8881. [CrossRef] [PubMed]
27. Elkington, P.; Shiomi, T.; Breen, R.; Nuttall, R.K.; Ugarte-Gil, C.A.; Walker, N.F.; Saraiva, L.; Pedersen, B.; Mauri, F.; Lipman, M.; et al. MMP-1 drives immunopathology in human tuberculosis and transgenic mice. *J. Clin. Investig.* **2011**, *121*, 1827–1833. [CrossRef]
28. Walker, N.F.; Clark, S.O.; Oni, T.; Andreu, N.; Tezera, L.; Singh, S.; Saraiva, L.; Pederson, B.; Kelly, D.L.; Tree, J.A.; et al. Doxycycline and HIV infection suppress tuberculosis-induced matrix metalloproteinases. *Am. J. Respir. Crit. Care Med.* **2012**, *185*, 989–997. [CrossRef] [PubMed]
29. Ganachari, M.; Guio, H.; Zhao, N.; Flores-Villanueva, P.O. Host gene-encoded severe lung TB: From genes to the potential pathways. *Genes Immun.* **2012**, *13*, 605–620. [CrossRef]
30. Seddon, J.; Kasprowicz, V.; Walker, N.F.; Yuen, H.M.; Sunpath, H.; Tezera, L.; Meintjes, G.; Wilkinson, R.J.; Bishai, W.R.; Friedland, J.S.; et al. Procollagen III N-terminal propeptide and desmosine are released by matrix destruction in pulmonary tuberculosis. *J. Infect. Dis.* **2013**, *208*, 1571–1579. [CrossRef]
31. Ugarte-Gil, C.A.; Elkington, P.; Gilman, R.H.; Coronel, J.; Tezera, L.B.; Bernabe-Oritz, A.; Friedland, J.S.; Moore, D.A. Induced sputum MMP-1, -3 & -8 concentrations during treatment of tuberculosis. *PLoS ONE* **2013**, *8*, e61333.
32. Kubler, A.; Luna, B.; Larsson, C.; Ammerman, N.C.; Andrade, B.B.; Orandle, M.; Bock, K.W.; Xu, Z.; Bagci, U.; Mollura, D.J.; et al. Mycobacterium tuberculosis dysregulates MMP/TIMP balance to drive rapid cavitation and unrestrained bacterial proliferation. *J. Pathol.* **2015**, *235*, 431–444. [CrossRef] [PubMed]
33. Singh, S.; Kubler, A.; Singh, U.K.; Singh, A.; Gardiner, H.; Prasad, R.; Elkington, P.T.; Firedland, J.S. Antimycobacterial drugs modulate immunopathogenic matrix metalloproteinases in a cellular model of pulmonary tuberculosis. *Antimicrob. Agents Chemother.* **2014**, *58*, 4657–4665. [CrossRef] [PubMed]
34. Chen, W.; Sheu, J.; Chen, R.; Hsiao, C.; Chou, Y.; Chung, C.; Hsiao, G. Mycobacterium tuberculosis upregulates TNF-α Expression via TLR2/ERK signaling and induces MMP-1 and MMP-9 production in human pleural mesothelial cells. *PLoS ONE* **2015**, *10*, e0137979. [CrossRef] [PubMed]
35. Sathyamoorthy, T.; Sandhu, G.; Tezera, L.B.; Thomas, R.; Singhania, A.; Woelk, C.H.; Dimitrov, B.D.; Agranoff, D.; Evans, C.A.; Friedland, J.S.; et al. Gender-dependent differences in plasma matrix metalloproteinase-8 elevated in pulmonary tuberculosis. *PLoS ONE* **2015**, *10*, e0117605. [CrossRef] [PubMed]
36. Lee, M.R.; Tsai, C.J.; Wang, W.J.; Chuang, T.Y.; Yang, C.M.; Chang, L.Y.; Lin, C.K.; Wang, J.Y.; Shu, C.C.; Lee, L.N.; et al. Plasma Biomarkers Can Predict Treatment Response in Tuberculosis Patients: A Prospective Observational Study. *Medicine (Baltimore)* **2015**, *94*, e1628. [CrossRef] [PubMed]
37. Andrade, B.B.; Pavan Kumar, N.; Amaral, E.P.; Riteau, N.; Mayer-Barber, K.D.; Tosh, K.W.; Maier, N.; Conceição, E.L.; Kubler, A.; Sridhar, R.; et al. Heme Oxygenase-1 Regulation of Matrix Metalloproteinase-1 Expression Underlies Distinct Disease Profiles in Tuberculosis. *J. Immunol.* **2015**, *195*, 2763–2773. [CrossRef] [PubMed]
38. Ong, C.W.; Elkington, P.T.; Brilha, S.; Ugarte-Gil, C.; Tome-Esteban, M.T.; Tezera, L.B.; Pabisiak, P.J.; Moores, R.C.; Sathyamoorthy, T.; Patel, V.; et al. Neutrophil-derived MMP-8 drives AMPK-dependent matrix destruction in human pulmonary tuberculosis. *PLoS Pathogens* **2015**, *11*, e1004917. [CrossRef]
39. Sathyamoorthy, T.; Tezera, L.B.; Walker, N.F.; Brilha, S.; Saraiva, L.; Mauri, F.; Wilkinson, R.J.; Friedland, J.S.; Elkington, P.T. Membrane Type 1 Matrix Metalloproteinase Regulates Monocyte Migration and Collagen Destruction in Tuberculosis. *J. Immunol.* **2015**, *195*, 882–891. [CrossRef]
40. Brilha, S.; Sathyamoorthy, T.; Stuttaford, L.H.; Walker, N.F.; Wilkinson, R.J.; Singh, S.; Moores, R.C.; Elkington, P.T.; Friedland, J.S. Early secretory antigenic target-6 drives matrix metalloproteinase-10 gene expression and secretion in tuberculosis. *Am. J. Respir. Cell Mol. Biol.* **2017**, *56*, 223–232.

41. Fox, K.A.; Kirwan, D.E.; Whittington, A.M.; Krishnan, N.; Robertson, B.D.; Gilman, R.H.; Lopez, J.W.; Singh, S.; Porter, J.C.; Friedland, J.S. Platelets regulate pulmonary inflammation and tissue destruction in tuberculosis. *Am. J. Respir. Crit. Care Med.* **2018**, *198*, 245–255. [CrossRef]
42. Singh, S.; Maniakis-Grivas, G.; Singh, U.K.; Asher, R.M.; Mauri, F.; Elkington, P.T.; Friedland, J.S. Interleukin-17 regulates matrix metalloproteinase activity in human pulmonary tuberculosis. *J. Pathol.* **2018**, *244*, 311–322. [CrossRef] [PubMed]
43. Matsuura, E.; Umehara, F.; Hashiguchi, T.; Fujimoto, N.; Okada, Y.; Osame, M. Marked increase of matrix metalloproteinase 9 in cerebrospinal fluid of patients with fungal or tuberculous meningoencephalitis. *J. Neurol. Sci.* **2000**, *173*, 45–52. [CrossRef]
44. Price, N.M.; Farrar, J.; Tran, T.T.; Nguyen, T.H.; Tran, T.H.; Friedland, J.S. Identification of a matrix-degrading phenotype in human tuberculosis in vitro and in vivo. *J. Immunol.* **2001**, *166*, 4223–4230. [CrossRef] [PubMed]
45. Thwaites, G.E.; Simmons, C.P.; Than Ha Quyen, N.; Thi Hong Chau, T.; Phuong Mai, P.; Thi Dung, N.; Hoan Phu, N.; White, N.P.; Tinh Hien, T.; Farrar, J.J. Pathophysiology and prognosis in vietnamese adults with tuberculous meningitis. *J. Infect. Dis.* **2003**, *188*, 1105–1115. [CrossRef] [PubMed]
46. Lee, K.Y.; Kim, E.H.; Yang, W.S.; Ryu, H.; Cho, S.N.; Lee, B.I.; Heo, J.H. Persistent increase of matrix metalloproteinases in cerebrospinal fluid of tuberculous meningitis. *J. Neurol. Sci.* **2004**, *220*, 73–78. [CrossRef] [PubMed]
47. Green, J.A.; Tran, C.T.; Farrar, J.J.; Nguyen, M.T.; Dinh, S.X.; Ho, N.D.; Ly, C.V.; Tran, H.T.; Friedland, J.S.; Thwaites, G.E. Dexamethasone, cerebrospinal fluid matrix metalloproteinase concentrations and clinical outcomes in tuberculous meningitis. *PLoS ONE* **2009**, *4*, e7277. [CrossRef] [PubMed]
48. Rai, D.; Garg, R.K.; Mahdi, A.A.; Jain, A.; Verma, R.; Tripathi, A.K.; Singh, M.K.; Malhotra, H.S.; Singh, G.P.; Ahmad, M.K. Cerebrospinal fluid cytokines and matrix metalloproteinases in human immunodeficiency seropositive and seronegative patients of tuberculous meningitis. *Ann. Indian Acad. Neurol.* **2014**, *17*, 171–178. [PubMed]
49. Marais, S.; Wilkinson, K.A.; Lesosky, M.; Coussens, A.K.; Deffur, A.; Pepper, D.J.; Schutz, C.; Ismail, Z.; Meintjes, G.; Wilkinson, R.J. Neutrophil-associated central nervous system inflammation in tuberculous meningitis immune reconstitution inflammatory syndrome. *Clin. Infect. Dis.* **2014**, *59*, 1638–1647. [CrossRef]
50. Majeed, S.; Singh, P.; Sharma, N.; Sharma, S. Role of matrix metalloproteinase-9 in progression of tuberculous meningitis: A pilot study in patients at different stages of the disease. *BMC Infect. Dis.* **2016**, *16*, 722. [CrossRef]
51. Marais, S.; Lai, R.P.; Wilkinson, K.A.; Meintjes, G.; O'garra, A.; Wilkinson, R.J. Inflammasome Activation Underlying Central Nervous System Deterioration in HIV-Associated Tuberculosis. *J. Infect. Dis.* **2017**, *215*, 677–686.
52. Mailankody, S.; Dangeti, G.V.; Soundravally, R.; Joseph, N.M.; Mandal, J.; Dutta, T.K.; Kadhiravan, T. Cerebrospinal fluid matrix metalloproteinase 9 levels, blood-brain barrier permeability, and treatment outcome in tuberculous meningitis. *PLoS ONE* **2017**, *12*, e0181262. [CrossRef]
53. Li, Y.J.; Wilkinson, K.A.; Wilkinson, R.J.; Figaji, A.A.; Rohlwink, U.K. Elevated matrix metalloproteinases offer novel insight into their role in paediatric tuberculous meningitis. *J. Pediatr. Infect. Dis. Soc.* **2019**. [Epub ahead of print]. [CrossRef]
54. Dheda, K.; Lenders, L.; Magombedze, G.; Srivastava, S.; Raj, P.; Arning, E.; Ashcraft, P.; Bottiglieri, T.; Wainwright, H.; Pennel, T.; et al. Drug-Penetration Gradients Associated with Acquired Drug Resistance in Patients with Tuberculosis. *Am. J. Respir. Crit. Care Med.* **2018**, *198*, 1208–1219. [CrossRef]
55. Ong, C.W.; Elkington, P.T.; Friedland, J.S. Tuberculosis, pulmonary cavitation, and matrix metalloproteinases. *Am. J. Respir. Crit. Care Med.* **2014**, *190*, 9–18. [CrossRef]
56. Chang, J.C.; Wysocki, A.; Tchou-Wong, K.M.; Moskowitz, N.; Zhang, Y.; Rom, W.N. Effect of Mycobacterium tuberculosis and its components on macrophages and the release of matrix metalloproteinases. *Thorax* **1996**, *51*, 306–311. [CrossRef]
57. Walker, N.F.; Wilkinson, K.A.; Meintjes, G.; Tezera, L.B.; Goliath, R.; Peyper, J.M.; Tadokera, R.; Opondo, C.; Coussens, A.K.; Wilkinson, R.J.; et al. Matrix Degradation in Human Immunodeficiency Virus Type 1–Associated Tuberculosis and Tuberculosis Immune Reconstitution Inflammatory Syndrome: A Prospective Observational Study. *Clin. Infect. Dis.* **2017**, *65*, 121–132. [CrossRef]

58. Al Shammari, B.; Shiomi, T.; Tezera, L.; Bielecka, M.K.; Workman, V.; Sathyamoorthy, T.; Mauri, F.; Jayasinghe, S.N.; Robertson, B.D.; D'Armiento, J.; et al. The extracellular matrix regulates granuloma necrosis in tuberculosis. *J. Infect. Dis.* **2015**, *212*, 463–473. [CrossRef]
59. Brilha, S.; Wysoczanski, R.; Whittington, A.M.; Friedland, J.S.; Porter, J.C. Monocyte Adhesion, Migration, and Extracellular Matrix Breakdown Are Regulated by Integrin alphaVbeta3 in Mycobacterium tuberculosis Infection. *J. Immunol.* **2017**, *199*, 982–991. [CrossRef]
60. Chen, Y.; Wang, J.; Ge, P.; Cao, D.; Miao, B.; Robertson, I.; Zhou, X.; Zhang, L.; Chen, H.; Guo, A. Tissue inhibitor of metalloproteinases 1, a novel biomarker of tuberculosis. *Mol. Med. Rep.* **2017**, *15*, 483–487. [CrossRef]
61. Pavan Kumar, N.; Anuradha, R.; Andrade, B.B.; Suresh, N.; Ganesh, R.; Shankar, J.; Kumaraswami, V.; Nutman, T.B.; Babu, S. Circulating biomarkers of pulmonary and extrapulmonary tuberculosis in children. *Clin. Vaccine Immunol.* **2013**, *20*, 704–711. [CrossRef]
62. Eum, S.Y.; Kong, J.H.; Hong, M.S.; Lee, Y.; Kim, J.; Hwang, S.; Cho, S.; Via, L.E.; Barry, C.E. Neutrophils are the predominant infected phagocytic cells in the airways of patients with active pulmonary TB. *Chest* **2010**, *137*, 122–128. [CrossRef]
63. Lowe, D.M.; Redford, P.S.; Wilkinson, R.J.; O'Garra, A.; Martineau, A.R. Neutrophils in tuberculosis: Friend or foe? *Trends Immunol.* **2012**, *33*, 14–25. [CrossRef]
64. Lowe, D.M.; Bandara, A.K.; Packe, G.E.; Barker, R.D.; Wilkinson, R.J.; Griffiths, C.J.; Martineau, A.R. Neutrophilia independently predicts death in tuberculosis. *Eur. Respir. J.* **2013**, *42*, 1752–1757. [CrossRef]
65. Ravimohan, S.; Tamuhla, N.; Kung, S.J.; Nfanyana, K.; Steenhoff, A.P.; Gross, R.; Weissman, D.; Bisson, G.P. Matrix Metalloproteinases in Tuberculosis-Immune Reconstitution Inflammatory Syndrome and Impaired Lung Function Among Advanced HIV/TB Co-infected Patients Initiating Antiretroviral Therapy. *EBioMedicine* **2015**, *3*, 100–107. [CrossRef]
66. Via, L.E.; Lin, P.L.; Ray, S.M.; Carrillo, J.; Allen, S.S.; Eum, S.Y.; Taylor, K.; Klein, E.; Manjunatha, U.; Gonzales, J.; et al. Tuberculous granulomas are hypoxic in guinea pigs, rabbits, and nonhuman primates. *Infect. Immunol.* **2008**, *76*, 2333–2340. [CrossRef]
67. Belton, M.; Brilha, S.; Manavaki, R.; Mauri, F.; Nijran, K.; Hong, Y.T.; Patel, N.H.; Dembek, M.; Tezera, L.; Green, J.; et al. Hypoxia and tissue destruction in pulmonary TB. *Thorax* **2016**, *71*, 1145–1153. [CrossRef]
68. Ong, C.W.; Fox, K.; Ettorre, A.; Elkington, P.T.; Friedland, J.S. Hypoxia increases neutrophil-driven matrix destruction after exposure to Mycobacterium tuberculosis. *Sci. Rep.* **2018**, *8*, 11475. [CrossRef]
69. Matthews, K.; Deffur, A.; Ntsekhe, M.; Syed, F.; Russell, J.B.W.; Tibazarawa, K.; Wolske, J.; Brink, J.; Mayosi, B.M.; Wilkinson, R.J.; et al. A compartmentalized profibrotic immune response characterizes pericardial tuberculosis, irrespective of HIV-1 infection. *Am. J. Respir. Crit. Care Med.* **2015**, *192*, 1518–1521. [CrossRef]
70. O'Kane, C.M.; Elkington, P.T.; Friedland, J.S. Monocyte-dependent oncostatin M and TNF-α synergize to stimulate unopposed matrix metalloproteinase-1/3 secretion from human lung fibroblasts in tuberculosis. *Eur. J. Immunol.* **2008**, *38*, 1321–1330. [CrossRef]
71. Comstock, G.W.; Livesay, V.T.; Woolpert, S.F. The prognosis of a positive tuberculin reaction in childhood and adolescence. *Am. J. Epidemiol.* **1974**, *99*, 131–138. [CrossRef]
72. Elkington, P.T.; Bateman, A.C.; Thomas, G.J.; Ottensmeier, C.H. Implications of Tuberculosis Reactivation after Immune Checkpoint Inhibition. *Am. J. Respir. Crit. Care Med.* **2018**, *198*, 1451–1453. [CrossRef]
73. Barber, D.L.; Sakai, S.; Kudchadkar, R.R.; Fling, S.P.; Day, T.A.; Vergara, J.A.; Ashkin, D.; Cheng, J.H.; Lundgren, L.M.; Raabe, V.N.; et al. Tuberculosis following PD-1 blockade for cancer immunotherapy. *Sci. Transl. Med.* **2019**, *11*. [CrossRef]
74. Singh, S.; Saraiva, L.; Elkington, P.T.; Friedland, J.S. Regulation of matrix metalloproteinase-1,-3, and-9 in Mycobacterium tuberculosis-dependent respiratory networks by the rapamycin-sensitive PI3K/p70S6K cascade. *FASEB J.* **2014**, *28*, 85–93. [CrossRef]
75. Moores, R.C.; Brilha, S.; Schutgens, F.; Elkington, P.T.; Friedland, J.S. Epigenetic regulation of Matrix Metalloproteinase-1 and-3 expression in Mycobacterium tuberculosis infection. *Front. Immunol.* **2017**, *8*, 602. [CrossRef]
76. Brace, P.T.; Tezera, L.B.; Bielecka, M.K.; Mellows, T.; Garay, D.; Tian, S.; Rand, L.; Green, J.; Jogai, S.; Steele, A.J.; et al. Mycobacterium tuberculosis subverts negative regulatory pathways in human macrophages to drive immunopathology. *PLoS Pathogens* **2017**, *13*, e1006367. [CrossRef]

77. Sebastian, V.P.; Salazar, G.A.; Coronado-Arrazola, I.; Schultz, B.M.; Vallejos, O.P.; Berkowitz, L.; Alvarez-Lobos, M.M.; Riedel, C.A.; Kalergis, A.M.; Bueno, S.M. Heme Oxygenase-1 as a Modulator of Intestinal Inflammation Development and Progression. *Front. Immunol.* **2018**, *9*, 1956. [CrossRef]
78. Coussens, A.K.; Wilkinson, R.J.; Nikolayevskyy, V.; Elkington, P.T.; Hanifa, Y.; Islam, K.; Timms, P.M.; Bothamley, G.H.; Claxton, A.P.; Packe, G.E.; et al. Ethnic variation in inflammatory profile in tuberculosis. *PLoS Pathogens* **2013**, *9*, e1003468. [CrossRef]
79. Esmail, H.; Riou, C.; du Bruyn, E.; Lai, R.P.; Harley, Y.X.R.; Meintjes, G.; Wilkinson, K.A.; Wilkinson, R.J. The immune response to Mycobacterium tuberculosis in HIV-1-coinfected persons. *Annu. Rev. Immunol.* **2018**, *36*, 603–638. [CrossRef]
80. Meintjes, G.; Lawn, S.D.; Scano, F.; Maartens, G.; French, M.A.; Worodria, W.; Elliott, J.H.; Murdoch, D.; Wilkinson, R.J.; Seyler, C.; et al. Tuberculosis-associated immune reconstitution inflammatory syndrome: Case definitions for use in resource-limited settings. *Lancet Infect. Dis.* **2008**, *8*, 516–523. [CrossRef]
81. Walker, N.F.; Stek, C.; Wasserman, S.; Wilkinson, R.J.; Meintjes, G. The tuberculosis-associated immune reconstitution inflammatory syndrome: Recent advances in clinical and pathogenesis research. *Curr. Opin. HIV AIDS* **2018**, *13*, 512–521. [CrossRef]
82. Meintjes, G.; Stek, C.; Blumenthal, L.; Thienemann, F.; Schutz, C.; Buyze, J.; Ravinetto, R.; van Loen, H.; Nair, A.; Jackson, A.; et al. Prednisone for the Prevention of Paradoxical Tuberculosis-Associated IRIS. *N. Engl. J. Med.* **2018**, *379*, 1915–1925. [CrossRef]
83. Tadokera, R.; Meintjes, G.A.; Wilkinson, K.A.; Skolimowska, K.H.; Walker, N.; Friedland, J.S.; Maartens, G.; Elkington, P.T.; Wilkinson, R.J. Matrix metalloproteinases and tissue damage in HIV-tuberculosis immune reconstitution inflammatory syndrome. *Eur. J. Immunol.* **2014**, *44*, 127–136. [CrossRef]
84. Meintjes, G.; Wilkinson, R.J.; Morroni, C.; Pepper, D.J.; Rebe, K.; Rangaka, M.X.; Oni, T.; Maartens, G. Randomized placebo-controlled trial of prednisone for paradoxical tuberculosis-associated immune reconstitution inflammatory syndrome. *AIDS* **2010**, *24*, 2381–2390. [CrossRef]
85. De Vries, H.E.; Kuiper, J.; de Boer, A.G.; Van Berkel, T.J.; Breimer, D.D. The blood-brain barrier in neuroinflammatory diseases. *Pharmacol. Rev.* **1997**, *49*, 143–155.
86. Rosenberg, G.A.; Mun-Bryce, S. Matrix metalloproteinases in neuroinflammation and cerebral ischemia. In Proceedings of the Ernst Schering Research Foundation Workshop, Berlin, Heidelberg, Germany, 2004; pp. 1–16.
87. Barr, T.L.; Latour, L.L.; Lee, K.Y.; Schaewe, T.L.; Luby, M.; Chang, G.S.; El-Zammer, Z.; Alam, S.; Hallenbeck, J.M.; Kidwell, C.S.; et al. Blood-brain barrier disruption in humans is independently associated with increased matrix metalloproteinase-9. *Stroke* **2010**, *41*, e123–e128. [CrossRef]
88. Tayebjee, M.H.; Nadar, S.; Blann, A.D.; Beevers, D.G.; MacFadyen, R.J.; Lip, G.Y. Matrix metalloproteinase-9 and tissue inhibitor of metalloproteinase-1 in hypertension and their relationship to cardiovascular risk and treatment: A substudy of the Anglo-Scandinavian Cardiac Outcomes Trial (ASCOT). *Am. J. Hicsypertens.* **2004**, *17*, 764–769. [CrossRef]
89. Leppert, D.; Leib, S.; Grygar, C.; Miller, K.; Schaad, U.; Holländer, G. Matrix metalloproteinase (MMP)-8 and MMP-9 in cerebrospinal fluid during bacterial meningitis: Association with blood-brain barrier damage and neurological sequelae. *Clin. Infect. Dis.* **2000**, *31*, 80–84. [CrossRef]
90. Green, R.C.; Berg, J.S.; Grody, W.W.; Kalia, S.S.; Korf, B.R.; Martin, C.L.; McGuire, A.L.; Nussbaum, R.L.; O-Daniel, J.M.; Ormond, K.E.; et al. ACMG recommendations for reporting of incidental findings in clinical exome and genome sequencing. *Genetics in Medicine.* **2013**, *15*, 565–574. [CrossRef]
91. Green, J.A.; Dholakia, S.; Janczar, K.; Ong, C.W.; Moores, R.; Fry, J.; Elkington, P.T.; Roncaroli, F.; Friedland, J.S. Mycobacterium tuberculosis-infected human monocytes down-regulate microglial MMP-2 secretion in CNS tuberculosis via TNFα, NFκB, p38 and caspase 8 dependent pathways. *J. Neuroinflam.* **2011**, *8*, 46. [CrossRef]
92. Gurney, K.J.; Estrada, E.Y.; Rosenberg, G.A. Blood–brain barrier disruption by stromelysin-1 facilitates neutrophil infiltration in neuroinflammation. *Neurobiol. Dis.* **2006**, *23*, 87–96. [CrossRef]
93. Asahi, M.; Wang, X.; Mori, T.; Sumii, T.; Jung, J.C.; Moskowitz, M.A.; Fin, M.E.; Lo, E.H. Effects of matrix metalloproteinase-9 gene knock-out on the proteolysis of blood-brain barrier and white matter components after cerebral ischemia. *J. Neurosci.* **2001**, *21*, 7724–7732. [CrossRef]
94. Price, N.M.; Gilman, R.H.; Uddin, J.; Recavarren, S.; Friedland, J.S. Unopposed matrix metalloproteinase-9 expression in human tuberculous granuloma and the role of TNF-alpha-dependent monocyte networks. *J. Immunol.* **2003**, *171*, 5579–5586. [CrossRef]

95. Coussens, L.M.; Fingleton, B.; Matrisian, L.M. Matrix metalloproteinase inhibitors and cancer: Trials and tribulations. *Science* **2002**, *295*, 2387–2392. [CrossRef]
96. Davis, A.G.; Rohlwink, U.K.; Proust, A.; Figaji, A.A.; Wilkinson, R.J. The pathogenesis of tuberculous meningitis. *J. Leukoc. Biol.* **2019**, *105*, 267–280. [CrossRef]
97. Marais, S.; Meintjes, G.; Pepper, D.J.; Dodd, L.E.; Schutz, C.; Ismail, Z.; Wilkinson, K.A.; Wilkinson, R.J. Frequency, severity, and prediction of tuberculous meningitis immune reconstitution inflammatory syndrome. *Clin. Infect. Dis.* **2013**, *56*, 450–460. [CrossRef]
98. Lai, R.P.; Meintjes, G.; Wilkinson, R.J. HIV-1 tuberculosis-associated immune reconstitution inflammatory syndrome. *Semin. Immunopathol.* **2016**, *38*, 185–198. [CrossRef]
99. Marais, S.; Meintjes, G.; Lesosky, M.; Wilkinson, K.A.; Wilkinson, R.J. Interleukin-17 mediated differences in the pathogenesis of HIV-1-associated tuberculous and cryptococcal meningitis. *AIDS* **2016**, *30*, 395–404. [CrossRef]
100. Van Well, G.T.; Paes, B.F.; Terwee, C.B.; Springer, P.; Roord, J.J.; Donald, P.R.; van Furth, A.M.; Schoeman, J.F. Twenty years of pediatric tuberculous meningitis: A retrospective cohort study in the Western Cape of South Africa. *Pediatrics* **2009**, *123*, e1–e8. [CrossRef]
101. Van den Bos, F.; Terken, M.; Ypma, L.; Kimpen, J.L.; Nel, E.D.; Schaaf, H.S.; Schoeman, J.F.; Donald, P.R. Tuberculous meningitis and miliary tuberculosis in young children. *Trop. Med. Int. Health* **2004**, *9*, 309–313.
102. Murase, S.; Lantz, C.L.; Kim, E.; Gupta, N.; Higgins, R.; Stopfer, M.; Hoffman, D.A.; Quinlan, E.M. Matrix metalloproteinase-9 regulates neuronal circuit development and excitability. *Mol. Neurobiol.* **2016**, *53*, 3477–3493. [CrossRef]
103. Gonthier, B.; Nasarre, C.; Roth, L.; Perraut, M.; Thomasset, N.; Roussel, G.; Aunis, D.; Bagnard, D. Functional interaction between matrix metalloproteinase-3 and semaphorin-3C during cortical axonal growth and guidance. *Cereb. Cortex* **2006**, *17*, 1712–1721. [CrossRef]
104. Van Hove, I.; Verslegers, M.; Buyens, T.; Delorme, N.; Lemmens, K.; Stroobants, S.; Gantois, I.; D'Hooge, R.; Moons, L. An aberrant cerebellar development in mice lacking matrix metalloproteinase-3. *Mol. Neurobiol.* **2012**, *45*, 17–29. [CrossRef]
105. Ordonez, A.A.; Maiga, M.; Gupta, S.; Weinstein, E.A.; Bishai, W.R.; Jain, S.K. Novel adjunctive therapies for the treatment of tuberculosis. *Curr. Mol. Med.* **2014**, *14*, 385–395. [CrossRef]
106. Stek, C.; Allwood, B.; Walker, N.F.; Wilkinson, R.J.; Lynen, L.; Meintjes, G. The Immune Mechanisms of Lung Parenchymal Damage in Tuberculosis and the Role of Host-Directed Therapy. *Front. Microbiol.* **2018**, *9*, 2603. [CrossRef]
107. Tiberi, S.; du Plessis, N.; Walzl, G.; Vjecha, M.J.; Rao, M.; Ntoumi, F.; Mfinanga, S.; Kapata, N.; Mwaba, P.; McHugh, T.D.; et al. Tuberculosis: Progress and advances in development of new drugs, treatment regimens, and host-directed therapies. *Lancet Infect. Dis.* **2018**, *18*, e183–e198. [CrossRef]
108. Schutz, C.; Davis, A.G.; Sossen, B.; Lai, R.P.; Ntsekhe, M.; Harley, Y.X.; Wilkinson, R.J. Corticosteroids as an adjunct to tuberculosis therapy. *Expert Rev. Respir. Med.* **2018**, *12*, 881–891. [CrossRef]
109. Hanemaaijer, R.; Visser, H.; Koolwijk, P.; Sorsa, T.; Salo, T.; Golub, L.M.; van Hinsbergh, V.W. Inhibition of MMP synthesis by doxycycline and chemically modified tetracyclines (CMTs) in human endothelial cells. *Adv. Dent. Res.* **1998**, *12*, 114–118. [CrossRef]
110. Henehan, M.; Montuno, M.; De Benedetto, A. Doxycycline as an anti-inflammatory agent: Updates in dermatology. *J. Eur. Acad. Dermatol. Venereol.* **2017**, *31*, 1800–1808. [CrossRef]
111. Subbian, S.; Tsenova, L.; O'Brien, P.; Yang, G.; Koo, M.S.; Peixoto, B.; Fallows, D.; Zeldis, J.B.; Muller, G.; Kaplan, G. Phosphodiesterase-4 inhibition combined with isoniazid treatment of rabbits with pulmonary tuberculosis reduces macrophage activation and lung pathology. *Am. J. Pathol.* **2011**, *179*, 289–301. [CrossRef]
112. Rothenberg, M.L.; Nelson, A.R.; Hande, K.R. New Drugs on the Horizon: Matrix Metalloproteinase Inhibitors. *Oncologist* **1998**, *3*, 271–274. [CrossRef]
113. Hernandez-Pando, R.; Orozco, H.; Arriaga, K.; Pavön, L.; Rook, G. Treatment with BB-94, a broad spectrum inhibitor of zinc-dependent metalloproteinases, causes deviation of the cytokine profile towards Type-2 in experimental pulmonary tuberculosis in Balb/c mice. *Int. J. Exp. Pathol.* **2000**, *81*, 199–209. [CrossRef]
114. Izzo, A.; Izzo, L.; Kasimos, J.; Majka, S. A matrix metalloproteinase inhibitor promotes granuloma formation during the early phase of Mycobacterium tuberculosis pulmonary infection. *Tuberculosis* **2004**, *84*, 387–396. [CrossRef] [PubMed]

115. Taylor, J.L.; Hattle, J.M.; Dreitz, S.A.; Troudt, J.M.; Izzo, L.S.; Basaraba, R.J.; Orme, I.M.; Matrisian, L.M.; Izzo, A.A. Role for matrix metalloproteinase 9 in granuloma formation during pulmonary Mycobacterium tuberculosis infection. *Infect. Immun.* **2006**, *74*, 6135–6144. [CrossRef] [PubMed]
116. Urbanowski, M.E.; Ihms, E.A.; Bigelow, K.; Kübler, A.; Elkington, P.T.; Bishai, W.R. Repetitive aerosol exposure promotes cavitary tuberculosis and enables screening for targeted inhibitors of extensive lung destruction. *J. Infect. Dis.* **2018**, *218*, 53–63. [CrossRef] [PubMed]
117. Ordonez, A.A.; Pokkali, S.; Kim, S.; Carr, B.; Klunk, M.H.; Tong, L.; Saini, V.; Chang, Y.S.; McKevitt, M.; Smith, V.; et al. Adjunct antibody administration with standard treatment reduces relapse rates in a murine tuberculosis model of necrotic granulomas. *PLoS ONE* **2018**, *13*, e0197474. [CrossRef] [PubMed]
118. Xu, Y.; Wang, L.; Zimmerman, M.D.; Chen, K.Y.; Huang, L.; Fu, D.J.; Kaya, F.; Rakhilin, N.; Nazarova, E.V.; Bu, P.; et al. Matrix metalloproteinase inhibitors enhance the efficacy of frontline drugs against Mycobacterium tuberculosis. *PLoS Pathogens* **2018**, *14*, e1006974. [CrossRef] [PubMed]
119. Ordonez, A.A.; Pokkali, S.; Sanchez-Bautista, J.; Klunk, M.H.; Urbanowski, M.E.; Kubler, A.; Bishai, W.R.; Elkington, P.T.; Jain, S.K. Matrix Metalloproteinase Inhibition in a Murine Model of Cavitary Tuberculosis Paradoxically Worsens Pathology. *J. Infect. Dis.* **2018**, *219*, 633–636. [CrossRef]
120. Paul, R.; Lorenzl, S.; Koedel, U.; Sporer, B.; Vogel, U.; Frosch, M.; Pfister, H.W. Matrix metalloproteinases contribute to the blood—brain barrier disruption during bacterial meningitis. *Ann. Neurol.* **1998**, *44*, 592–600. [CrossRef]
121. Majeed, S.; Radotra, B.D.; Sharma, S. Adjunctive role of MMP-9 inhibition along with conventional anti-tubercular drugs against experimental tuberculous meningitis. *Int. J.Exp. Pathol.* **2016**, *97*, 230–237. [CrossRef]

International Journal of
Molecular Sciences

MDPI

Review

MT4-MMP: The GPI-Anchored Membrane-Type Matrix Metalloprotease with Multiple Functions in Diseases

Cassandre Yip, Pierre Foidart, Agnès Noël and Nor Eddine Sounni *

Laboratory of Tumor and Development Biology, Groupe Interdisciplinaire de Génoprotéomique Appliqué-Cancer (GIGA-Cancer), University of Liège, Liège 4000, Belgium; cassandre.yip@uliege.be (C.Y.); pierre.foidart@chuliege.be (P.F.); agnes.noel@uliege.be (A.N.)

* Correspondence: nesounni@uliege.be; Tel.: +3243662570

Received: 24 December 2018; Accepted: 12 January 2019; Published: 16 January 2019

Abstract: MT4-MMP (or MMP17) belongs to the Membrane-Type Matrix Metalloproteinase (MT-MMP) family. This family of proteases contributes to extracellular matrix remodeling during several physiological processes, including embryogenesis, organogenesis, tissue regeneration, angiogenesis, wound healing, and inflammation. MT4-MMP (MMP17) presents unique characteristics compared to other members of the family in terms of sequence homology, substrate specificity, and internalization mode, suggesting distinct physiological and pathological functions. While the physiological functions of MT4-MMP are poorly understood, it has been involved in different pathological processes such as arthritis, cardiovascular disease, and cancer progression. The *mt4-mmp* transcript has been detected in a large diversity of cancers. The contribution of MT4-MMP to tumor development has been further investigated in gastric cancer, colon cancer, head and neck cancer, and more deeply in breast cancer. Given its contribution to different pathologies, particularly cancers, MT4-MMP represents an interesting therapeutic target. In this review, we examine its biological and structural properties, and we propose an overview of its physiological and pathological functions.

Keywords: MT4-MMP; cancer; diseases

1. Introduction

The integrity of interstitial compartments is crucial for tissue homeostasis. The perturbation of the extracellular matrix and its related components destabilizes this balance, leading to pathogenesis. Matrix Metalloproteinases (MMPs) are the main remodeling enzymes of the extracellular matrix. This protease family counts more than 20 members, and most of them are secreted in the extracellular microenvironment. The membrane-type MMP (MT-MMPs) are associated to the membrane by a transmembrane domain, an amino-terminal link, or a glycosylphosphatidylinositol anchor (GPI). Together, secreted or attached to the membrane, MMPs can directly cleave almost all components of the extracellular matrix (ECM). However, the GPI anchor confers to MMPs a unique location in the lipid raft, giving access to a specific set of substrates. Only two MT-MMPs display this anchor: MT4-MMP (MMP-17) and MT6-MMP (MMP-25). In this review, we focus on MT4-MMP. Discovered more than 20 years ago, this protease aroused interest only a decade ago [1–4]. MT4-MMP exhibits unique characteristics, which distinguishes it from other MMPs. Unlike the others, it is unable to activate pro-MMP2 and cleaves only a few ECM components [5,6]. Its sensitivity to tissue inhibitors of metalloproteinases (TIMPs) is also different, with MT4-MMP being more sensitive to TIMP1 than TIMP2 [6,7]. These differences are probably due to the least degree of sequence identity [1]. MT4-MMP has been involved in inflammation and angiogenesis, contributing to associated pathologies such as osteoarthritis and atherosclerosis, as well as thoracic aortic aneurysms and dissection (TAAD) [2,8,9].

Interestingly, inflammation and angiogenesis are two pillars of tumor development. The *mt4-mmp* transcript has been detected in prostate carcinomas, oral carcinomas, osteosarcomas, gastrointestinal adenocarcinomas, embryonal carcinomas, leukemias, lung carcinomas, glioblastomas, cervical carcinomas, melanomas, adrenal adenocarcinomas, and thyroid carcinomas [3,10–12]. MT4-MMP was first described in breast cancers [1], in which it has been more widely investigated compared to the other cancers. The pro-angiogenic and pro-metastatic functions of MT4-MMP have been highlighted in breast cancer [4,13]. MT4-MMP-mediated metastatic dissemination has been also pointed out in colon cancer and head and neck cancer [3,11]. All these data propel MT4-MMP onto the stage of future potential therapeutic treatments.

2. Characteristics of the MMP family

The MMPs are endopeptidases characterized by the presence of a zinc ion in the catalytic domain. Twenty-four members have been identified and are separated into two different groups: The soluble MMPs (MMP-1, -2, -3, -7, -8, -9, -10, -11, -12, -13, -19, -20, -21, -22, -27, and -28) and the MMPs linked to the membrane by a transmembrane domain (MMP-14, -15, -16, and -24), a glycosylphosphatidylinositol (GPI) anchor (MMP-17 and -25), or an amino-terminal signal peptide (MMP-23A and -23B). The groups are shown in Figure 1.

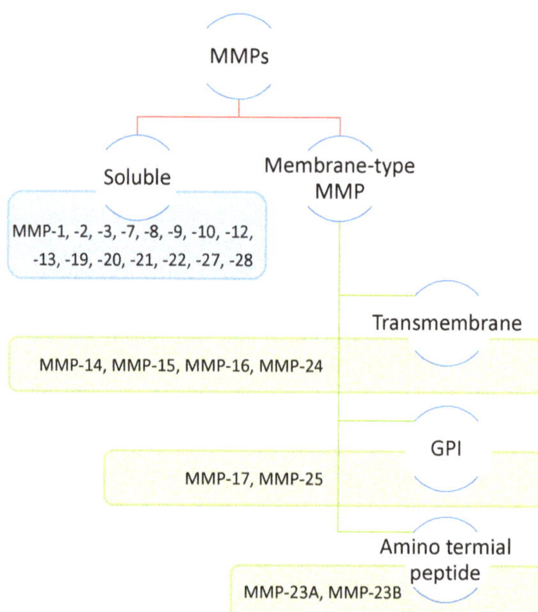

Figure 1. Classification of different Matrix Metalloproteinases (MMPs) according to their structure. Matrix Metalloproteases are either soluble (MMPs) or membrane-tethered (MT-MMPs). MMP14 (MT1-MMP), MMP15 (MT2-MMP), MMP16 (MT3-MMP), and MMP24 (MT5-MMP) are attached to the cell membrane by a transmembrane domain. MMP17 (MT4-MMP) and MMP25 (MT6-MMP) are linked to the cell membrane by a glycosylphosphatidylinositol anchor (GPI).

The MMPs share common structures including: (1) The pre-domain, an N-terminal sequence driving the MMP to the endoplasmic reticulum (ER); (2) the pro-domain, keeping enzymes in an inactive form; and (3) the catalytic domain, implicated in the recognition and cleavage of substrates. These are shown in Figure 2.

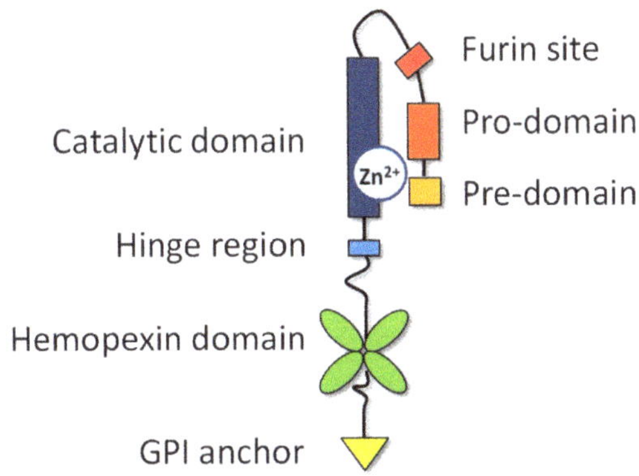

Figure 2. Structural domains of MT4-MMP, including the pre-domain or signal peptide (amino acids 1 to 41), the pro-domain (42–128), the catalytic domain with a zinc ion (129–297), a linker (298–333) containing the furin site (R–X–K/R–R), the hemopexin domain (334–535), and the glycosylphosphatidylinositol (GPI) anchored to the membrane (572–605).

The catalytic domain is characterized by a consensus sequence "HEXXHXXGXXH", which allows the linking of a zinc ion. The presence of a zinc ion facilitates the binding of H2O molecules, thus providing the hydrolytic reactions of peptides and substrates [14]. Except for MMP-7, -26, and -23, all MMP family members display an hemopexin domain known to play a role in substrate recognition, proteolytic activity, and inhibitor binding. The GPI-anchored MT4-MMP displays unique features as compared to other MT-MMP members [15]. First, MT4-MMP is distantly related in its amino acid sequence to the other members. The catalytic domain displays less than 40% sequence identity, while the sequence identity is more than 65% among the other MMP members [1]. Second, MT4-MMP is unable to process pro-MMP2 into its active form, in contrast with MT1-, MT2-, MT3-, and MT5-MMP [5,6,16]. The pro-MMP2-activating MT-MMPs contain eight amino acids located in the catalytic domain, the so-called "MT-loop", which are lacking in MT4-MMP [17]. It has been reported that the pro-MMP2 activation is impaired when the MT-loop of MT1-MMP is deleted or inhibited by neutralizing antibodies [18]. These results are consistent with the capacity of the MT-Loop of MT1-MMP to interact with the fibronectin-like domain of pro-MMP2. Furthermore, a mutation in the MT-Loop of MT1-MMP impairs pro-MMP2 activation [19]. Thirdly, unlike other MMPs, MT4-MMP has a small repertoire of substrates among the ECM, with the exception of weak hydrolyzing capacities against fibrinogen, fibrin, and gelatin [5,6]. However, MT4-MMP is efficient in the cleavage of proTNF, ADAMTS4, α-2-macroglobulin, low density lipoprotein receptor-related protein, osteopontin, and cartilage oligomeric matrix protein, as detailed in Table 1 [2,5,6,8,20,21]. Interestingly, a soluble MT4-MMP form has been abundantly detected in the media of cells overexpressing active MT4-MMP, but not in the media of cells overexpressing its inactive form [22]. These data suggest that MT4-MMP is able to process itself by an active autocleavage. Finally, MT4-MMP is most potently inhibited by TIMP1 rather than TIMP2 or TIMP3, unlike the pro-MMP2-activating MT-MMPs that are efficiently inhibited by TIMP-2, -3, and -4, but not by TIMP-1 [6,7].

Table 1. Overview of identified substrates of MT4-MMP.

Substrates	References
ECM substrates:	
Gelatin	Wang et al., 1999 [5]
Fibrin	English et al., 2000 [6]
Fibrinogen	English et al., 2000 [6]
Other substrates:	
proTNF	Wang et al., 1999 [5]
	English et al., 2000 [6]
COMP	English et al., 2000 [6]
α-2-macroglobulin	English et al., 2000 [6]
LRP1	Rozanov et al., 2004 [20]
ADAMTS4	Gao et al., 2004 [2]
	Patwari et al., 2005 [23]
	Clements et al., 2011 [24]
MT4-MMP	Host et al., 2012 [22]
Osteopontin	Martin-Alonso et al., 2015 [8]
	Papke et al., 2015 [25]
Thrombospondin 4	Martin-Alonso et al., 2015 [8]
αM integrin	Clemente et al., 2018 [9]

3. Biosynthesis and Trafficking

MMPs are synthetized as inactive zymogens. The inactive form is maintained by the interaction between the cysteine sulfhydryl group in the pro-domain and the zinc ion bound to the catalytic domain. The activation of MMPs requires a proteolytic cleavage of their pro-domain [26]. The MT4-MMP contains a furin consensus sequence (R–X–K/R–R) and can be activated by furin [7,27]. The precursor form (69 kDa) is found in the ER and in the Golgi compartment, while the processed form (58 kDa) is present at the membrane [12]. The biosynthesis of GPI proteins such as MT4-MMP follows a unique pathway that starts in the ER and finishes in the Golgi [28]. The mature GPI protein is finally transported to the cytoplasmic membrane by endoplasmic vesicles. In the lumen of the ER, the nascent protein is attached to a preformed GPI present in the inner membrane of the ER by a GPI transamidase. The preformed GPI is then modified following different steps to reach a mature form. The acyl chain is removed by PGAP1 (Post-GPI attachment to proteins 1) and the ethanolamine phosphate chain is released by PGAP5. After arrival in the Golgi, the unsaturated acid is replaced by a saturated fatty acid under the actions of PGAP3 and PGAP2, as illustrated in Figure 3 [28]. The fatty acid remodeling is essential for the incorporation of GPI-anchored proteins into lipid rafts, conferring to them the ability to interact with and degrade specific components of the raft environment. During this process, MT4-MMP is also N-glycosylated at the Asn318 site [12].

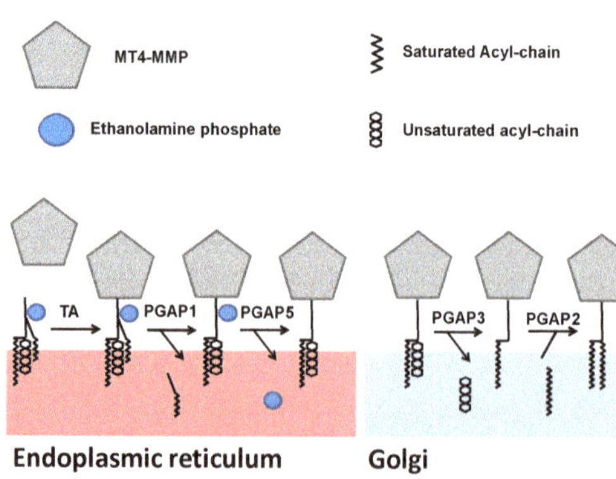

Figure 3. The biosynthesis of the glycosylphosphatidylinositol anchor on the MT4-MMP.

While MT4-MMP has been shown to colocalize with caveolin-1, a major structural protein associated with lipid rafts in mammalian cells, MT4-MMP internalization is independent of the caveolin-1 pathway [29]. Indeed, its internalization is not disturbed by filipin III, a caveolae pathway inhibitor. Similarly, treatment with chlorpromazine, an inhibitor of the clathrin pathway, does not block MT4-MMP internalization. Moreover, no colocalization of MT4-MMP and clathrin has been observed. MT4-MMP internalization involves a clathrin-independent carriers/GPI-enriched early endosomal compartments (CLIC/GEEC) pathway commonly used for GPI-anchored protein endocytosis, which is shown in Figure 4 [29]. The silencing of CDC42, Rac1, and RhoA, three regulators of the CLIC/GEEC endocytic pathway, disturbs MT4-MMP internalization. This implication of the CLIC/GEEC pathway in MT4-MMP internalization is up till today unique in the MMP family, further underlying the specific features of this membrane-associated enzyme. MT4-MMP present at the cell surface is internalized in early endosomes, and part of the enzyme is intracellularly autodegraded or recycled to the cell surface [29]. We and others have demonstrated the key role of the cysteine residue C564 in MT4-MMP dimerization, whereas the effect of its demonization on its activity is still unknown [29,30].

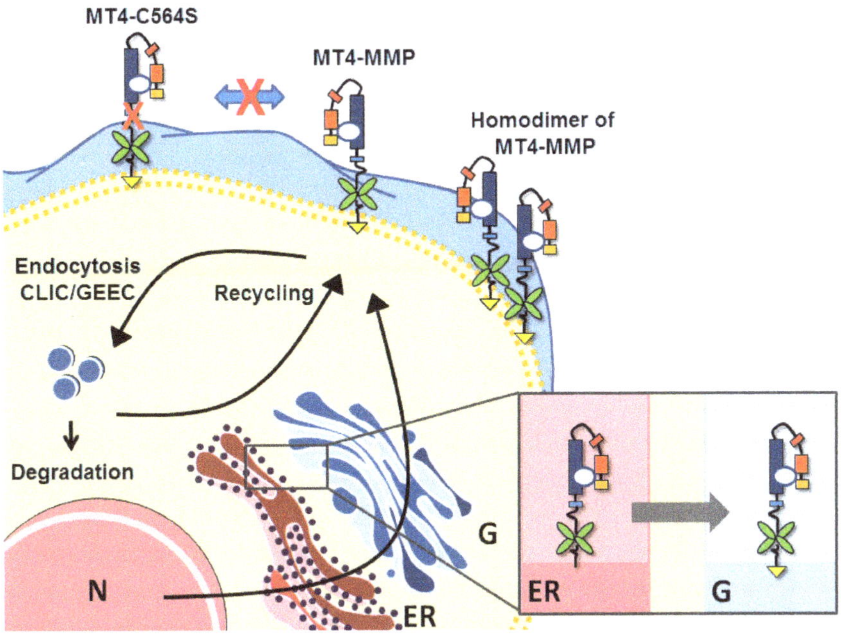

Figure 4. Biosynthesis and trafficking of MT4-MMP. MT4-MMP is associated to the glycosylphosphatidylinositol anchor in the endoplasmic reticulum (ER) and the Golgi (G) before being brought to the membrane. MT4-MMP can form homodimers, but the switch of cysteine with a serine prevents the dimerization. The protease is internalized by the clathrin-independent carriers/GPI-enriched early endosomal compartments pathway (CLIC/GEEC). In the cytoplasm, the protease is degraded or recycled at the membrane. N: Nucleus.

4. Physiological Expression and Functions of MT4-MMP

The spatiotemporal expression of *mt4-mmp* during murine embryonic development has been linked to its key role in angiogenesis, limb development, and brain formation [31]. The implication of MT4-MMP in the neuronal system is not limited only to the brain, but also involved in the migration of neural crest cells during embryogenesis of the zebrafish. Indeed, its inhibition by morpholino or by broad spectrum MMP inhibitors (Marimastat and ONO-4817) results in aberrant neural crest cell migration, with minimal changes in cell proliferation or apoptosis [32]. In transgenic mouse models expressing different types of retinal ganglion neurons with distinct directional migration, *mt4-mmp* transcripts have been identified in retinal ganglion cells with preferences for nasal motion, but not for ventral motion [33]. These data suggest a role of MT4-MMP in the orientation of the retinal ganglion neuron migration [34]. Overall, MT4-MMP-deficient mice grow normally and display a normal appearance, behaviour, life span, and fertility [21]. The unique observable effect in this model is a hypodipsia with a decreased daily urine output. While the kidney function is normal and MT4-MMP is expressed in the hypothalamus, which regulates thirst, it has been proposed that MT4-MMP may play a role in thirst regulation [35]. The expression of MT4-MMP is prominent in the brain, lungs, and uterus, and is transitional in the spleen, stomach, intestines, and reproductive organs [1,21]. MT4-MMP is also detected in human eosinophils, lymphocytes, monocytes, and macrophages, suggesting a role of this protease in inflammation [1,6,9,21,36,37]. Its expression in eosinophils is increased upon stimulation with TNF-α [37]. While the expression of MT4-MMP has been linked to different physiological processes, the mechanisms underlying these processes are still unknown. In contrast, more investigations were dedicated to understanding the role of this protease in pathologies, and several mechanisms were unravelled and will be described in this review and summarized in Table 2. The mechanism of action of MT4-MMP is not only related to its proteolytic activity, but also to its non-proteolytic function. We previously demonstrated that MT4-MMP is a key precursor and partner of Epidermal Growth Factor Receptor (EGFR), and enhances its activation leading to cancer cell proliferation in a non-proteolytic activity [38]. Future studies aiming at targeting these two functions will further define its relevance in cancer and diseases.

Table 2. Proteolytic and non-proteolytic functions of MT4-MMP in different pathologies.

Diseases	Proteolytic Functions	Non-Proteolytic Functions	Unknown
Breast cancer	Angiogenesis [1,2], Metastatic dissemination [1,2]	Cell proliferation [3]	
Colon cancer			Metastatic dissemination [4]
Head and neck cancer			Metastatic dissemination [5], Hypoxia [5]
Osteoarthritis	Processing of ADAMTS4 p68 isoform [6,7,8]		
Thoracic aortic aneurysms and dissections	Processing of osteopontin [9,10]		
Atherosclerosis	Processing of αMintegrin [11]		

[1] Chabottaux et al., 2009 [4], [2] Host et al., 2012 [22], [3] Paye et al., 2014 [38], [4] Nimri et al., 2013 [11], [5] Huang et al., 2009 [3], [6] Gao et al., 2004 [2], [7] Patwari et al., 2005 [23], [8] Clements et al., 2011 [24], [9] Martin-Alonso et al., 2015 [8], [10] Papke et al., 2015 [25], [11] Clemente et al., 2018 [9].

5. Osteoarthritis

Articular cartilage function depends on ECM homeostasis that is dependent on a fine-tuned equilibrium between synthetic and degradative processes. Osteoarthritis (OA) is characterized by inflammation and a degradation of cartilage resulting from an increase of aggrecanase activity. Accordingly, MT4-MMP has been described with an aggrecanase activity, suggesting a role in cartilage homeostasis [2]. In OA, both *mt4-mmp* mRNA and the protein were found upregulated in this pathology [24,39]. *Mt4-mmp* transcripts are detected in human OA cartilage, but not in intact control cartilage [24]. Patwari et al. demonstrated that interleukin-1 (IL-1) considerably increases MT4-MMP expression in articular cartilage disk explants obtained from the femoropatellar groove of calves and in the conditioned medium [23]. Similar results were observed in C57BL6/Jax mice injected with IL-1 [24]. MT4-MMP is able to cleave the ADAMTS4 p68 isoform, which possesses relatively poor aggrecanase activity, to generate the ADAMTS4 p53 isoform, which is the highly active form [2]. In inflammatory conditions, the amount of total ADAMTS4 protein is not modulated. However, the ADAMTS4 p53 isoform is more abundant than the p68 isoform, reflecting an increaseof aggrecanase activity.

The generation of the p53 isoform is directly linked to MT4-MMP expression. Indeed, MT4-MMP null mice are protected from IL-1-induced cartilage aggrecanolysis [24]. All these data imply the role of MT4-MMP in the regulation of aggrecan processing under inflammatory contexts, and open new therapeutic perspectives in osteoarthritis treatment.

6. Thoracic Aortic Aneurysms and Dissections

The role of MT4-MMP in the regulation of vessel stability in cancer and vascular diseases is well established [4,22,40]. More recently, its role in the thoracic aortic aneurysms and dissections (TAAD) has been reported in an elegant study by Martin-Alonso and colleagues, who performed screening of 58 patients with inherited predispositions to TAAD and identified a missense mutation R373H in the *mt4-mmp* gene that prevents the expression of the protease [8]. In a genetic mouse models of MT4-MMP loss of function, dilated aortas are observed, with dysfunctional vascular smooth muscle cells (VSMCs) and ECM changes [8,21]. Moreover, these mice are hypotensive and display an adventitial fibrosis as described in other mouse models of TAAD [41]. Furthermore, MT4-MMP expression is detected in periaortic progenitors during embryogenesis, suggesting a role of MT4-MMP in the early construction of the aortic wall and VSMC maturation. While no TAAD are observed spontaneously in MT4-MMP-deficient mice, its incidence is increased in MT4-MMP-null mice compared to the wild type mice when vessel wall stress is induced by angiotensin-II treatment. In another model of vascular injury triggered by ligation of the carotid artery, MT4-MMP-null mice display an alteration of vascular remodelling, characterized by a neointima more prominent in the carotid arteries and increased VSMC proliferation. Among the vascular substrates of MT4-MMP, osteopontin has been identified. This protein is expressed in the vessel wall, including in the embryonic aorta, and is associated with VSMC migration and differentiation, and more interestingly with aortic aneurysm [42–45]. Parallel to the increase of osteopontin fragments mediated by MT4-MMP cleavage in the aorta, an increase of c-Jun N-terminal kinase (JNK) phosphorylation is observed, highlighting a signalling pathway activated by MT4-MMP during VSMC maturation. Finally, the restoration of MT4-MMP expression by lentivirus partially rescues the vessel-wall phenotype, providing the way for new therapeutic strategies [8,25]. Atherosclerosis can be associated with TAAD [46,47]. Interestingly, the absence of MT4-MMP expression in mice is correlated with the recruitment of patrolling monocytes and lipid deposits in atherosclerotic plaques. Indeed, MT4-MMP cleaves αM integrin at the surface of crawling monocytes, inducing their detachment and the loss of their function in the regulation of atherosclerosis [9].

7. Cancer

7.1. Gastric Cancer

Wang et al. investigated the expression of *mt4-mmp* transcripts and proteins in 42 cases of gastric cancer and normal tissues, and 40 cases of atrophic gastritis [48]. Interestingly, no difference in MT4-MMP expression is observed between normal tissues and atrophic gastritis cases. However, its expression is higher in gastric cancer patients than in normal and atrophic gastritis tissues. This study highlighted an association of MT4-MMP expression with the depth of tumor invasion, lymph node metastasis and serosal involvement of gastric cancer patients [48]. In an experimental model of gastric cancer cell lines exposed to oxidative stress in vitro, the expression of MT4-MMP was increased among other MMPs and β-catenin, suggesting a potential role of this enzyme in the pathogenesis of gastric cancer, dependent on the continuous exposure of the mucosa to oxidative stress [49].

7.2. Colon Cancer

MT4-MMP is expressed in lipid rafts of highly metastatic colon cancer cell line HM-7, but not in the parental lower metastatic cell line LS174T, suggesting a role of MT4-MMP in the metastatic dissemination of colon cancer [11]. Inversely, caveolin-1 is not expressed in metastatic HM-7 cells and weakly expressed in the cytosolic fraction of parental LS174T cells. Interestingly, the restoration of caveolin-1 expression in metastatic HM-7 cells inhibits MT4-MMP expression in the lipid rafts, suppressing the metastatic phenotype of colon cancer cells. While the role of caveolin-1 in MT4-MMP trafficking has been excluded [29], the impact of caveolin-1 on MT4-MMP expression could be explained by other mechanisms, including the regulation of *mt4-mmp* transcription or translation, or the release of MT4-MMP from the membrane by proteases or phospholipases, which all can be regulated by caveolin-1 activity [11].

7.3. Head and Neck Cancer

Huang et al. described an unprecedented link between hypoxia and the regulation of MT4-MMP expression in head and neck cancer [3]. The experiments on hypopharyngeal squamous cell carcinoma (FADU) and tongue squamous cell carcinoma (SAS) showed an increase of *mt4-mmp* transcripts and proteins in hypoxic conditions or under constitutive expression of HIF-1. In reverse, HIF-1 silencing decreases MT4-MMP expression. Among the key regulators of hypoxia, SLUG has been identified as the major factor responsible for hypoxia and HIF-1α-induced MT4-MMP expression. Indeed, SLUG activates the transcription of *mt4-mmp* through direct interaction with the E-box located in the promoter of the *mt4-mmp* gene. Interestingly, MT4-MMP silencing reduces HIF-1α- or SLUG-induced invasiveness and lung dissemination of cancer cells. In a retrospective clinical study, MT4-MMP and HIF-1α colocalized in 20 out of 68 patients, and the colocalization of the both proteins was associated with poor overall survival [3].

7.4. Breast Cancer

Chabottaux et al. investigated the expression of MT4-MMP in 21 samples of healthy breast tissue and 63 breast adenocarcinomas [13]. Immunohistochemistry staining revealed a higher expression of MT4-MMP in breast adenocarcinomas, while healthy breast tissues presented a negative to moderate positivity [13]. Interestingly, the overexpression of MT4-MMP in MDA-MB-231 cells (triple-negative breast cancer cell line) promotes cell proliferation in the 3D matrix in vitro and in subcutaneous xenografts [13,38]. Moreover, MT4-MMP expression induces lung metastases by destabilizing blood vasculature, characterized by an enlargement of blood vessel lumens and pericyte detachment [4]. While no differences in the production of key angiogenic modulators (VEGF, PDGFR, FGF, and their receptors) were detected, expression of human thrombospondin-2 (TSP-2) is decreased in MT4-MMP xenografts. This result fits with the decrease of this anti-angiogenic factor that has been associated with impaired vascular integrity and permeability in mouse models [50]. Moreover, genetic deletion of TSP-2 in knock-out mice favors tumor growth, metastasis, and angiogenesis [51,52]. Interestingly, Salz et al. found that the transcription of different MMPs, including *mt4-mmp*, is regulated by hSETD1A, a methyltransferase overexpressed in metastatic human breast cancer cell lines and patients [53]. The silencing of hSETD1A decreases H3K4 methylation in promoters, decreasing MMP transcription and leading to decreased cell migration and invasion in vitro, and a reduction of lung metastases in mice [53]. Host et al. provided evidence for the requirement of proteolytic activity for MT4-MMP-mediated proangiogenic and prometastatic effects [22]. In sharp contrast, MT4-MMP has been reported to exert mitogenic effects on triple-negative breast cancer cells that are independent of its proteolytic activity. Indeed, MT4-MMP stimulates cell proliferation by interacting with EGFR and enhancing its activation in response to its ligands, Epidermal Growth Factor (EGF) and Tumor Growth Factor (TGF) [38]. Moreover, superimposition of EGFR and MT4-MMP has been observed in human triple-negative breast cancer (TNBC) patients [54]. Interestingly, MT4-MMP expression has been recently identified as a biomarker for TNBC patient responses to chemotherapy and to a combination of anti-EGFR drugs such as Erlotinib and Palbociclib, an inhibitor of Cyclin-dependent kinases 4 and 6, which are involved in the cell cycle [54,55]. These recent data highlight the clinical relevance of using the MT4-MMP/EGFR axis to select patients who could benefit from specific combinations of targeted therapies.

8. Clinical Inhibitors of MMPs

Almost thirty years ago, MMPs were viewed as interesting targets for therapeutic compounds. Rationally, the TIMPs, endogenous inhibitors of MMPs, have been considered as potential treatments. Unfortunately, the development of clinical drugs has been aborted due to technical difficulties with the production and use of these proteins [56]. Also, studies on TIMPs have revealed dual functions in inhibiting or promoting cancer progression through several mechanisms involving intracellular signalling that are independent from their inhibitory activity of MMPs [57–59]. Other natural inhibitors have been taken into consideration, such as Neovastat, a molecule extracted from shark cartilage, which prevents angiogenesis and metastases. These anti-tumor effects are not only due to the inhibition of MMPs, but also the inhibition of VEGF [60]. Another natural compound is Genistein, a soy isoflavonoïd that blocks tumor growth and invasions by altering, among other things, the expression and the activity of MMPs and TIMPs [61,62]. Rapidly, synthetic inhibitors of MMPs have been produced and tested in different clinical trials [63]. However, those clinical trials using broad-spectrum MMP inhibitors were disappointing despite the promising preclinical studies. The clinical benefit was not convincing, and the secondary effects were intolerable [64]. The new generation of MMP inhibitors are designed to be more selective to decrease the secondary effects [65]. Monoclonal antibodies specifically directed against the catalytic domains of MT1-MMP or MMP-9 were promising in vitro and in vivo [66,67]. To date, only monoclonal anti-MMP-9 (GS-5745; Gilead Science) is being investigated in clinical trials. More recently, an antibody directed towards active MMP-13 has been produced and characterized [68]. Interestingly, MT4-MMP is emerging as a marker of interest to select patients who could benefit from specific combinations of existing treatments [55]. These findings hold new promise in the field of MMPs.

9. Discussion and Perspectives

The specific roles of MMPs and their inhibitors in cancer progression are widely recognized [69]. Although the family of human MMPs is composed of 25 enzymes, only five MMPs, including four soluble forms (MMPs-1, -2, -9, and -13) and a membrane form (MT1-MMP), were intensively studied. This interest in these MMPs is linked to their overexpression identified on the basis of the genomic and transcriptomic data collected on different types of human cancers [70]. Because of its membrane localization and its pericellular proteolytic action, MT1-MMP has been the subject of intensive studies that have elucidated its primordial role in the migration of cancer cells [71,72]. The search for scientific publications in the national center for biotechnology information (NCBI) database by combining the terms "MT1-MMP" and "cancer" reveals 340 articles from 1998 to the present. On the other hand, only 60 scientific articles are proposed for MT4-MMP in cancer. The low interest for this protease is due to the fact that the overexpression of MT4-MMP was not detected in transcriptomic studies. However, analyses by immunohistochemistry showed that the protein is very abundant in the tumor compartment of human breast cancers [13,38,54]. Several hypotheses can be envisaged, such as: (1) The existence of several transcripts resulting from alternative splicing that would not be detected in the transcriptomic analyses, (2) post-transcriptional regulation via miRNAs, (3) translation control, and/or (4) epigenetic mRNA changes [73]. Since 2006, special attention has been brought to the study of MT4-MMP, which has been identified in experimental models of mammary cancers as a driver of metastasis [4,13]. In addition, this protease has unique biochemical characteristics that distinguish it from other MMPs, suggesting different functions [15]. First, MT4-MMP is anchored to the membrane by a GPI anchor, a feature only shared with MT6-MMP, and second, unlike most MT-MMPs, MT4-MMP is inefficient in the activation of pro-MMP2 and hydrolyses very few ECM components [5,6]. Moreover, its sensitivity to TIMPs is different from that of other MMPs [6,7]. One of the plausible explanations for this specificity of MT4-MMP is the low-percentage sequence homology of its catalytic domain compared to other MMPs [1]. All these data suggest that MT4-MMP presents unique functions and different substrates, promoting different pathologies, particularly in tumor progression. All these data support the development of blocking molecules to counter the effects of MT4-MMP. The pathological functions of this protease depend mainly on its proteolytic activity. Interestingly, the catalytic domain of MT4-MMP possesses a distinct sequence to the other MMPs, allowing the development of specific inhibitors as function-blocking antibodies.

Funding: This research received no external funding.

Conflicts of Interest: The authors declare no conflict of interest.

References

1. Puente, X.S.; Pendas, A.M.; Llano, E.; Velasco, G.; Lopez-Otin, C. Molecular cloning of a novel membrane-type matrix metalloproteinase from a human breast carcinoma. *Cancer Res.* **1996**, *56*, 944–949. [PubMed]
2. Gao, G.; Plaas, A.; Thompson, V.P.; Jin, S.; Zuo, F.; Sandy, J.D. ADAMTS4 (aggrecanase-1) activation on the cell surface involves C-terminal cleavage by glycosylphosphatidyl inositol-anchored membrane type 4-matrix metalloproteinase and binding of the activated proteinase to chondroitin sulfate and heparan sulfate on syndecan-1. *J. Biol. Chem.* **2004**, *279*, 10042–10051. [PubMed]
3. Huang, C.H.; Yang, W.H.; Chang, S.Y.; Tai, S.K.; Tzeng, C.H.; Kao, J.Y.; Wu, K.J.; Yang, M.H. Regulation of membrane-type 4 matrix metalloproteinase by SLUG contributes to hypoxia-mediated metastasis. *Neoplasia* **2009**, *11*, 1371–1382. [CrossRef]
4. Chabottaux, V.; Ricaud, S.; Host, L.; Blacher, S.; Paye, A.; Thiry, M.; Garofalakis, A.; Pestourie, C.; Gombert, K.; Bruyere, F.; et al. Membrane-type 4 matrix metalloproteinase (MT4-MMP) induces lung metastasis by alteration of primary breast tumour vascular architecture. *J. Cell. Mol. Med.* **2009**, *13*, 4002–4013. [CrossRef] [PubMed]
5. Wang, Y.; Johnson, A.R.; Ye, Q.Z.; Dyer, R.D. Catalytic activities and substrate specificity of the human membrane type 4 matrix metalloproteinase catalytic domain. *J. Biol. Chem.* **1999**, *274*, 33043–33049. [CrossRef]
6. English, W.R.; Puente, X.S.; Freije, J.M.; Knauper, V.; Amour, A.; Merryweather, A.; Lopez-Otin, C.; Murphy, G. Membrane type 4 matrix metalloproteinase (MMP17) has tumor necrosis factor-alpha convertase activity but does not activate pro-MMP2. *J. Biol. Chem.* **2000**, *275*, 14046–14055. [CrossRef]
7. Sohail, A.; Sun, Q.; Zhao, H.; Bernardo, M.M.; Cho, J.A.; Fridman, R. MT4-(MMP17) and MT6-MMP (MMP25), A unique set of membrane-anchored matrix metalloproteinases: Properties and expression in cancer. *Cancer Metast. Rev.* **2008**, *27*, 289–302. [CrossRef]
8. Martin-Alonso, M.; Garcia-Redondo, A.B.; Guo, D.; Camafeita, E.; Martinez, F.; Alfranca, A.; Mendez-Barbero, N.; Pollan, A.; Sanchez-Camacho, C.; Denhardt, D.T.; et al. Deficiency of MMP17/MT4-MMP proteolytic activity predisposes to aortic aneurysm in mice. *Circ. Res.* **2015**, *117*, e13–e26. [CrossRef]
9. Clemente, C.; Rius, C.; Alonso-Herranz, L.; Martin-Alonso, M.; Pollan, A.; Camafeita, E.; Martinez, F.; Mota, R.A.; Nunez, V.; Rodriguez, C.; et al. MT4-MMP deficiency increases patrolling monocyte recruitment to early lesions and accelerates atherosclerosis. *Nat. Commun.* **2018**, *9*, 910. [CrossRef]
10. Grant, G.M.; Giambernardi, T.A.; Grant, A.M.; Klebe, R.J. Overview of expression of matrix metalloproteinases (MMP-17, MMP-18, and MMP-20) in cultured human cells. *Matrix Biol.* **1999**, *18*, 145–148. [CrossRef]
11. Nimri, L.; Barak, H.; Graeve, L.; Schwartz, B. Restoration of caveolin-1 expression suppresses growth, membrane-type-4 metalloproteinase expression and metastasis-associated activities in colon cancer cells. *Mol. Carcinog.* **2013**, *52*, 859–870. [CrossRef] [PubMed]
12. Hieronimus, B.; Pfohl, J.; Busch, C.; Graeve, L. Expression and Characterization of Membrane-Type 4 Matrix Metalloproteinase (MT4-MMP) and its Different Forms in Melanoma. *Cell. Physiol. Biochem.* **2017**, *42*, 198–210. [CrossRef] [PubMed]
13. Chabottaux, V.; Sounni, N.E.; Pennington, C.J.; English, W.R.; van den Brule, F.; Blacher, S.; Gilles, C.; Munaut, C.; Maquoi, E.; Lopez-Otin, C.; et al. Membrane-type 4 matrix metalloproteinase promotes breast cancer growth and metastases. *Cancer Res.* **2006**, *66*, 5165–5172. [CrossRef] [PubMed]
14. Manzetti, S.; McCulloch, D.R.; Herington, A.C.; van der Spoel, D. Modeling of enzyme-substrate complexes for the metalloproteases MMP-3, ADAM-9 and ADAM-10. *J. Comput. Aided Mol. Des.* **2003**, *17*, 551–565. [CrossRef] [PubMed]
15. Itoh, Y.; Kajita, M.; Kinoh, H.; Mori, H.; Okada, A.; Seiki, M. Membrane type 4 matrix metalloproteinase (MT4-MMP, MMP-17) is a glycosylphosphatidylinositol-anchored proteinase. *J. Biol. Chem.* **1999**, *274*, 34260–34266. [CrossRef] [PubMed]

16. Kolkenbrock, H.; Essers, L.; Ulbrich, N.; Will, H. Biochemical characterization of the catalytic domain of membrane-type 4 matrix metalloproteinase. *Biol. Chem.* **1999**, *380*, 1103–1108. [CrossRef]
17. Sounni, N.E.; Janssen, M.; Foidart, J.M.; Noel, A. Membrane type-1 matrix metalloproteinase and TIMP-2 in tumor angiogenesis. *Matrix Biol.* **2003**, *22*, 55–61. [CrossRef]
18. Woskowicz, A.M.; Weaver, S.A.; Shitomi, Y.; Ito, N.; Itoh, Y. MT-LOOP-dependent localization of membrane type I matrix metalloproteinase (MT1-MMP) to the cell adhesion complexes promotes cancer cell invasion. *J. Biol. Chem.* **2013**, *288*, 35126–35137. [CrossRef]
19. English, W.R.; Holtz, B.; Vogt, G.; Knauper, V.; Murphy, G. Characterization of the role of the "MT-loop": An eight-amino acid insertion specific to progelatinase A (MMP2) activating membrane-type matrix metalloproteinases. *J. Biol. Chem.* **2001**, *276*, 42018–42026. [CrossRef]
20. Rozanov, D.V.; Hahn-Dantona, E.; Strickland, D.K.; Strongin, A.Y. The low density lipoprotein receptor-related protein LRP is regulated by membrane type-1 matrix metalloproteinase (MT1-MMP) proteolysis in malignant cells. *J. Biol. Chem.* **2004**, *279*, 4260–4268. [CrossRef]
21. Rikimaru, A.; Komori, K.; Sakamoto, T.; Ichise, H.; Yoshida, N.; Yana, I.; Seiki, M. Establishment of an MT4-MMP-deficient mouse strain representing an efficient tracking system for MT4-MMP/MMP-17 expression in vivo using beta-galactosidase. *Genes Cells* **2007**, *12*, 1091–1100. [CrossRef] [PubMed]
22. Host, L.; Paye, A.; Detry, B.; Blacher, S.; Munaut, C.; Foidart, J.M.; Seiki, M.; Sounni, N.E.; Noel, A. The proteolytic activity of MT4-MMP is required for its pro-angiogenic and pro-metastatic promoting effects. *Int. J. Cancer* **2012**, *131*, 1537–1548. [CrossRef]
23. Patwari, P.; Gao, G.; Lee, J.H.; Grodzinsky, A.J.; Sandy, J.D. Analysis of ADAMTS4 and MT4-MMP indicates that both are involved in aggrecanolysis in interleukin-1-treated bovine cartilage. *Osteoarthritis Cartilage* **2005**, *13*, 269–277. [CrossRef] [PubMed]
24. Clements, K.M.; Flannelly, J.K.; Tart, J.; Brockbank, S.M.; Wardale, J.; Freeth, J.; Parker, A.E.; Newham, P. Matrix metalloproteinase 17 is necessary for cartilage aggrecan degradation in an inflammatory environment. *Ann. Rheum. Dis.* **2011**, *70*, 683–689. [CrossRef]
25. Papke, C.L.; Yamashiro, Y.; Yanagisawa, H. MMP17/MT4-MMP and thoracic aortic aneurysms: OPNing new potential for effective treatment. *Circ. Res.* **2015**, *117*, 109–112. [CrossRef] [PubMed]
26. Sternlicht, M.D.; Werb, Z. How matrix metalloproteinases regulate cell behavior. *Annu. Rev. Cell Dev. Biol.* **2001**, *17*, 463–516. [CrossRef] [PubMed]
27. Pei, D. Leukolysin/MMP25/MT6-MMP: A novel matrix metalloproteinase specifically expressed in the leukocyte lineage. *Cell Res.* **1999**, *9*, 291–303. [CrossRef]
28. Udenfriend, S.; Kodukula, K. How glycosylphosphatidylinositol-anchored membrane proteins are made. *Annu. Rev. Biochem.* **1995**, *64*, 563–591. [CrossRef]
29. Truong, A.; Yip, C.; Paye, A.; Blacher, S.; Munaut, C.; Deroanne, C.; Noel, A.; Sounni, N.E. Dynamics of internalization and recycling of the prometastatic membrane type 4 matrix metalloproteinase (MT4-MMP) in breast cancer cells. *FEBS J.* **2016**, *283*, 704–722. [CrossRef]
30. Sohail, A.; Marco, M.; Zhao, H.; Shi, Q.; Merriman, S.; Mobashery, S.; Fridman, R. Characterization of the dimerization interface of membrane type 4 (MT4)-matrix metalloproteinase. *J. Biol. Chem.* **2011**, *286*, 33178–33189. [CrossRef]
31. Blanco, M.J.; Rodriguez-Martin, I.; Learte, A.I.R.; Clemente, C.; Montalvo, M.G.; Seiki, M.; Arroyo, A.G.; Sanchez-Camacho, C. Developmental expression of membrane type 4-matrix metalloproteinase (Mt4-mmp/Mmp17) in the mouse embryo. *PLoS ONE* **2017**, *12*, e0184767. [CrossRef] [PubMed]
32. Leigh, N.R.; Schupp, M.O.; Li, K.; Padmanabhan, V.; Gastonguay, A.; Wang, L.; Chun, C.Z.; Wilkinson, G.A.; Ramchandran, R. Mmp17b is essential for proper neural crest cell migration in vivo. *PLoS ONE* **2013**, *8*, e76484. [CrossRef] [PubMed]
33. Kim, I.J.; Zhang, Y.; Meister, M.; Sanes, J.R. Laminar restriction of retinal ganglion cell dendrites and axons: Subtype-specific developmental patterns revealed with transgenic markers. *J. Neurosci.* **2010**, *30*, 1452–1462. [CrossRef]
34. Kay, J.N.; de la Huerta, I.; Kim, I.J.; Zhang, Y.; Yamagata, M.; Chu, M.W.; Meister, M.; Sanes, J.R. Retinal ganglion cells with distinct directional preferences differ in molecular identity, structure, and central projections. *J. Neurosci.* **2011**, *31*, 7753–7762. [CrossRef] [PubMed]

35. Srichai, M.B.; Colleta, H.; Gewin, L.; Matrisian, L.; Abel, T.W.; Koshikawa, N.; Seiki, M.; Pozzi, A.; Harris, R.C.; Zent, R. Membrane-type 4 matrix metalloproteinase (MT4-MMP) modulates water homeostasis in mice. *PLoS ONE* **2011**, *6*, e17099. [CrossRef] [PubMed]
36. Kajita, M.; Kinoh, H.; Ito, N.; Takamura, A.; Itoh, Y.; Okada, A.; Sato, H.; Seiki, M. Human membrane type-4 matrix metalloproteinase (MT4-MMP) is encoded by a novel major transcript: Isolation of complementary DNA clones for human and mouse mt4-mmp transcripts. *FEBS Lett.* **1999**, *457*, 353–356. [CrossRef]
37. Gauthier, M.C.; Racine, C.; Ferland, C.; Flamand, N.; Chakir, J.; Tremblay, G.M.; Laviolette, M. Expression of membrane type-4 matrix metalloproteinase (metalloproteinase-17) by human eosinophils. *Int. J. Biochem. Cell Biol.* **2003**, *35*, 1667–1673. [CrossRef]
38. Paye, A.; Truong, A.; Yip, C.; Cimino, J.; Blacher, S.; Munaut, C.; Cataldo, D.; Foidart, J.M.; Maquoi, E.; Collignon, J.; et al. EGFR activation and signaling in cancer cells are enhanced by the membrane-bound metalloprotease MT4-MMP. *Cancer Res.* **2014**, *74*, 6758–6770. [CrossRef]
39. Xiao, Y.; Li, B.; Liu, J. miRNA27a regulates arthritis via PPARgamma in vivo and in vitro. *Mol. Med. Rep.* **2018**, *17*, 5454–5462.
40. Sounni, N.E.; Paye, A.; Host, L.; Noel, A. MT-MMPS as Regulators of Vessel Stability Associated with Angiogenesis. *Front. Pharmacol.* **2011**, *2*, 111. [CrossRef] [PubMed]
41. Li, W.; Li, Q.; Jiao, Y.; Qin, L.; Ali, R.; Zhou, J.; Ferruzzi, J.; Kim, R.W.; Geirsson, A.; Dietz, H.C.; et al. Tgfbr2 disruption in postnatal smooth muscle impairs aortic wall homeostasis. *J. Clin. Investig.* **2014**, *124*, 755–767. [CrossRef] [PubMed]
42. Scatena, M.; Liaw, L.; Giachelli, C.M. Osteopontin: A multifunctional molecule regulating chronic inflammation and vascular disease. *Arterioscler. Thromb. Vasc. Biol.* **2007**, *27*, 2302–2309. [CrossRef] [PubMed]
43. Isoda, K.; Nishikawa, K.; Kamezawa, Y.; Yoshida, M.; Kusuhara, M.; Moroi, M.; Tada, N.; Ohsuzu, F. Osteopontin plays an important role in the development of medial thickening and neointimal formation. *Circ. Res.* **2002**, *91*, 77–82. [CrossRef] [PubMed]
44. Gao, H.; Steffen, M.C.; Ramos, K.S. Osteopontin regulates alpha-smooth muscle actin and calponin in vascular smooth muscle cells. *Cell Biol. Int.* **2012**, *36*, 155–161. [CrossRef] [PubMed]
45. Giachelli, C.; Bae, N.; Lombardi, D.; Majesky, M.; Schwartz, S. Molecular cloning and characterization of 2B7, a rat mRNA which distinguishes smooth muscle cell phenotypes in vitro and is identical to osteopontin (secreted phosphoprotein I, 2aR). *Biochem. Biophys. Res. Commun.* **1991**, *177*, 867–873. [CrossRef]
46. Frischhertz, B.P.; Shamszad, P.; Pedroza, C.; Milewicz, D.M.; Morris, S.A. Thoracic aortic dissection and rupture in conotruncal cardiac defects: A population-based study. *Int. J. Cardiol.* **2015**, *184*, 521–527. [CrossRef] [PubMed]
47. Ahmad, M.M.; Kiani, I.A.; Ammar, K.A.; Ahmad, M.N.; Khandheria, B.K.; Paterick, T.E.; Jain, R.; Tajik, A.J. Ascending Aortic Aneurysm Is an Inherited Disease: A Contemporary Literature Review Based on Hill's Criteria of Specificity, Strength of Association, and Biological Coherence. *Cardiol. Rev.* **2017**, *25*, 268–278. [CrossRef]
48. Wang, Y.; Yu, S.J.; Li, Y.X.; Luo, H.S. Expression and clinical significance of matrix metalloproteinase-17 and -25 in gastric cancer. *Oncol. Lett.* **2015**, *9*, 671–676. [CrossRef]
49. Gencer, S.; Cebeci, A.; Irmak-Yazicioglu, M.B. Matrix metalloproteinase gene expressions might be oxidative stress targets in gastric cancer cell lines. *Chin. J. Cancer Res.* **2013**, *25*, 322–333.
50. Chen, J.; Somanath, P.R.; Razorenova, O.; Chen, W.S.; Hay, N.; Bornstein, P.; Byzova, T.V. Akt1 regulates pathological angiogenesis, vascular maturation and permeability in vivo. *Nat. Med.* **2005**, *11*, 1188–1196. [CrossRef]
51. Hawighorst, T.; Velasco, P.; Streit, M.; Hong, Y.K.; Kyriakides, T.R.; Brown, L.F.; Bornstein, P.; Detmar, M. Thrombospondin-2 plays a protective role in multistep carcinogenesis: A novel host anti-tumor defense mechanism. *EMBO J.* **2001**, *20*, 2631–2640. [CrossRef]
52. Fears, C.Y.; Grammer, J.R.; Stewart, J.E., Jr.; Annis, D.S.; Mosher, D.F.; Bornstein, P.; Gladson, C.L. Low-density lipoprotein receptor-related protein contributes to the antiangiogenic activity of thrombospondin-2 in a murine glioma model. *Cancer Res.* **2005**, *65*, 9338–9346. [CrossRef]
53. Salz, T.; Deng, C.; Pampo, C.; Siemann, D.; Qiu, Y.; Brown, K.; Huang, S. Histone Methyltransferase hSETD1A Is a Novel Regulator of Metastasis in Breast Cancer. *Mol. Cancer Res.* **2015**, *13*, 461–469. [CrossRef]

54. Yip, C.; Foidart, P.; Somja, J.; Truong, A.; Lienard, M.; Feyereisen, E.; Schroeder, H.; Gofflot, S.; Donneau, A.F.; Collignon, J.; et al. MT4-MMP and EGFR expression levels are key biomarkers for breast cancer patient response to chemotherapy and erlotinib. *Br. J. Cancer* **2017**, *116*, 742–751. [CrossRef] [PubMed]
55. Foidart, P.; Yip, C.; Radermacher, J.; Lienard, M.; Blacher, S.; Montero-Ruiz, L.; Maquoi, E.; Montaudon, E.; Chateau-Joubert, S.; Collignon, J.; et al. Expression of MT4-MMP, EGFR and RB in triple negative breast cancer strongly sensitizes tumors to erlotinib and palbociclib combination therapy. *Clin. Cancer Res.* **2018**. [CrossRef] [PubMed]
56. Whittaker, M.; Floyd, C.D.; Brown, P.; Gearing, A.J. Design and therapeutic application of matrix metalloproteinase inhibitors. *Chem. Rev.* **1999**, *99*, 2735–2776. [CrossRef]
57. Sun, J.; Stetler-Stevenson, W.G. Overexpression of tissue inhibitors of metalloproteinase 2 up-regulates NF-kappaB activity in melanoma cells. *J. Mol. Signal.* **2009**, *4*, 4. [CrossRef] [PubMed]
58. Sounni, N.E.; Rozanov, D.V.; Remacle, A.G.; Golubkov, V.S.; Noel, A.; Strongin, A.Y. Timp-2 binding with cellular MT1-MMP stimulates invasion-promoting MEK/ERK signaling in cancer cells. *Int. J. Cancer* **2010**, *126*, 1067–1078. [CrossRef]
59. Emonard, H.; Duca, L.; Dedieu, S. Editorial: Matricellular Receptors As Potential Targets in Anti-Cancer Therapeutic Strategies. *Front. Pharmacol.* **2016**, *7*, 95. [CrossRef]
60. Falardeau, P.; Champagne, P.; Poyet, P.; Hariton, C.; Dupont, E. Neovastat, a naturally occurring multifunctional antiangiogenic drug, in phase III clinical trials. *Semin. Oncol.* **2001**, *28*, 620–625. [CrossRef]
61. Huang, X.; Chen, S.; Xu, L.; Liu, Y.; Deb, D.K.; Platanias, L.C.; Bergan, R.C. Genistein inhibits p38 map kinase activation, matrix metalloproteinase type 2, and cell invasion in human prostate epithelial cells. *Cancer Res.* **2005**, *65*, 3470–3478. [CrossRef]
62. Kousidou, O.C.; Mitropoulou, T.N.; Roussidis, A.E.; Kletsas, D.; Theocharis, A.D.; Karamanos, N.K. Genistein suppresses the invasive potential of human breast cancer cells through transcriptional regulation of metalloproteinases and their tissue inhibitors. *Int. J. Oncol.* **2005**, *26*, 1101–1109. [CrossRef] [PubMed]
63. Coussens, L.M.; Fingleton, B.; Matrisian, L.M. Matrix metalloproteinase inhibitors and cancer: Trials and tribulations. *Science* **2002**, *295*, 2387–2392. [CrossRef] [PubMed]
64. Cathcart, J.; Pulkoski-Gross, A.; Cao, J. Targeting Matrix Metalloproteinases in Cancer: Bringing New Life to Old Ideas. *Genes Dis.* **2015**, *2*, 26–34. [CrossRef] [PubMed]
65. Amar, S.; Fields, G.B. Potential clinical implications of recent matrix metalloproteinase inhibitor design strategies. *Expert Rev. Proteomics* **2015**, *12*, 445–447. [CrossRef]
66. Zhou, N.; Paemen, L.; Opdenakker, G.; Froyen, G. Cloning and expression in Escherichia coli of a human gelatinase B-inhibitory single-chain immunoglobulin variable fragment (scFv). *FEBS Lett.* **1997**, *414*, 562–566. [CrossRef]
67. Devy, L.; Huang, L.; Naa, L.; Yanamandra, N.; Pieters, H.; Frans, N.; Chang, E.; Tao, Q.; Vanhove, M.; Lejeune, A.; et al. Selective inhibition of matrix metalloproteinase-14 blocks tumor growth, invasion, and angiogenesis. *Cancer Res.* **2009**, *69*, 1517–1526. [CrossRef]
68. Naito, S.; Takahashi, T.; Onoda, J.; Yamauchi, A.; Kawai, T.; Kishino, J.; Yamane, S.; Fujii, I.; Fukui, N.; Numata, Y. Development of a neutralizing antibody specific for the active form of matrix metalloproteinase-13. *Biochemistry* **2012**, *51*, 8877–8884. [CrossRef]
69. Kessenbrock, K.; Plaks, V.; Werb, Z. Matrix metalloproteinases: Regulators of the tumor microenvironment. *Cell* **2010**, *141*, 52–67. [CrossRef] [PubMed]
70. Mason, S.D.; Joyce, J.A. Proteolytic networks in cancer. *Trends Cell Biol.* **2011**, *21*, 228–237. [CrossRef]
71. Itoh, Y. MT1-MMP: A key regulator of cell migration in tissue. *IUBMB Life* **2006**, *58*, 589–596. [CrossRef] [PubMed]
72. Pahwa, S.; Stawikowski, M.J.; Fields, G.B. Monitoring and Inhibiting MT1-MMP during Cancer Initiation and Progression. *Cancers* **2014**, *6*, 416–435. [CrossRef] [PubMed]
73. Delatte, B.; Wang, F.; Ngoc, L.V.; Collignon, E.; Bonvin, E.; Deplus, R.; Calonne, E.; Hassabi, B.; Putmans, P.; Awe, S.; et al. RNA biochemistry. Transcriptome-wide distribution and function of RNA hydroxymethylcytosine. *Science* **2016**, *351*, 282–285. [CrossRef] [PubMed]

International Journal of *Molecular Sciences*

MDPI

Review

Matrix Metalloproteases as Influencers of the Cells' Social Media

Daniel Young [1,2,†], Nabangshu Das [2,3,†], Anthonia Anowai [1,2,4,†] and Antoine Dufour [1,2,3,4,*]

1 Department of Physiology and Pharmacology, University of Calgary, Calgary, AB T2N 4N1, Canada
2 McCaig Institute for Bone and Joint Health, University of Calgary, Calgary, AB T2N 4N1, Canada
3 Faculty of Kinesiology, University of Calgary, Calgary, AB T2N 4N1, Canada
4 Department of Biochemistry and Molecular Biology, University of Calgary, Calgary, AB T2N 4N1, Canada
* Correspondence: antoine.dufour@ucalgary.ca; Tel.: +1-403-210-7426
† These authors contributed equally to this work.

Received: 8 July 2019; Accepted: 2 August 2019; Published: 7 August 2019

Abstract: Matrix metalloproteinases (MMPs) have been studied in the context of cancer due to their ability to increase cell invasion, and were initially thought to facilitate metastasis solely through the degradation of the extracellular matrix (ECM). MMPs have also been investigated in the context of their ECM remodeling activity in several acute and chronic inflammatory diseases. However, after several MMP inhibitors failed in phase III clinical trials, a global reassessment of their biological functions was undertaken, which has revealed multiple unanticipated functions including the processing of chemokines, cytokines, and cell surface receptors. Despite what their name suggests, the matrix aspect of MMPs could contribute to a lesser part of their physiological functions in inflammatory diseases, as originally anticipated. Here, we present examples of MMP substrates implicated in cell signaling, independent of their ECM functions, and discuss the impact for the use of MMP inhibitors.

Keywords: matrix metalloproteinases (MMPs); protease; inflammation; signaling; invasion; apoptosis; chemokine; cytokine; proteomics; interferon

---

1. Introduction: MMPs Act as Emojis in Cell–Cell Communication

Matrix metalloproteinases (MMPs) have been associated with multiple inflammatory diseases [1–3], but the initial hypothesis that their proteolytic functions were solely linked to extracellular matrix (ECM) remodeling is outdated and needs to be extended to include additional substrates such as chemokines, cytokines, and cell surface receptors [4–9]. The catalytic activity of MMPs is tightly regulated by the tissue inhibitors of metalloproteinases (TIMPs), and is a key contributor in the overall outcome of substrate processing [10]. Interestingly, MMPs can also induce cell–cell communications through a non-proteolytic and non-ECM manner. For example, MMP2 was demonstrated to induce cell migration and signaling under static mechanical strain [11]. Increased mechanosensing properties in osteocytes was observed through decreased MT1–MMP expression [12]. Thus, MMPs are multi-tasking proteins that play key roles in cellular interactions and signaling; similar to emojis in our messages and social media posts, MMPs are able to tune and modulate cellular communications. Unfortunately, both emojis and MMPs are often misunderstood. Here, we describe examples of MMP substrates that are implicated in cell-to-cell communication and their physiological connection to inflammatory diseases.

2. MMPs Process the Gatekeepers of Cell Signaling

Cell surface receptors act as gatekeepers of cell signaling networks, as they provide the integral component that allows extracellular messages to be converted to intracellular signals. Proteolytic regulation is a key physiological process, as the cleavage of surface receptors can impact their activation state and half-life, and alter normal downstream transduction of receptor signals. Protease-activated receptors (PARs) are a family of G-protein coupled receptors (GPCRs) that are activated via a proteolytic mechanism [13]. They exist on most cell types, including macrophages, smooth muscle cells, and endothelial cells [14], and play roles in platelet aggregation, adhesion, cytokine production, and migration through the activation of different G proteins (G_{s}, G_{q}, and $G_{12/13}$) [14,15]. PAR1 is cleaved into its active form by thrombin, but also by other proteases including tryptase, neutrophil elastase, and MMP1 through its N-terminal extracellular exodomain (Figure 1) [13]. This occurs at a unique peptide sequence between $^{41}R{\downarrow}S^{42}$ (Table 1), liberating an N-terminal fragment that functionally interacts with the C-terminal region of the exodomain on PAR1 and triggers signal transduction from the GPCR (Figure 1). In cancer cells, thrombin cleavage of PAR1 increases tumor cell adhesion to matrix components, including platelets and vascular epithelial cells [16]. In platelets, PAR1 forms a heterodimer with PAR4, which is an isoform with lower affinity for thrombin. Thrombin cleavage of the surface receptor activates G_{q}-driven and G_{13}-driven pathways, leading to increased intracellular calcium and platelet aggregation [16]. MMP1 and MMP13 were found to be highly expressed in platelets, and are converted to their catalytically active form in the presence of fibrillar collagen [15]. MMP1 can also cleave membrane-tethered PAR1 at the *N*-terminal exodomain to generate signal transduction, albeit in a non-classical manner compared to thrombin–PAR1 cleavage [17]. Functional analysis involving the replacement of single amino acids around the cut site of PAR1 determined that MMP1's cleavage site is unique and specific to MMP1's active site [15]. Using a synthetic peptide matched to the exodomain sequence, the cleavage site was identified between $^{39}D{\downarrow}P^{40}$, which corresponds to a fragment of two amino acids that is distal to the thrombin cleavage site. MMP1 processing of PAR1 induces signaling, resulting in changes in morphology, activation via an alternate pathway, and phosphorylation of the p38 protein [13]. Overall, MMP1 can modulate PAR signaling in a cell-dependent manner, leading to various outcomes in downstream cell signaling.

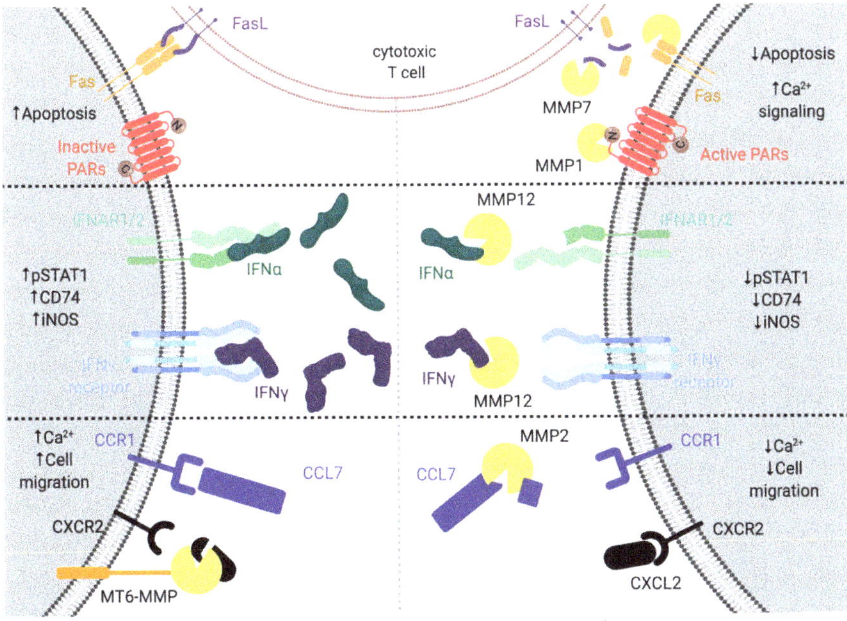

Figure 1. *Upper*, matrix metalloproteinase 1 (MMP1) cleavage of protease-activated receptors (PARs) result in increased Ca^{2+} signaling. MMP7 processing of tumor necrosis factor ligand superfamily member 6 (FASL) leads to decreased apoptosis. The right side displays MMP7 proteolysis on tumor necrosis factor receptor superfamily member 6 (FAS), the CD95 ligand (FasL), and PARs, leading to decreased apoptosis and increased Ca^{2+} activation. *Middle*, MMP12 can cleave interferon-alpha (IFNα) and interferon-gamma (IFNγ) at the C-terminus, leading to the termination of the tyrosine-protein kinase JAK1-signal transducer and activator of transcription 1-alpha/beta (JAK–STAT1) signaling pathway. The right side displays MMP12 proteolysis on IFNα and IFNγ, leading to the decreased phosphorylation of STAT1, HLA class II histocompatibility antigen gamma chain (CD74), and inducible nitric oxide synthase (iNOS). *Lower*, MMP2 cleaves C-C motif chemokine 7 (CCL7), and MT6-MMP cleaves C-X-C motif chemokine 2 (CXCL2), resulting in decreased migration and Ca^{2+} activation. The left side displays MT6-MMP proteolysis on CXCL2, leading to increased cell migration and Ca^{2+} activation. The right side displays MMP2 proteolysis on CCL7, leading to decreased cell migration and Ca^{2+} activation. MMPs are depicted as yellow Pacman.

Table 1. Cell surface proteins and receptors cleaved by various MMPs. The cleavage site is indicated for each MMP.

Protein Name	MMP1	MMP2	MMP3	MMP7	MMP9	MMP11	MMP12	MMP13	MMP14/ MT1–MMP	MMP16/ MT3–MMP	MMP25/ MT6–MMP	References
Amyloid protein precursor (APP)					^{687}K↓L^{689} ^{691}F↓A^{692} ^{694}D↓V^{695} ^{701}A↓I^{702}				^{579}N↓M^{580} ^{687}K↓L^{689}	^{463}A↓M^{464} ^{622}H↓S^{623} ^{579}N↓M^{580} ^{685}H↓Q^{686}		[18–21]
CD44 antigen (CD44)									^{162}R↓T^{163} ^{186}R↓S^{187} ^{192}G↓Y^{193}			[22]
C-type lectin domain family 3 member A (CLEC3A)				^{57}A↓L^{58} ^{63}A↓L^{64} ^{151}F↓L^{152}								[23]
CX3CL1 (fractalkine)		^{71}A↓L^{72} ^{4}G↓M^{5}										[24]
β-dystroglycan					^{715}H↓L^{716}							[25]
EMMPRIN/CD147									^{209}P↓M^{210} ^{214}N↓I^{215}			[26,27]
Ephrin B2 receptor		^{394}N↓I^{395} ^{432}D↓L^{433}			^{394}N↓I^{395} ^{432}D↓L^{433}							[28]
Fas Receptor (FAS)				^{19}E↓L^{20} ^{32}N↓L^{33}								[29,30]
Fibroblast growth factor receptor 1 (FGFR1)		^{368}V↓M^{369}										[31]
Integrin αV (CD51)									^{891}D↓L^{892}			[32]
Integrin β2 (CD18)					^{705}A↓I^{706}							[33]
Integrin-associated protein (IAP/CD47)												[34]
Intercellular adhesion molecule (ICAM)-1								^{60}I↓E^{61} ^{97}D↓G^{98}				[35–37]
Laminin receptor						^{115}A↓F^{116} ^{133}P↓I^{134}						[38]
Glutamate receptor ionotropic, NMDA 1 (NMDA receptor)				^{516}E↓K^{517}								[39]

Table 1. *Cont.*

Protein Name	MMP1	MMP2	MMP3	MMP7	MMP9	MMP11	MMP12	MMP13	MMP14/ MT1–MMP	MMP16/ MT3–MMP	MMP25/ MT6–MMP	References
Myelin-associated glycoprotein		$^{233}S{\downarrow}M^{234}$ $^{508}R{\downarrow}L^{509}$		$^{233}S{\downarrow}M^{234}$ $^{508}R{\downarrow}L^{509}$	$^{233}S{\downarrow}M^{234}$ $^{508}R{\downarrow}L^{509}$							[40]
Protease-activated receptor-1 (PAR-1)	$^{41}R{\downarrow}S^{42}$											[17]
Protein-tyrosine phosphatase receptor type Z (Ptprz)					$^{1625}R{\downarrow}I^{1626}$ $^{1627}G{\downarrow}L^{1628}$							[41]
Tumor necrosis factor ligand superfamily member 11(RANKL)				$^{145}M{\downarrow}M^{146}$								[42]
Semaphorin 4D												[43]
Tissue transglutaminase		$^{375}P{\downarrow}V^{376}$ $^{458}R{\downarrow}A^{459}$ $^{461}H{\downarrow}L^{462}$							$^{375}P{\downarrow}V^{376}$ $^{458}R{\downarrow}A^{459}$ $^{461}H{\downarrow}L^{462}$	$^{375}P{\downarrow}V^{376}$ $^{458}R{\downarrow}A^{459}$ $^{461}H{\downarrow}L^{462}$		[44,45]
TRANCE/OPGL (TNF-related activation-induced cytokine/osteoprotegrin ligand)									$^{138}R{\downarrow}F^{139}$ $^{145}M{\downarrow}M^{146}$			[46]
Urokinase plasminogen activator surface receptor (uPAR/CD87)			$^{108}T{\downarrow}Y^{109}$				$^{108}T{\downarrow}Y^{109}$				$^{108}T{\downarrow}Y^{109}$ $^{109}Y{\downarrow}S^{110}$ $^{111}R{\downarrow}S^{112}$	[47]

3. MMP7 Modulates CD95/Fas Signaling in Cell Death

CD95, apoptosis-mediating surface antigen tumor necrosis factor receptor superfamily member 6, (FAS)-mediated apoptosis is an extrinsic contact-dependent cell death pathway that is used by the adaptive immune system to eradicate neoplastic or infected cells [48]. The CD95 ligand (tumor necrosis factor ligand superfamily member 6/FASL) is expressed on the surface of cytotoxic T lymphocytes, which seek out target cells expressing the CD95 receptor [49]. Upon the association of a T cell with its cognate Fas receptor, apoptosis is initiated via recruitment of the adapter protein, Fas-associated death domain (FADD), procaspases, and a caspase regulator to form the death-inducing signaling complex (DISC), resulting in cell death [30,48,49]. Inhibitors that increase FasL expression at cell surfaces are able to terminate tumor cell proliferation [50]. Subverting Fas-mediated cell death signals allows cancer cells to develop resistance to chemotherapy drugs and increases chances of tumor survival. Some methods of Fas-related apoptosis evasion include occupation of the Fas receptor by a soluble FasL antagonistic ligand, the downregulation of Fas receptor expression, the cleavage of FasL ligand by MMP7 [51], and cleavage of the Fas receptor by MMP7 [29] (Figure 1). The extent to which MMP7 contributes to cell death escape via cleavage of the Fas/CD95 receptor was investigated in a previous study using the recombinant CD95 protein [29]. MMP7 was shown to cleave CD95 when on the surface of HT-29 colon cancer cells and prevented apoptosis, while treatment with a broad spectrum MMP inhibitor increased cell sensitivity to CD95-mediated death [29]. MMP7 has been shown to cleave FasL, producing a fragment from the soluble FasL ligand that is able to initiate apoptosis via Fas [52]. In contrast, others have suggested that the MMP7 cleavage of FasL inhibits the tumor killing action of cytotoxic T cells [51]. Importantly, Fas/CD95 processing is cell type-dependent, and further examination is needed to clarify the mechanism. Thus, MMP7 activity could be responsible for multiple mechanisms of immune defense evasion, in a cell-dependent manner, via the Fas/FasL system.

4. Co-Expression of MMP14/MT1–MMP and DR6 in Cell Death

Death receptor 6 (DR6) is a member of the tumor necrosis factor receptor (TNFR) family, which is ubiquitously expressed in humans, with higher occurrence in the heart, brain, and lymphoid organs [53,54]. Among its structural features are four highly conserved extracellular cysteine-rich domains that form the ligand binding site. This site is an intracellular death domain that is homologous to that of other members of the death receptor subfamily, as well as numerous post-translational modifications, including glycosylation, which can influence DR6 trafficking and interactions [53,55]. DR6 expression can be increased via activation of the NF-κB pathway in tumor cells; however, the overexpression of ectopic DR6 can activate the NF-κB and JNK pathways, and lead to apoptosis [54]. A cognate ligand is yet to be identified for DR6, and has generated speculations leading to discoveries regarding the role of DR6 in healthy tissues and pathologic conditions. Interestingly, the amyloid precursor protein (APP) was recently suggested as a major agonist for DR6 based on the ability of cleaved N-terminal APP fragments to bind DR6 [56]. DR6 is implicated in the degeneration of neurons in Alzheimer's disease via caspase-mediated apoptosis [56], the development and differentiation of hematopoietic cells [54,57], and immune system modulation [54,57]. In addition, DR6 is widely expressed on the surface of tumor cells, particularly in prostate and breast cancer [57]. This suggests unconventional interactions of pro-apoptotic DR6 in neuronal inflammation in comparison to cancer cells that may alter apoptosis, but no clear mechanism has been described yet. Explorative proteomics experiments have identified DR6 as a substrate of MMP14/membrane type 1 matrix metalloproteinase (MT1–MMP) [58]. The ectodomain of DR6 spans amino acids 67 to 211, and MT1–MMP cleavage releases the entire extracellular portion to create a soluble DR6 fragment [53,58]. MT1–MMP cleavage of DR6 ectodomain diverted T cell differentiation away from Th1, induced monocyte cell death, and affected the cytokine profiles of immature dendritic cells [57]. The extent to which DR6 cleavage influences physiological conditions in other tissues remains unclear; however, this suggests interesting functions of DR6 and MT1–MMP in boosting the innate and adaptive immune response in anti-tumor therapy.

5. Processing of Ephrin B2 Receptor Mediates Cell Motility

Receptor tyrosine kinase ephrin type-B receptor 2 (EPHB2) and its associated ligand ephrin-B2 (EFNB2) are both transmembrane proteins whose interactions guide embryonic neuronal development [59], specifically the spatial organization of cellular networks in the nervous system [60]. They are part of the larger Eph–ephrin family, whose roles also include directing cell boundary division, the morphogenesis of vasculature, and axon guidance [61]. There are two groups within this family: ephrin A-type ligands that connect to the cell membrane via a glycosylphosphatidylinositol anchor (GPI) to which Ephrin A receptors bind, and ephrin B-type ligands that are transmembrane proteins that associate with Eph B receptors [59–61]. The responses of receptor–ligand are enhanced when a large cluster of proteins interact at the contact site [61]. Activation at the contact site occurs in the ligand-bearing and receptor-bearing cells, which are termed reverse and forward signaling respectively, and is thus bi-directional [62]. Forward signaling (Eph-expressing cell) activates tyrosine kinase activity and the downstream phosphorylation of members of the Rho family GTPases that modulate intracellular morphological changes [61]. Mutations or the loss of ephrin B ligands impacts the reverse signaling capabilities, and in mice, it may alter the migration of capillary, lymphatic, and neuronal networks [61]. Ephrin-B2 activates the EphB2 receptor on the surface of neighboring cells to trigger cell adhesion and repulsion, thereby directing cell movements. The processes that allow repulsion and attraction between cells involves the remodeling of the cytoskeleton as well as actin polymerization [63]; however, the mechanism by which high-affinity binding of the surface-bound receptor and ligand is translated into repulsion remains uncharacterized. One proposed mechanism is that the specific cleavage of the EphB2 extracellular domain by MMP7 and MMP9 alter ephrin-B2-induced activation [28]. Two distinct large and short fragments of the EphB2 ectodomain were isolated from murine hippocampal neurons and 14-day-old mice embryonic brains. This cleavage was inhibited by an MMP2/MMP9 inhibitor, suggesting the involvement of MMPs in ephrin-B2-directed shedding of the EphB2 ectodomain [28]. MMP7 and MMP9 can cleave endogenous and recombinant EphB2 in vitro and on the surface of HEK-293 cells and hippocampal neurons to generate two fragments of lengths matching those originally discovered [28]. Further investigation revealed two sites at which MMPs targeted the EphB2 receptor: between $^{394}N{\downarrow}I^{395}$ and $^{432}D{\downarrow}L^{433}$ (Table 1). Amino acid substitution at these sites rendered mutant Eph2 insusceptible to MMP cleavage and thus failed to illicit cell–cell repulsion, although ephrin-B2 could still bind to the receptor [28]. One of these sites was conserved amongst all the Eph receptors, while the other was only expressed in EphB2 and EphB3, suggesting that repulsive signaling from Eph receptors depends on the predisposition of the receptor to MMP cleavage [28].

6. Processing of CD44 Decreases Cell Adhesion

Tumor invasion and metastasis are key processes of cancer progression that require cells to enhance their migratory ability [64–66]. CD44, a ubiquitous transmembrane receptor, is necessary for cell adhesion, wound healing, tumor invasion, and cell migration due to its broad-spectrum ligand interactions. CD44 interacts with ECM components, cytokines, and growth factors [22,67,68]. Its ability to facilitate multiple binding interactions with ECM components may promote the establishment of the cellular niche [69]. CD44 is the primary cognate receptor for hyaluronate, which is a molecule that aids in migration by inducing changes in the ECM to promote cell movement [70]. CD44 is derived from a single gene, but has multiple variants due to alternative splicing in 10 of its 19 exons [67]. Common in the structure of CD44 are post-translational N-glycosylations and O- glycosylations, and phosphorylation within the ectodomain, which might determine CD44 receptor potency [71]. Multiple CD44 variants are markers of stem cells and cancer stem cells, and have been linked to angiogenesis and increased metastatic activity in tumors [22,67,72]. However, findings of increased soluble CD44 (sCD44) in patients with aggressive cancers suggest that additional post-translational modifications, such as proteolytic processing, may play essential roles in CD44's mechanism of action [22,72]. Additionally, CD44 maintains adhesive interactions to other ECM components that have to be detached in order for migration to occur [67]. MT1–MMP is a key protease that is implicated in increasing cell migration and invasion and interacts with CD44 at the cell surface in the direction of migration [72]. Kajita et al. analyzed and confirmed the processing of CD44 by MT1–MMP in cancer cell lines [22]. The co-expression of MMP-14 and CD44 in human breast carcinoma cells revealed measurable amounts of 70-kDa fragments and a significant decrease in hyaluronate binding ability [22,70]. Three ~26-kDa sized CD44 fragments were generated at ^{162}R↓T^{163}, ^{186}R↓S^{187}, and ^{192}G↓Y^{193} by MT1–MMP [22]. The blade I of the MT1–MMP hemopexin domain was shown to interact with CD44, resulting in the phosphorylation of the epidermal growth factor receptor (EGFR) [72]. This interaction was blocked with peptides mimicking blade I of the MT1–MMP hemopexin domain, leading to decreased cancer cell metastasis in a mouse model of breast cancer and a reduction in new blood vessel formation in a chorioallantoic membrane assay in chickens [72]. Thus, CD44–MT1–MMP interactions may be responsible for migration, invasion, metastasis, and angiogenesis through proteolytic modulation and impact on downstream signaling.

7. MMP Processing of Chemokines

Chemokines are essential regulators in the immune system that act on G protein-coupled receptors (GPCRs) [73–75]. There are four main types, which are classified by the structure of a conserved cysteine-containing motif at the amine terminus, including the C, CC, CXC, and the CX3C motifs (C represents a cysteine and X represents any amino acid) [76]. Chemokine receptors are involved in multiple biological processes, including angiogenesis, wound healing, virus sensing, cell signaling, calcium activation, and hematopoiesis (Figure 1) [76,77]. Immune cells respond to multiple chemokines, and are thus subjected to this complex landscape of immune responses and signaling networks. For example, the binding of CXCL12 to CXCR4 results in the development and maturation of several immune cell linages including B cells, monocytes, natural killer cells, neutrophils, and macrophages [77]. Antagonists to CXCR4 generate abnormal numbers of neutrophils in peripheral circulation [78]. CCL7 attracts monocytes and eosinophils, but not neutrophils, via interactions with CCR1, CCR2, or CCR3 [79]. CCL7 acts as a ligand for CCR2, which is required for monocyte mobilization out of the bone marrow [80]. CCL13 attracts lymphocytes, monocytes, basophils, and eosinophils, but not neutrophils via CCR2B or CCR3, resulting in downstream signaling and chemotaxis [79,81]. To date, little is known about the post-translational modification of chemokines. Several proteases, including MMPs, are capable of cleaving chemokines at their *N*-terminus or *C*-terminus to alter their function, resulting in the formation of haptotactic gradients that help direct cell migration [6,9]. The establishment of potent chemical gradients occurs through interactions between peptide sequences of various chemokines and unique glycoproteins known as glycosaminoglycans (GAGs) [76]. GAGs are present in the ECM and on the surface of endothelial cells where they can facilitate the efficient infiltration of inflammatory cells into tissues from the blood stream [82]. MMPs modulate the functions of CCL7 and CCL13 through precise processing (Table 2). The truncated form of both CCL7 (5–76) and CCL13 (8–75) lose agonist activity in comparison to their full-length counterparts, as demonstrated using cell migration assays [83]. Proteolytic truncations of CCL7 and CCL13 results in a dampening of the inflammatory response, as demonstrated in a mouse model of inflammatory edema [83].

Table 2. Chemokine and cytokine cleaved by various MMPs. The cleavage site is indicated for each MMP. ND = not determined.

Chemokine Name	MMP1	MMP2	MMP3	MMP7	MMP8	MMP9	MMP12	MMP13	MMP14/ MT1–MMP	MMP25/ MT6–MMP	References
CCL2	$^{4}S\downarrow A^{5}$ $^{27}A\downarrow I^{28}$		$^{4}S\downarrow A^{5}$ $^{27}A\downarrow I^{28}$	$^{4}S\downarrow A^{5}$ $^{67}K\downarrow T^{68}$	$^{4}S\downarrow A^{5}$	$^{4}S\downarrow A^{5}$	$^{4}S\downarrow A^{5}$	$^{4}S\downarrow A^{5}$		$^{4}S\downarrow A^{5}$	[6,9,83,84]
CCL3	$^{47}I\downarrow A^{48}$			$^{8}L\downarrow V^{9}$ $^{47}I\downarrow A^{48}$ $^{63}I\downarrow F^{64}$	$^{15}M\downarrow A^{16}$	$^{15}M\downarrow A^{16}$ $^{64}F\downarrow L^{65}$		$^{47}I\downarrow A^{48}$			[6]
CCL4	$^{15}A\downarrow A^{16}$ $^{61}P\downarrow A^{62}$	$^{5}V\downarrow T^{6}$ $^{6}T\downarrow V^{7}$ $^{44}P\downarrow R^{45}$	$^{44}P\downarrow R^{45}$	$^{5}V\downarrow T^{6}$ $^{61}P\downarrow A^{62}$	$^{6}T\downarrow V^{7}$	$^{9}L\downarrow V^{14}$ $^{61}P\downarrow A^{62}$	$^{6}T\downarrow V^{7}$		$^{5}V\downarrow T^{6}$ $^{6}T\downarrow V^{7}$ $^{44}P\downarrow R^{45}$	$^{6}T\downarrow V^{7}$	[6,84]
CCL5				$^{65}V\downarrow T^{66}$				$^{4}S\downarrow A^{5}$			[6]
CCL7	$^{4}S\downarrow A^{5}$ $^{27}G\downarrow I^{28}$	$^{4}S\downarrow A^{5}$ $^{27}G\downarrow I^{28}$	$^{4}S\downarrow A^{5}$ $^{27}G\downarrow I^{28}$	$^{4}S\downarrow A^{5}$ $^{6}A\downarrow L^{7}$		$^{4}S\downarrow A^{5}$ $^{27}G\downarrow I^{28}$	$^{4}S\downarrow A^{5}$	$^{4}S\downarrow A^{5}$ $^{27}G\downarrow I^{28}$	$^{6}A\downarrow L^{7}$ $^{8}L\downarrow C^{9}$ $^{27}G\downarrow I^{28}$	$^{4}S\downarrow A^{5}$	[6,9,83–85]
CCL8	$^{4}S\downarrow A^{5}$ $^{27}G\downarrow I^{28}$	$^{4}S\downarrow A^{5}$ $^{27}G\downarrow I^{28}$	$^{4}S\downarrow A^{5}$ $^{27}G\downarrow I^{28}$	$^{4}S\downarrow A^{5}$ $^{6}A\downarrow L^{7}$		$^{4}S\downarrow A^{5}$ $^{27}G\downarrow I^{28}$	$^{4}S\downarrow A^{5}$	$^{27}G\downarrow I^{28}$	$^{6}A\downarrow L^{7}$ $^{8}L\downarrow C^{9}$ $^{27}G\downarrow I^{28}$	$^{4}S\downarrow A^{5}$	[6,9,83,84]
CCL11		$^{9}W\downarrow L^{10}$		$^{9}W\downarrow L^{10}$				$^{9}W\downarrow L^{10}$			[6]
CCL13	$^{4}S\downarrow A^{5}$ $^{26}D\downarrow A^{27}$ $A^{27}\downarrow L^{28}$ $^{30}V\downarrow P^{31}$ $^{72}E\downarrow I^{73}$	$^{3}V\downarrow S^{4}$ $^{4}S\downarrow A^{5}$ $^{9}C\downarrow L^{10}$ $^{74}C\downarrow A^{75}$	$^{3}V\downarrow S^{4}$ $^{4}S\downarrow A^{5}$ $^{26}D\downarrow A^{27}$ $A^{27}\downarrow L^{28}$	$^{4}S\downarrow A^{5}$ $^{9}C\downarrow L^{10}$ $^{72}E\downarrow I^{73}$ $^{74}C\downarrow A^{75}$	$^{4}S\downarrow A^{5}$ $^{72}E\downarrow I^{73}$	$^{4}S\downarrow A^{5}$ $^{72}E\downarrow I^{73}$	$^{4}S\downarrow A^{5}$ $^{72}E\downarrow I^{73}$	$^{4}S\downarrow A^{5}$ $^{72}E\downarrow I^{73}$	$^{3}V\downarrow S^{4}$	$^{4}S\downarrow A^{5}$ $^{72}E\downarrow I^{73}$	[6,9,84]
CCL14				$^{66}F\downarrow I^{67}$ $^{70}R\downarrow G^{71}$			$^{3}I\downarrow S^{4}$ $^{6}A\downarrow L^{7}$ $^{71}G\downarrow H^{72}$ $^{72}H\downarrow S^{73}$				[6]
CCL15	$^{24}I\downarrow N^{25}$	$^{13}L\downarrow V^{14}$ $^{24}I\downarrow N^{25}$ $^{26}D\downarrow A^{27}$ $^{27}A\downarrow E^{28}$	$^{16}V\downarrow L^{17}$ $^{24}I\downarrow N^{25}$ $^{27}A\downarrow E^{28}$	$^{16}V\downarrow L^{17}$ $^{24}I\downarrow N^{25}$ $^{88}K\downarrow G^{89}$	$^{24}I\downarrow N^{25}$ $^{27}A\downarrow E^{28}$ $^{42}V\downarrow V^{43}$	$^{27}A\downarrow E^{28}$	$^{24}I\downarrow N^{25}$ $^{27}A\downarrow E^{28}$	$^{13}L\downarrow V^{14}$ $^{24}I\downarrow N^{25}$ $^{26}D\downarrow A^{27}$ $^{27}A\downarrow E^{28}$	$^{13}L\downarrow V^{14}$ $^{26}D\downarrow A^{27}$	$^{24}I\downarrow N^{25}$ $^{26}D\downarrow A^{27}$	[6,84]

Table 2. *Cont.*

Chemokine Name	MMP1	MMP2	MMP3	MMP7	MMP8	MMP9	MMP12	MMP13	MMP14/ MT1–MMP	MMP25/ MT6–MMP	References
CCL16	^{7}A↓L^{8} ^{85}Q↓E^{86}	^{4}S↓E^{5} ^{7}A↓L^{8} ^{77}T↓N^{78}	^{7}A↓L^{8} ^{85}Q↓E^{86}	^{7}A↓L^{885}Q↓E^{86}	^{4}S↓E^{5} ^{7}A↓L^{8} ^{85}Q↓E^{86}	^{7}A↓L^{8} ^{77}T↓N^{78} ^{85}Q↓E^{86}	^{7}A↓L^{8} ^{77}T↓N^{78} ^{85}Q↓E^{86}	^{7}A↓L^{8} ^{77}T↓N^{78}	^{4}S↓E^{5} ^{7}A↓L^{8} ^{85}Q↓E^{86}	^{4}S↓E^{5}	[6,84]
CCL17				^{3}P↓L^{4} ^{69}G↓R^{70}	^{8}A↓L^{9}		^{3}P↓L^{4} ^{69}G↓R^{70}	^{3}P↓L^{4}			[6]
CCL23	^{10}C↓L^{11} ^{16}A↓L^{17} ^{22}R↓V^{23} ^{25}K↓D^{26} ^{27}A↓E^{28} ^{29}T↓E^{30}	^{10}C↓L^{11} ^{16}A↓L^{17} ^{25}K↓D^{26} ^{27}A↓E^{28}	^{20}Q↓A^{21} ^{25}K↓D^{26} ^{90}G↓R^{91}	^{10}C↓L^{11} ^{16}A↓L^{17} ^{22}R↓V^{23} ^{25}K↓D^{26} ^{90}G↓R^{91}	^{10}C↓L^{11} ^{16}A↓L^{17} ^{25}K↓D^{26}	^{10}C↓L^{11} ^{27}A↓E^{28}	^{10}C↓L^{11} ^{16}A↓L^{17} ^{22}R↓V^{23} ^{25}K↓D^{26}	^{90}G↓R^{91}	^{13}L↓V^{14} ^{29}T↓E^{30}	^{10}C↓L^{11} ^{20}Q↓A^{21} ^{25}K↓D^{26}	[6,84]
IFNα							^{160}L↓Q^{161} ^{157}N↓L^{158}				[8]
IFNβ						^{25}N↓L^{26} ^{29}F↓L^{30} ^{30}L↓Q^{31} ^{107}N↓L^{108} ^{114}N↓L^{115}					[86]
IFNγ							^{136}E↓L^{115} ^{157}M↓L^{158}				[7]
Tumor necrosis factor (TNF)	ND	ND	ND	ND		ND	ND		^{69}L↓I^{70} ^{72}P↓L^{73}		[58,87]
IL-1β		^{141}E↓L^{142}									[88]
TGFβ		ND				ND			ND		[89,90]

Chemokines also regulate T-cell development by inducing several developmental stages that occur in the thymus. Progenitor T cells (thymocytes) are double negatives, meaning that they lack markers denoting any specialized role ($CD4^-$, $CD8^-$). Upon entering the thymus, they may be pushed toward a $CD4^+/CD8^-$ or $CD4^-/CD8^+$ phenotype [77]. Progression through these stages is well coordinated and occurs in various microenvironments within the thymus [91]. Initial migration to the thymus is CCR7/CCR9-dependent, and activated by CCL21 or CCL19 [92]. Once in the thymus, thymocytes move toward the cortex in response to CCL19 and CXCL12, which is mediated by CCR9 [93]. There, they receive further cues for T-cell maturation, including signals that cause the apoptosis of autoreactive T cells. $CD4^+/CD8^-$ or $CD4^-/CD8^+$ T cells also begin to express CCR7 via CCL19 signaling [94]. MMPs were demonstrated to process multiple chemokines, including CCL5, CCL19, CCL21, and CXCL12 (Table 2); however, the proteolytic effect of MMPs on these chemokines during T-cell development and adaptation has yet to be characterized, and remains unexplored.

Membrane-type matrix metalloproteinase 6 (MT6–MMP or MMP25) is predominantly expressed by neutrophils, and has been shown to cleave 14 chemokines that are implicated in the recruitment of macrophages and monocytes during inflammatory processes (Table 2). Of these 14 chemokine substrates, CXCL2 and CXCL5 are involved in the recruitment of macrophages and monocytes [95,96]. The cleavage of CXCL2 at $^4A\downarrow T^5$ and CXCL5 at $^7R\downarrow A^8$ increases their chemotactic activity [84,97]. Ten MMPs can cleave CCL15 at six different sites (Table 2), resulting either in a decrease or increase of T cells and monocytes recruitment via interactions with chemokine receptor 1 (CCR1) [98]. MMP12 was identified as the most kinetically efficient MMP responsible for CCL15 proteolysis [6]. In addition to CCL15, several MMPs can cleave CCL23, which is a chemokine that is involved in the recruitment of monocytes, T lymphocytes, and neutrophils (Table 2). Cleaved CCL15 and CCL23 have increased binding to GAGs and induce chemotaxis [6]. The cleavage of CCL7 (MCP-3) by MMP2 dampens the immune response by preventing the additional recruitment of macrophages and lymphocytes [99]. CCL7 binds to CCR-1, CCR-2, and CCR-3, which are cell surface receptors on leukocytes, and when activated, these receptors cause an influx of calcium and leukocyte migration. MMP2 cleaves the first four amino acids in CCL7 (Table 1); cleaved CCL7 (CCL 5–76) is unable to induce calcium via either CCR-1 or CCR-2, and acts as an antagonist to the CCR-3 receptor [99]. Mice treated with cleaved CCL7 (5–76) suffered a reduction in mononuclear inflammatory cell infiltration as compared to full-length CCL7 [99]. Together, these results demonstrate that MMPs are key regulators of immune cell recruitment during various inflammatory processes.

8. MMPs–Orchestrator of the Fine Tuning of Cytokine Signaling

Cytokines are key regulators and messengers modulating the social landscape of cell-to-cell communication. Cytokine-mediated communication among lymphocytes, inflammatory cells, and hematopoietic cells initiate and organize an effective immune response [100]. Cytokine–receptor engagement triggers an intracellular signaling cascade; interestingly, most if not all cytokines can be post-translationally modified, altering their ability to bind to their respective cell surface receptors [7,8,86]. Traditionally, MMPs have been defined as ECM degrading proteases, but recent evidence has demonstrated that the majority of their substrates are not ECM-related proteins; rather, they are related to chemokine/cytokine networks, cell migration, kinase signaling, or transcription factors [4,5,101]. The altered expression of proteases is typically present in autoimmune and inflammatory diseases wherein an imbalance in homeostatic pro-inflammatory and anti-inflammatory cytokine expression and signaling is observed. For example, MMP12 was shown to affect both the expression and secretion of interferon-alpha (IFNα), which is a key cytokine implicated in regulating autoimmune diseases and viral infection [8,102]. When cells were infected with either coxsackievirus type B3 (CVB3) or respiratory syncytial virus (RSV), MMP12 was demonstrated to control INFα secretion indirectly by regulating the transcription of IκBα in the nucleus [8]. Extracellularly, MMP12 prevented aberrant IFNα-mediated signaling through proteolytic processing of the C-terminus, which is known to bind IFNα/β receptor 1, and terminated the phosphorylation of STAT1 (Figure 1) [8]. By the injecting a membrane-impermeable MMP12 inhibitor in CVB3-infected mice, extracellular MMP12 activity was inhibited. This resulted in an elevation of systemic IFN-α levels and a reduction of viral replication in the pancreas, without loss of the beneficial roles of intracellular MMP12 [8]. During a CVB3 murine infection, MMP9 was found to prevent virus propagation in the heart, promote immune infiltration and remodeling, and preserve cardiac output–product of the heart rate and the stroke volume [103]. *Mmp9*$^{-/-}$ mice infected with CVB3 showed more severe myocardial injury, elevated the foci of infection and immune infiltrates along with increased levels of IFNβ1 and IFNγ in comparison to wild types [103]. Similar to IFNα, MMP12 was demonstrated to truncate the C-terminus of IFNγ, causing a loss of receptor binding ability and a decrease in JAK–STAT1 signaling [7]. In a murine model of acute peritonitis, IFNγ levels were elevated, leading to an increase in IFNγ pro-inflammatory protein signature (S100A8, S100A9, inducible nitric oxide synthase (iNOS), and STAT1) in *Mmp12*$^{-/-}$ as compared to the wild-type counterparts [7]. In two autoimmune murine arthritis models using two different genetic backgrounds (MRL/*lpr* and B10.RIII), increased disease severity along with elevated IFNγ-dependent pro-inflammatory protein signatures were observed in *Mmp12*$^{-/-}$ as compared to wild type [7].

Tumor necrosis factor alpha (TNFα) is a pro-inflammatory cytokine released by activated macrophages, CD4^{+} lymphocytes, natural killer cells, neutrophils, mast cells, and eosinophils. It is involved in the perpetuation of systemic inflammation in diseases such as in rheumatoid arthritis, Crohn's disease, multiple sclerosis, and cancer [104]. TNFα is produced as a trimeric membrane-anchored precursor and secreted through a disintegrin and metalloproteinase domain-containing protein 17 (ADAM17)-regulated proteolytic pathway, resulting in the release of an active, soluble TNFα mediator [101,104,105]. The processing of TNFα by MMPs was demonstrated to modulate inflammatory signaling responses [87]. In a model of macrophage-mediated herniated disc resorption, macrophage MMP7 was shown to cleave TNFα and hindered macrophage infiltration into the disc [106]. Multiple MMPs (-1, -2, -3, -7, -9, -12, -14, and -17) have been shown to cleave active TNFα on the cell surface, further supporting their roles as influencers of cell signaling [58,87,107–109].

Interleukin-1 beta (IL-1β) is a cytokine that is cleaved into its active form by cellular proteases. Mature IL-1β is secreted from activated cells following caspase-1-dependent processing [110,111]. In addition to cleavage by cytosolic caspase-1, IL-1β can also be cleaved by MMP2, -3, and -9 in order to achieve a biologically mature form [112]. In a feed-forward mechanism, IL-1β induces MMP synthesis, which in turn dampens IL-1β activity. For example, the addition of MMP-1 reduced the IL-1β-dependent synthesis of prostaglandin E2 in human fibroblasts [88]. MMPs can regulate IL-1β activity by processing the soluble type II IL-1 decoy receptor (sIL-1R II) [113,114].

MMPs promote inflammation, but are also implicated in the anti-inflammatory regulation of immune responses. Transforming growth factor beta (TGF)-β, an anti-inflammatory cytokine, is processed by MMPs [90,115,116]. Initially, TGF-β is produced as a latent precursor containing two disulfide-linked short carboxy-terminal and large amino-terminal homodimers, which act as a latency-associated protein. Non-covalent association of these homodimers in the latent complex prevents receptor-ligand binding and the induction of TGF-β-mediated responses [115]. Following secretion, the latent TGF-β complex is cross-linked to the cellular membrane, forming a reservoir of latent TGF-β in the extracellular environment [115]. Several mechanisms including MMP proteolysis have been implicated in the release of mature TGF-β from the latent complex. MMP2 [90], MMP3 [116], MMP9 [90], and MT1–MMP [117] have demonstrated cleavage of the latent TGF-β complex, untethering TGF-β from the cell surface [118]. Furthermore, MMPs can release TGF-β by degrading decorin, which is a small collagen-associated proteoglycan that is known to act as a depot for TGF-β in the ECM [119]. In a feed-forward mechanism, TGF-β and IL-1β stimulate the production of MMP9 in rabbit corneal epithelial cells [120]. This induction of MMP9 by IL-1β occurs via NF-kB and AP-1-dependent pathways [121]. The characterization of signaling cascades regulated by MMP processing has been demonstrated in multiple systems and tissues, but more validation is needed to better understand these effects in vivo and their links to disease pathology. This could help identify new means to develop therapeutic modalities that restore normal cytokine interactions in inflammatory diseases.

9. How Can We Stop MMPs *Influencing* the Cells' Social Media Communications?

After the failure of broad-spectrum MMP inhibitors for the treatment of cancer and arthritic diseases, a global reassessment of their physiological functions was undertaken. We now know that (1) more than 10 MMPs can be beneficial in inflammatory diseases; (2) tissue and cellular localization are key in dictating MMP functions and substrate processing; (3) MMPs cleave more substrates than only ECM proteins; and (4) MMPs have non-proteolytic functions [3,5,10,64,65,122–124]. The question is now: would a selective MMP inhibitor have better success? Sela-Passwell et al. [125] demonstrated that by immunizing mice with a synthetic molecule that mimics the conserved structure of MMP2 and MMP9 catalytic zinc-histidine complex, they were able to protect mice from dextran sulfate sodium (DSS)-induced colitis both prophylactically and therapeutically. Several small molecules that selectively target the hemopexin domain of MMP9 were designed and were demonstrated to inhibit lung metastasis of MDA-MD-435 cells in a xenograft mouse model of breast cancer [65]. A selective MT1–MMP monoclonal antibody was demonstrated to be effective when used in combination with oseltamivir phosphate (Tamiflu) in an influenza virus model composed of PR8 influenza and *Streptococcus pneumonia* intra-nasal infection at sublethal doses [126]. Combination therapy demonstrated a synergistic therapeutic effect of the two inhibitors by both reducing the viral load and keeping ECM remodeling in homeostasis. A similar approach using a human Fab display phage library led to the development of DX-2400, which is a selective MT1–MMP monoclonal antibody that only inhibited the active form of the protease [127]. Efficacy was demonstrated in xenograft murine models of breast cancer using MDA-MB-231 cells and a syngeneic model using murine 4T1 cells by reducing tumor growth, metastasis, and angiogenesis; however, so far, little efficacy was demonstrated in human clinical trials [127,128]. It appears that broad-spectrum inhibition may not be ideal for long-term usage, but could be an alternative for short-term life-threatening conditions such as viral infections [8] or sepsis [129–132]. A better understanding of the specific roles of each MMP in a specific tissue and their communicating roles in diseases will pave the way for either selective MMP inhibition or targeting MMP substrates. Despite the long history of MMP inhibitors, we still have limited knowledge about these inflammatory proteases, and how they influence of the cells' social media.

Conflicts of Interest: The authors declare no conflict of interest.

References

1. Dufour, A. Degradomics of matrix metalloproteinases in inflammatory diseases. *Front. Biosci.* **2015**, *7*, 150–167. [CrossRef]
2. Butler, G.S.; Overall, C.M. Proteomic identification of multitasking proteins in unexpected locations complicates drug targeting. *Nat. Rev. Drug Discov.* **2009**, *8*, 935–948. [CrossRef]
3. Hu, J.; Van den Steen, P.E.; Sang, Q.-X.A.; Opdenakker, G. Matrix metalloproteinase inhibitors as therapy for inflammatory and vascular diseases. *Nat. Rev. Drug Discov.* **2007**, *6*, 480–498. [CrossRef] [PubMed]
4. Dufour, A.; Overall, C.M. Subtracting Matrix Out of the Equation: New Key Roles of Matrix Metalloproteinases in Innate Immunity and Disease. In *Matrix Metalloproteinase Biology, Sagi/Matrix Metalloproteinase Biology*; John Wiley & Sons, Inc.: Hoboken, NJ, USA, 2015; Volume 48, pp. 131–152.
5. Cauwe, B.; Opdenakker, G. Intracellular substrate cleavage: A novel dimension in the biochemistry, biology and pathology of matrix metalloproteinases. *Crit. Rev. Biochem. Mol. Biol.* **2010**, *45*, 351–423. [CrossRef] [PubMed]
6. Starr, A.E.; Dufour, A.; Maier, J.; Overall, C.M. Biochemical analysis of matrix metalloproteinase activation of chemokines CCL15 and CCL23 and increased glycosaminoglycan binding of CCL16. *J. Biol. Chem.* **2012**, *287*, 5848–5860. [CrossRef]

7. Dufour, A.; Bellac, C.L.; Eckhard, U.; Solis, N.; Klein, T.; Kappelhoff, R.; Fortelny, N.; Jobin, P.; Rozmus, J.; Mark, J.; et al. C-terminal truncation of IFN-γ inhibits proinflammatory macrophage responses and is deficient in autoimmune disease. *Nat. Commun.* **2018**, *9*, 2416. [CrossRef] [PubMed]
8. Marchant, D.J.; Bellac, C.L.; Moraes, T.J.; Wadsworth, S.J.; Dufour, A.; Butler, G.S.; Bilawchuk, L.M.; Hendry, R.G.; Robertson, A.G.; Cheung, C.T.; et al. A new transcriptional role for matrix metalloproteinase-12 in antiviral immunity. *Nat. Med.* **2014**, *20*, 493–502. [CrossRef]
9. Dean, R.A.; Cox, J.H.; Bellac, C.L.; Doucet, A.; Starr, A.E.; Overall, C.M. Macrophage-specific metalloelastase (MMP-12) truncates and inactivates ELR+ CXC chemokines and generates CCL2, -7, -8, and -13 antagonists: Potential role of the macrophage in terminating polymorphonuclear leukocyte influx. *Blood* **2008**, *112*, 3455–3464. [CrossRef]
10. Khokha, R.; Murthy, A.; Weiss, A. Metalloproteinases and their natural inhibitors in inflammation and immunity. *Nat. Rev. Immunol.* **2013**, *13*, 649–665. [CrossRef]
11. Yang, Z.; Wu, B.; Jia, S.; Zhao, Y.; Hou, R.; Liu, X.; Wang, X.; Chen, L.; Yang, X.; Lei, D.; et al. The mechanically activated p38/MMP-2 signaling pathway promotes bone marrow mesenchymal stem cell migration in rats. *Arch. Oral Biol.* **2017**, *76*, 55–60. [CrossRef]
12. Kulkarni, R.N.; Bakker, A.D.; Gruber, E.V.; Chae, T.D.; Veldkamp, J.B.B.; Klein-Nulend, J.; Everts, V. MT1-MMP modulates the mechanosensitivity of osteocytes. *Biochem. Biophys. Res. Commun.* **2012**, *417*, 824–829. [CrossRef] [PubMed]
13. Gieseler, F.; Ungefroren, H.; Settmacher, U.; Hollenberg, M.D.; Kaufmann, R. Proteinase-activated receptors (PARs)-focus on receptor-receptor-interactions and their physiological and pathophysiological impact. *Cell Commun. Signal.* **2013**, *11*, 86. [CrossRef] [PubMed]
14. Austin, K.M.; Covic, L.; Kuliopulos, A. Matrix metalloproteases and PAR1 activation. *Blood* **2013**, *121*, 431–439. [CrossRef] [PubMed]
15. Trivedi, V.; Boire, A.; Tchernychev, B.; Kaneider, N.C.; Leger, A.J.; O'Callaghan, K.; Covic, L.; Kuliopulos, A. Platelet matrix metalloprotease-1 mediates thrombogenesis by activating PAR1 at a cryptic ligand site. *Cell* **2009**, *137*, 332–343. [CrossRef]
16. Wojtukiewicz, M.Z.; Hempel, D.; Sierko, E.; Tucker, S.C.; Honn, K.V. Protease-activated receptors (PARs)–biology and role in cancer invasion and metastasis. *Cancer Metastasis Rev.* **2015**, *34*, 775–796. [CrossRef] [PubMed]
17. Boire, A.; Covic, L.; Agarwal, A.; Jacques, S.; Sherifi, S.; Kuliopulos, A. PAR1 is a matrix metalloprotease-1 receptor that promotes invasion and tumorigenesis of breast cancer cells. *Cell* **2005**, *120*, 303–313. [CrossRef] [PubMed]
18. Amour, A.; Knight, C.G.; English, W.R.; Webster, A.; Slocombe, P.M.; Knäuper, V.; Docherty, A.J.P.; Becherer, J.D.; Blobel, C.P.; Murphy, G. The enzymatic activity of ADAM8 and ADAM9 is not regulated by TIMPs. *FEBS Lett.* **2002**, *524*, 154–158. [CrossRef]
19. Yan, P.; Hu, X.; Song, H.; Yin, K.; Bateman, R.J.; Cirrito, J.R.; Xiao, Q.; Hsu, F.F.; Turk, J.W.; Xu, J.; et al. Matrix metalloproteinase-9 degrades amyloid-beta fibrils in vitro and compact plaques in situ. *J. Biol. Chem.* **2006**, *281*, 24566–24574. [CrossRef] [PubMed]
20. Ahmad, M.; Takino, T.; Miyamori, H.; Yoshizaki, T.; Furukawa, M.; Sato, H. Cleavage of amyloid-beta precursor protein (APP) by membrane-type matrix metalloproteinases. *J. Biochem.* **2006**, *139*, 517–526. [CrossRef]
21. LePage, R.N.; Fosang, A.J.; Fuller, S.J.; Murphy, G.; Evin, G.; Beyreuther, K.; Masters, C.L.; Small, D.H. Gelatinase A possesses a beta-secretase-like activity in cleaving the amyloid protein precursor of Alzheimer's disease. *FEBS Lett.* **1995**, *377*, 267–270. [CrossRef]
22. Kajita, M.; Itoh, Y.; Chiba, T.; Mori, H.; Okada, A.; Kinoh, H.; Seiki, M. Membrane-type 1 matrix metalloproteinase cleaves CD44 and promotes cell migration. *J. Cell Biol.* **2001**, *153*, 893–904. [CrossRef] [PubMed]
23. Tsunezumi, J.; Higashi, S.; Miyazaki, K. Matrilysin (MMP-7) cleaves C-type lectin domain family 3 member A (CLEC3A) on tumor cell surface and modulates its cell adhesion activity. *J. Cell. Biochem.* **2009**, *106*, 693–702. [CrossRef] [PubMed]
24. Dean, R.A.; Overall, C.M. Proteomics discovery of metalloproteinase substrates in the cellular context by iTRAQ labeling reveals a diverse MMP-2 substrate degradome. *Mol. Cell Proteom.* **2007**, *6*, 611–623. [CrossRef] [PubMed]

25. Bozzi, M.; Inzitari, R.; Sbardell, D.; Monaco, S.; Pavoni, E.; Gioia, M.; Marini, S.; Morlacchi, S.; Sciandra, F.; Castagnola, M.; et al. Enzymatic processing of beta-dystroglycan recombinant ectodomain by MMP-9: Identification of the main cleavage site. *IUBMB Life* **2009**, *61*, 1143–1152. [CrossRef] [PubMed]
26. Tang, Y.; Kesavan, P.; Nakada, M.T.; Yan, L. Tumor-stroma interaction: Positive feedback regulation of extracellular matrix metalloproteinase inducer (EMMPRIN) expression and matrix metalloproteinase-dependent generation of soluble EMMPRIN. *Mol. Cancer Res.* **2004**, *2*, 73–80. [PubMed]
27. Haug, C.; Lenz, C.; Díaz, F.; Bachem, M.G. Oxidized low-density lipoproteins stimulate extracellular matrix metalloproteinase Inducer (EMMPRIN) release by coronary smooth muscle cells. *Arterioscler. Thromb. Vasc. Biol.* **2004**, *24*, 1823–1829. [CrossRef] [PubMed]
28. Lin, K.-T.; Sloniowski, S.; Ethell, D.W.; Ethell, I.M. Ephrin-B2-induced cleavage of EphB2 receptor is mediated by matrix metalloproteinases to trigger cell repulsion. *J. Biol. Chem.* **2008**, *283*, 28969–28979. [CrossRef]
29. Strand, S.; Vollmer, P.; van den Abeelen, L.; Gottfried, D.; Alla, V.; Heid, H.; Kuball, J.; Theobald, M.; Galle, P.R.; Strand, D. Cleavage of CD95 by matrix metalloproteinase-7 induces apoptosis resistance in tumour cells. *Oncogene* **2004**, *23*, 3732–3736. [CrossRef]
30. Almendro, V.; Ametller, E.; García-Recio, S.; Collazo, O.; Casas, I.; Augé, J.M.; Maurel, J.; Gascón, P. The role of MMP7 and its cross-talk with the FAS/FASL system during the acquisition of chemoresistance to oxaliplatin. *PLoS ONE* **2009**, *4*, e4728. [CrossRef]
31. Levi, E.; Fridman, R.; Miao, H.Q.; Ma, Y.S.; Yayon, A.; Vlodavsky, I. Matrix metalloproteinase 2 releases active soluble ectodomain of fibroblast growth factor receptor 1. *Proc. Natl. Acad. Sci. USA* **1996**, *93*, 7069–7074. [CrossRef]
32. Ratnikov, B.I.; Rozanov, D.V.; Postnova, T.I.; Baciu, P.G.; Zhang, H.; DiScipio, R.G.; Chestukhina, G.G.; Smith, J.W.; Deryugina, E.I.; Strongin, A.Y. An alternative processing of integrin alpha(v) subunit in tumor cells by membrane type-1 matrix metalloproteinase. *J. Biol. Chem.* **2002**, *277*, 7377–7385. [CrossRef]
33. Vaisar, T.; Kassim, S.Y.; Gomez, I.G.; Green, P.S.; Hargarten, S.; Gough, P.J.; Parks, W.C.; Wilson, C.L.; Raines, E.W.; Heinecke, J.W. MMP-9 sheds the beta2 integrin subunit (CD18) from macrophages. *Mol. Cell Proteom.* **2009**, *8*, 1044–1060. [CrossRef]
34. Maile, L.A.; Capps, B.E.; Miller, E.C.; Allen, L.B.; Veluvolu, U.; Aday, A.W.; Clemmons, D.R. Glucose regulation of integrin-associated protein cleavage controls the response of vascular smooth muscle cells to insulin-like growth factor-I. *Mol. Endocrinol.* **2008**, *22*, 1226–1237. [CrossRef]
35. Sithu, S.D.; English, W.R.; Olson, P.; Krubasik, D.; Baker, A.H.; Murphy, G.; D'Souza, S.E. Membrane-type 1-matrix metalloproteinase regulates intracellular adhesion molecule-1 (ICAM-1)-mediated monocyte transmigration. *J. Biol. Chem.* **2007**, *282*, 25010–25019. [CrossRef]
36. Tarín, C.; Gomez, M.; Calvo, E.; López, J.A.; Zaragoza, C. Endothelial nitric oxide deficiency reduces MMP-13-mediated cleavage of ICAM-1 in vascular endothelium: A role in atherosclerosis. *Arterioscler. Thromb. Vasc. Biol.* **2009**, *29*, 27–32. [CrossRef]
37. Essick, E.; Sithu, S.; Dean, W.; D'Souza, S. Pervanadate-induced shedding of the intercellular adhesion molecule (ICAM)-1 ectodomain is mediated by membrane type-1 matrix metalloproteinase (MT1-MMP). *Mol. Cell. Biochem.* **2008**, *314*, 151–159. [CrossRef]
38. Amano, T.; Kwak, O.; Fu, L.; Marshak, A.; Shi, Y.-B. The matrix metalloproteinase stromelysin-3 cleaves laminin receptor at two distinct sites between the transmembrane domain and laminin binding sequence within the extracellular domain. *Cell Res.* **2005**, *15*, 150–159. [CrossRef]
39. Szklarczyk, A.; Ewaleifoh, O.; Beique, J.-C.; Wang, Y.; Knorr, D.; Haughey, N.; Malpica, T.; Mattson, M.P.; Huganir, R.; Conant, K. MMP-7 cleaves the NR1 NMDA receptor subunit and modifies NMDA receptor function. *FASEB J.* **2008**, *22*, 3757–3767. [CrossRef]
40. Milward, E.; Kim, K.J.; Szklarczyk, A.; Nguyen, T.; Melli, G.; Nayak, M.; Deshpande, D.; Fitzsimmons, C.; Hoke, A.; Kerr, D.; et al. Cleavage of myelin associated glycoprotein by matrix metalloproteinases. *J. Neuroimmunol.* **2008**, *193*, 140–148. [CrossRef]
41. Chow, J.P.H.; Fujikawa, A.; Shimizu, H.; Suzuki, R.; Noda, M. Metalloproteinase- and gamma-secretase-mediated cleavage of protein-tyrosine phosphatase receptor type Z. *J. Biol. Chem.* **2008**, *283*, 30879–30889. [CrossRef]
42. Lynch, C.C.; Hikosaka, A.; Acuff, H.B.; Martin, M.D.; Kawai, N.; Singh, R.K.; Vargo-Gogola, T.C.; Begtrup, J.L.; Peterson, T.E.; Fingleton, B.; et al. MMP-7 promotes prostate cancer-induced osteolysis via the solubilization of RANKL. *Cancer Cell* **2005**, *7*, 485–496. [CrossRef]

43. Basile, J.R.; Holmbeck, K.; Bugge, T.H.; Gutkind, J.S. MT1-MMP controls tumor-induced angiogenesis through the release of semaphorin 4D. *J. Biol. Chem.* **2007**, *282*, 6899–6905. [CrossRef]
44. Belkin, A.M.; Zemskov, E.A.; Hang, J.; Akimov, S.S.; Sikora, S.; Strongin, A.Y. Cell-surface-associated tissue transglutaminase is a target of MMP-2 proteolysis. *Biochemistry* **2004**, *43*, 11760–11769. [CrossRef]
45. Belkin, A.M.; Akimov, S.S.; Zaritskaya, L.S.; Ratnikov, B.I.; Deryugina, E.I.; Strongin, A.Y. Matrix-dependent proteolysis of surface transglutaminase by membrane-type metalloproteinase regulates cancer cell adhesion and locomotion. *J. Biol. Chem.* **2001**, *276*, 18415–18422. [CrossRef]
46. Schlöndorff, J.; Lum, L.; Blobel, C.P. Biochemical and pharmacological criteria define two shedding activities for TRANCE/OPGL that are distinct from the tumor necrosis factor alpha convertase. *J. Biol. Chem.* **2001**, *276*, 14665–14674. [CrossRef]
47. Andolfo, A.; English, W.R.; Resnati, M.; Murphy, G.; Blasi, F.; Sidenius, N. Metalloproteases cleave the urokinase-type plasminogen activator receptor in the D1-D2 linker region and expose epitopes not present in the intact soluble receptor. *Thromb. Haemost.* **2002**, *88*, 298–306. [CrossRef]
48. Peter, M.E.; Legembre, P.; Barnhart, B.C. Does CD95 have tumor promoting activities? *Biochim. Biophys. Acta* **2005**, *1755*, 25–36. [CrossRef]
49. Shresta, S.; Pham, C.T.; Thomas, D.A.; Graubert, T.A.; Ley, T.J. How do cytotoxic lymphocytes kill their targets? *Curr. Opin. Immunol.* **1998**, *10*, 581–587. [CrossRef]
50. Mollinedo, F.; Gajate, C. Fas/CD95 death receptor and lipid rafts: New targets for apoptosis-directed cancer therapy. *Drug Resist. Updat.* **2006**, *9*, 51–73. [CrossRef]
51. Mitsiades, N.; Yu, W.H.; Poulaki, V.; Tsokos, M.; Stamenkovic, I. Matrix metalloproteinase-7-mediated cleavage of Fas ligand protects tumor cells from chemotherapeutic drug cytotoxicity. *Cancer Res.* **2001**, *61*, 577–581.
52. Kayagaki, N.; Kawasaki, A.; Ebata, T.; Ohmoto, H.; Ikeda, S.; Inoue, S.; Yoshino, K.; Okumura, K.; Yagita, H. Metalloproteinase-mediated release of human Fas ligand. *J. Exp. Med.* **1995**, *182*, 1777–1783. [CrossRef]
53. Pan, G.; Bauer, J.H.; Haridas, V.; Wang, S.; Liu, D.; Yu, G.; Vincenz, C.; Aggarwal, B.B.; Ni, J.; Dixit, V.M. Identification and functional characterization of DR6, a novel death domain-containing TNF receptor. *FEBS Lett.* **1998**, *431*, 351–356. [CrossRef]
54. Benschop, R.; Wei, T.; Na, S. Tumor necrosis factor receptor superfamily member 21: TNFR-related death receptor-6, DR6. *Adv. Exp. Med. Biol.* **2009**, *647*, 186–194.
55. Klíma, M.; Zájedová, J.; Doubravská, L.; Andera, L. Functional analysis of the posttranslational modifications of the death receptor 6. *Biochim. Biophys. Acta* **2009**, *1793*, 1579–1587. [CrossRef]
56. Nikolaev, A.; McLaughlin, T.; O'Leary, D.D.M.; Tessier-Lavigne, M. APP binds DR6 to trigger axon pruning and neuron death via distinct caspases. *Nature* **2009**, *457*, 981–989. [CrossRef]
57. DeRosa, D.C.; Ryan, P.J.; Okragly, A.; Witcher, D.R.; Benschop, R.J. Tumor-derived death receptor 6 modulates dendritic cell development. *Cancer Immunol. Immunother.* **2008**, *57*, 777–787. [CrossRef]
58. Tam, E.M.; Morrison, C.J.; Wu, Y.I.; Stack, M.S.; Overall, C.M. Membrane protease proteomics: Isotope-coded affinity tag MS identification of undescribed MT1-matrix metalloproteinase substrates. *Proc. Natl. Acad. Sci. USA* **2004**, *101*, 6917–6922. [CrossRef]
59. Brambilla, R.; Br uuml ckner, K.; Orioli, D.; Bergemann, A.; Flanagan, J.; Klein, R. Similarities and Differences in the Way Transmembrane-Type Ligands Interact with the Elk Subclass of Eph Receptors. *Mol. Cell. Neurosci.* **1996**, *8*, 199–209. [CrossRef]
60. Flanagan, J.G.; Vanderhaeghen, P. The ephrins and Eph receptors in neural development. *Annu. Rev. Neurosci.* **1998**, *21*, 309–345. [CrossRef]
61. Klein, R. Eph/ephrin signalling during development. *Development* **2012**, *139*, 4105–4109. [CrossRef]
62. Zimmer, M.; Palmer, A.; Köhler, J.; Klein, R. EphB-ephrinB bi-directional endocytosis terminates adhesion allowing contact mediated repulsion. *Nat. Cell Biol.* **2003**, *5*, 869–878. [CrossRef]
63. Marston, D.J.; Dickinson, S.; Nobes, C.D. Rac-dependent trans-endocytosis of ephrinBs regulates Eph-ephrin contact repulsion. *Nat. Cell Biol.* **2003**, *5*, 879–888. [CrossRef]
64. Dufour, A.; Zucker, S.; Sampson, N.S.; Kuscu, C.; Cao, J. Role of matrix metalloproteinase-9 dimers in cell migration: Design of inhibitory peptides. *J. Biol. Chem.* **2010**, *285*, 35944–35956. [CrossRef]
65. Dufour, A.; Sampson, N.S.; Li, J.; Kuscu, C.; Rizzo, R.C.; Deleon, J.L.; Zhi, J.; Jaber, N.; Liu, E.; Zucker, S.; et al. Small-molecule anticancer compounds selectively target the hemopexin domain of matrix metalloproteinase-9. *Cancer Res.* **2011**, *71*, 4977–4988. [CrossRef]

66. Hanahan, D.; Weinberg, R.A. Hallmarks of cancer: The next generation. *Cell* **2011**, *144*, 646–674. [CrossRef]
67. Morath, I.; Hartmann, T.N.; Orian-Rousseau, V. CD44: More than a mere stem cell marker. *Int. J. Biochem. Cell Biol.* **2016**, *81*, 166–173. [CrossRef]
68. Nakamura, H.; Suenaga, N.; Taniwaki, K.; Matsuki, H.; Yonezawa, K.; Fujii, M.; Okada, Y.; Seiki, M. Constitutive and induced CD44 shedding by ADAM-like proteases and membrane-type 1 matrix metalloproteinase. *Cancer Res.* **2004**, *64*, 876–882. [CrossRef]
69. Zöller, M. CD44, Hyaluronan, the Hematopoietic Stem Cell, and Leukemia-Initiating Cells. *Front. Immunol.* **2015**, *6*, 235.
70. Aruffo, A.; Stamenkovic, I.; Melnick, M.; Underhill, C.B.; Seed, B. CD44 is the principal cell surface receptor for hyaluronate. *Cell* **1990**, *61*, 1303–1313. [CrossRef]
71. Naor, D.; Sionov, R.V.; Ish-Shalom, D. CD44: Structure, function, and association with the malignant process. *Adv. Cancer Res.* **1997**, *71*, 241–319.
72. Zarrabi, K.; Dufour, A.; Li, J.; Kuscu, C.; Pulkoski-Gross, A.; Zhi, J.; Hu, Y.; Sampson, N.S.; Zucker, S.; Cao, J. Inhibition of matrix metalloproteinase 14 (MMP-14)-mediated cancer cell migration. *J. Biol. Chem.* **2011**, *286*, 33167–33177. [CrossRef]
73. Hauser, A.S.; Attwood, M.M.; Rask-Andersen, M.; Schiöth, H.B.; Gloriam, D.E. Trends in GPCR drug discovery: New agents, targets and indications. *Nat. Rev. Drug Discov.* **2017**, *16*, 829–842. [CrossRef]
74. Hilger, D.; Masureel, M.; Kobilka, B.K. Structure and dynamics of GPCR signaling complexes. *Nat. Struct. Mol. Biol.* **2018**, *25*, 4–12. [CrossRef]
75. Latorraca, N.R.; Venkatakrishnan, A.J.; Dror, R.O. GPCR Dynamics: Structures in Motion. *Chem. Rev.* **2017**, *117*, 139–155. [CrossRef]
76. Laguri, C.; Arenzana-Seisdedos, F.; Lortat-Jacob, H. Relationships between glycosaminoglycan and receptor binding sites in chemokines-the CXCL12 example. *Carbohydr. Res.* **2008**, *343*, 2018–2023. [CrossRef]
77. Griffith, J.W.; Sokol, C.L.; Luster, A.D. Chemokines and chemokine receptors: Positioning cells for host defense and immunity. *Annu. Rev. Immunol.* **2014**, *32*, 659–702. [CrossRef]
78. Eash, K.J.; Greenbaum, A.M.; Gopalan, P.K.; Link, D.C. CXCR2 and CXCR4 antagonistically regulate neutrophil trafficking from murine bone marrow. *J. Clin. Invest.* **2010**, *120*, 2423–2431. [CrossRef]
79. Sabroe, I.; Hartnell, A.; Jopling, L.A.; Bel, S.; Ponath, P.D.; Pease, J.E.; Collins, P.D.; Williams, T.J. Differential regulation of eosinophil chemokine signaling via CCR3 and non-CCR3 pathways. *J. Immunol.* **1999**, *162*, 2946–2955.
80. Serbina, N.V.; Pamer, E.G. Monocyte emigration from bone marrow during bacterial infection requires signals mediated by chemokine receptor CCR2. *Nat. Immunol.* **2006**, *7*, 311–317. [CrossRef]
81. Berkhout, T.A.; Sarau, H.M.; Moores, K.; White, J.R.; Elshourbagy, N.; Appelbaum, E.; Reape, R.J.; Brawner, M.; Makwana, J.; Foley, J.J.; et al. Cloning, in vitro expression, and functional characterization of a novel human CC chemokine of the monocyte chemotactic protein (MCP) family (MCP-4) that binds and signals through the CC chemokine receptor 2B. *J. Biol. Chem.* **1997**, *272*, 16404–16413. [CrossRef]
82. Thompson, S.; Martínez-Burgo, B.; Sepuru, K.M.; Rajarathnam, K.; Kirby, J.A.; Sheerin, N.S.; Ali, S. Regulation of Chemokine Function: The Roles of GAG-Binding and Post-Translational Nitration. *Int. J. Mol. Sci.* **2017**, *18*, 1692. [CrossRef]
83. McQuibban, G.A.; Gong, J.-H.; Wong, J.P.; Wallace, J.L.; Clark-Lewis, I.; Overall, C.M. Matrix metalloproteinase processing of monocyte chemoattractant proteins generates CC chemokine receptor antagonists with anti-inflammatory properties in vivo. *Blood* **2002**, *100*, 1160–1167.
84. Starr, A.E.; Bellac, C.L.; Dufour, A.; Goebeler, V.; Overall, C.M. Biochemical characterization and N-terminomics analysis of leukolysin, the membrane-type 6 matrix metalloprotease (MMP25): Chemokine and vimentin cleavages enhance cell migration and macrophage phagocytic activities. *J. Biol. Chem.* **2012**, *287*, 13382–13395. [CrossRef]
85. Prudova, A.; Auf dem Keller, U.; Butler, G.S.; Overall, C.M. Multiplex N-terminome analysis of MMP-2 and MMP-9 substrate degradomes by iTRAQ-TAILS quantitative proteomics. *Mol. Cell Proteom.* **2010**, *9*, 894–911. [CrossRef]
86. Nelissen, I.; Martens, E.; Van den Steen, P.E.; Proost, P.; Ronsse, I.; Opdenakker, G. Gelatinase B/matrix metalloproteinase-9 cleaves interferon-beta and is a target for immunotherapy. *Brain* **2003**, *126*, 1371–1381. [CrossRef]

87. Gearing, A.J.; Beckett, P.; Christodoulou, M.; Churchill, M.; Clements, J.; Davidson, A.H.; Drummond, A.H.; Galloway, W.A.; Gilbert, R.; Gordon, J.L. Processing of tumour necrosis factor-alpha precursor by metalloproteinases. *Nature* **1994**, *370*, 555–557. [CrossRef]
88. Ito, A.; Mukaiyama, A.; Itoh, Y.; Nagase, H.; Thogersen, I.B.; Enghild, J.J.; Sasaguri, Y.; Mori, Y. Degradation of interleukin 1beta by matrix metalloproteinases. *J. Biol. Chem.* **1996**, *271*, 14657–14660. [CrossRef]
89. D'Angelo, M.; Billings, P.C.; Pacifici, M.; Leboy, P.S.; Kirsch, T. Authentic matrix vesicles contain active metalloproteases (MMP). a role for matrix vesicle-associated MMP-13 in activation of transforming growth factor-beta. *J. Biol. Chem.* **2001**, *276*, 11347–11353.
90. Yu, Q.; Stamenkovic, I. Cell surface-localized matrix metalloproteinase-9 proteolytically activates TGF-beta and promotes tumor invasion and angiogenesis. *Genes Dev.* **2000**, *14*, 163–176.
91. Shah, D.K.; Zúñiga-Pflücker, J.C. An overview of the intrathymic intricacies of T cell development. *J. Immunol.* **2014**, *192*, 4017–4023. [CrossRef]
92. Koch, U.; Radtke, F. Mechanisms of T cell development and transformation. *Annu. Rev. Cell Dev. Biol.* **2011**, *27*, 539–562. [CrossRef]
93. Uehara, S.; Grinberg, A.; Farber, J.M.; Love, P.E. A role for CCR9 in T lymphocyte development and migration. *J. Immunol.* **2002**, *168*, 2811–2819. [CrossRef]
94. Ueno, T.; Hara, K.; Willis, M.S.; Malin, M.A.; Höpken, U.E.; Gray, D.H.D.; Matsushima, K.; Lipp, M.; Springer, T.A.; Boyd, R.L.; et al. Role for CCR7 ligands in the emigration of newly generated T lymphocytes from the neonatal thymus. *Immunity* **2002**, *16*, 205–218. [CrossRef]
95. Al-Alwan, L.A.; Chang, Y.; Mogas, A.; Halayko, A.J.; Baglole, C.J.; Martin, J.G.; Rousseau, S.; Eidelman, D.H.; Hamid, Q. Differential roles of CXCL2 and CXCL3 and their receptors in regulating normal and asthmatic airway smooth muscle cell migration. *J. Immunol.* **2013**, *191*, 2731–2741. [CrossRef]
96. Wuyts, A.; Govaerts, C.; Struyf, S.; Lenaerts, J.P.; Put, W.; Conings, R.; Proost, P.; van Damme, J. Isolation of the CXC chemokines ENA-78, GRO alpha and GRO gamma from tumor cells and leukocytes reveals NH2-terminal heterogeneity. Functional comparison of different natural isoforms. *Eur. J. Biochem.* **1999**, *260*, 421–429. [CrossRef]
97. Van den Steen, P.E.; Wuyts, A.; Husson, S.J.; Proost, P.; Van Damme, J.; Opdenakker, G. Gelatinase B/MMP-9 and neutrophil collagenase/MMP-8 process the chemokines human GCP-2/CXCL6, ENA-78/CXCL5 and mouse GCP-2/LIX and modulate their physiological activities. *Eur. J. Biochem.* **2003**, *270*, 3739–3749. [CrossRef]
98. Berahovich, R.D.; Miao, Z.; Wang, Y.; Premack, B.; Howard, M.C.; Schall, T.J. Proteolytic activation of alternative CCR1 ligands in inflammation. *J. Immunol.* **2005**, *174*, 7341–7351. [CrossRef]
99. McQuibban, G.A.; Gong, J.H.; Tam, E.M.; McCulloch, C.A.; Clark-Lewis, I.; Overall, C.M. Inflammation dampened by gelatinase A cleavage of monocyte chemoattractant protein-3. *Science* **2000**, *289*, 1202–1206. [CrossRef]
100. Altan-Bonnet, G.; Mukherjee, R. Cytokine-mediated communication: A quantitative appraisal of immune complexity. *Nat. Rev. Immunol.* **2019**, *19*, 205–217. [CrossRef]
101. Van Lint, P.; Libert, C. Chemokine and cytokine processing by matrix metalloproteinases and its effect on leukocyte migration and inflammation. *J. Leukoc. Biol.* **2007**, *82*, 1375–1381. [CrossRef]
102. Crow, M.K. Autoimmunity: Interferon α or β: Which is the culprit in autoimmune disease? *Nat. Rev. Rheumatol.* **2016**, *12*, 439–440. [CrossRef]
103. Cheung, C.; Marchant, D.; Walker, E.K.-Y.; Luo, Z.; Zhang, J.; Yanagawa, B.; Rahmani, M.; Cox, J.; Overall, C.; Senior, R.M.; et al. Ablation of matrix metalloproteinase-9 increases severity of viral myocarditis in mice. *Circulation* **2008**, *117*, 1574–1582. [CrossRef]
104. Wajant, H.; Pfizenmaier, K.; Scheurich, P. Tumor necrosis factor signaling. *Cell Death Differ.* **2003**, *10*, 45–65. [CrossRef]
105. Le Gall, S.M.; Maretzky, T.; Issuree, P.D.A.; Niu, X.-D.; Reiss, K.; Saftig, P.; Khokha, R.; Lundell, D.; Blobel, C.P. ADAM17 is regulated by a rapid and reversible mechanism that controls access to its catalytic site. *J. Cell. Sci.* **2010**, *123*, 3913–3922. [CrossRef]
106. Haro, H.; Crawford, H.C.; Fingleton, B.; Shinomiya, K.; Spengler, D.M.; Matrisian, L.M. Matrix metalloproteinase-7-dependent release of tumor necrosis factor-alpha in a model of herniated disc resorption. *J. Clin. Investig.* **2000**, *105*, 143–150. [CrossRef]

107. English, W.R.; Puente, X.S.; Freije, J.M.; Knauper, V.; Amour, A.; Merryweather, A.; Lopez-Otin, C.; Murphy, G. Membrane type 4 matrix metalloproteinase (MMP17) has tumor necrosis factor-alpha convertase activity but does not activate pro-MMP2. *J. Biol. Chem.* **2000**, *275*, 14046–14055. [CrossRef]
108. Chandler, S.; Cossins, J.; Lury, J.; Wells, G. Macrophage metalloelastase degrades matrix and myelin proteins and processes a tumour necrosis factor-alpha fusion protein. *Biochem. Biophys. Res. Commun.* **1996**, *228*, 421–429. [CrossRef]
109. d'Ortho, M.P.; Will, H.; Atkinson, S.; Butler, G.; Messent, A.; Gavrilovic, J.; Smith, B.; Timpl, R.; Zardi, L.; Murphy, G. Membrane-type matrix metalloproteinases 1 and 2 exhibit broad-spectrum proteolytic capacities comparable to many matrix metalloproteinases. *Eur. J. Biochem.* **1997**, *250*, 751–757. [CrossRef]
110. Brough, D.; Rothwell, N.J. Caspase-1-dependent processing of pro-interleukin-1beta is cytosolic and precedes cell death. *J. Cell. Sci.* **2007**, *120*, 772–781. [CrossRef]
111. Thornberry, N.A.; Bull, H.G.; Calaycay, J.R.; Chapman, K.T.; Howard, A.D.; Kostura, M.J.; Miller, D.K.; Molineaux, S.M.; Weidner, J.R.; Aunins, J. A novel heterodimeric cysteine protease is required for interleukin-1 beta processing in monocytes. *Nature* **1992**, *356*, 768–774. [CrossRef]
112. Schönbeck, U.; Mach, F.; Libby, P. Generation of biologically active IL-1 beta by matrix metalloproteinases: A novel caspase-1-independent pathway of IL-1 beta processing. *J. Immunol.* **1998**, *161*, 3340–3346.
113. Bellehumeur, C.; Collette, T.; Maheux, R.; Mailloux, J.; Villeneuve, M.; Akoum, A. Increased soluble interleukin-1 receptor type II proteolysis in the endometrium of women with endometriosis. *Hum. Reprod.* **2005**, *20*, 1177–1184. [CrossRef]
114. Orlando, S.; Sironi, M.; Bianchi, G.; Drummond, A.H.; Boraschi, D.; Yabes, D.; Mantovani, A. Role of metalloproteases in the release of the IL-1 type II decoy receptor. *J. Biol. Chem.* **1997**, *272*, 31764–31769. [CrossRef]
115. Sheppard, D. Transforming growth factor beta: A central modulator of pulmonary and airway inflammation and fibrosis. *Proc. Am. Thorac. Soc.* **2006**, *3*, 413–417. [CrossRef]
116. Maeda, S.; Dean, D.D.; Gomez, R.; Schwartz, Z.; Boyan, B.D. The first stage of transforming growth factor beta1 activation is release of the large latent complex from the extracellular matrix of growth plate chondrocytes by matrix vesicle stromelysin-1 (MMP-3). *Calcif. Tissue Int.* **2002**, *70*, 54–65. [CrossRef]
117. Karsdal, M.A.; Larsen, L.; Engsig, M.T.; Lou, H.; Ferreras, M.; Lochter, A.; Delaissé, J.-M.; Foged, N.T. Matrix metalloproteinase-dependent activation of latent transforming growth factor-beta controls the conversion of osteoblasts into osteocytes by blocking osteoblast apoptosis. *J. Biol. Chem.* **2002**, *277*, 44061–44067. [CrossRef]
118. Dallas, S.L.; Rosser, J.L.; Mundy, G.R.; Bonewald, L.F. Proteolysis of latent transforming growth factor-beta (TGF-beta)-binding protein-1 by osteoclasts. A cellular mechanism for release of TGF-beta from bone matrix. *J. Biol. Chem.* **2002**, *277*, 21352–21360. [CrossRef]
119. Imai, K.; Hiramatsu, A.; Fukushima, D.; Pierschbacher, M.D.; Okada, Y. Degradation of decorin by matrix metalloproteinases: Identification of the cleavage sites, kinetic analyses and transforming growth factor-beta1 release. *Biochem. J.* **1997**, *322 (Pt 3)*, 809–814. [CrossRef]
120. Gordon, G.M.; Ledee, D.R.; Feuer, W.J.; Fini, M.E. Cytokines and signaling pathways regulating matrix metalloproteinase-9 (MMP-9) expression in corneal epithelial cells. *J. Cell. Physiol.* **2009**, *221*, 402–411. [CrossRef]
121. Tseng, H.-C.; Lee, I.-T.; Lin, C.-C.; Chi, P.-L.; Cheng, S.-E.; Shih, R.-H.; Hsiao, L.-D.; Yang, C.-M. IL-1β promotes corneal epithelial cell migration by increasing MMP-9 expression through NF-κB- and AP-1-dependent pathways. *PLoS ONE* **2013**, *8*, e57955. [CrossRef]
122. Dufour, A.; Sampson, N.S.; Zucker, S.; Cao, J. Role of the hemopexin domain of matrix metalloproteinases in cell migration. *J. Cell. Physiol.* **2008**, *217*, 643–651. [CrossRef]
123. Zucker, S.; Schmidt, C.E.; Dufour, A.; Kaplan, R.C.; Park, H.I.; Jiang, W. ProMMP-2: TIMP-1 complexes identified in plasma of healthy individuals. *Connect. Tissue Res.* **2009**, *50*, 223–231. [CrossRef]
124. Pavlaki, M.; Zucker, S.; Dufour, A.; Calabrese, N.; Bahou, W.; Cao, J. Furin Functions as a Nonproteolytic Chaperone for Matrix Metalloproteinase-28: MMP-28 Propeptide Sequence Requirement. *Biochem. Res. Int.* **2011**, *2011*, 630319. [CrossRef]
125. Sela-Passwell, N.; Kikkeri, R.; Dym, O.; Rozenberg, H.; Margalit, R.; Arad-Yellin, R.; Eisenstein, M.; Brenner, O.; Shoham, T.; Danon, T.; et al. Antibodies targeting the catalytic zinc complex of activated matrix metalloproteinases show therapeutic potential. *Nat. Med.* **2011**, *18*, 143–147. [CrossRef]

126. Talmi-Frank, D.; Altboum, Z.; Solomonov, I.; Udi, Y.; Jaitin, D.A.; Klepfish, M.; David, E.; Zhuravlev, A.; Keren-Shaul, H.; Winter, D.R.; et al. Extracellular Matrix Proteolysis by MT1-MMP Contributes to Influenza-Related Tissue Damage and Mortality. *Cell Host Microbe* **2016**, *20*, 458–470. [CrossRef]
127. Devy, L.; Huang, L.; Naa, L.; Yanamandra, N.; Pieters, H.; Frans, N.; Chang, E.; Tao, Q.; Vanhove, M.; Lejeune, A.; et al. Selective inhibition of matrix metalloproteinase-14 blocks tumor growth, invasion, and angiogenesis. *Cancer Res.* **2009**, *69*, 1517–1526. [CrossRef]
128. Ager, E.I.; Kozin, S.V.; Kirkpatrick, N.D.; Seano, G.; Kodack, D.P.; Askoxylakis, V.; Huang, Y.; Goel, S.; Snuderl, M.; Muzikansky, A.; et al. Blockade of MMP14 activity in murine breast carcinomas: Implications for macrophages, vessels, and radiotherapy. *J. Natl. Cancer Inst.* **2015**, *107*, 52. [CrossRef]
129. Vandenbroucke, R.E.; Dejager, L.; Libert, C. The first MMP in sepsis. *EMBO Mol. Med.* **2011**, *3*, 367–369. [CrossRef]
130. Vanlaere, I.; Libert, C. Matrix metalloproteinases as drug targets in infections caused by gram-negative bacteria and in septic shock. *Clin. Microbiol. Rev.* **2009**, *22*, 224–239. [CrossRef]
131. Vandenbroucke, R.E.; Dejonckheere, E.; Van Hauwermeiren, F.; Lodens, S.; De Rycke, R.; Van Wonterghem, E.; Staes, A.; Gevaert, K.; López-Otín, C.; Libert, C. Matrix metalloproteinase 13 modulates intestinal epithelial barrier integrity in inflammatory diseases by activating TNF. *EMBO Mol. Med.* **2013**, *5*, 1000–1016. [CrossRef]
132. Vandenbroucke, R.E.; Libert, C. Is there new hope for therapeutic matrix metalloproteinase inhibition? *Nat. Rev. Drug Discov.* **2014**, *13*, 904–927. [CrossRef]

International Journal of
Molecular Sciences

MDPI

Article

Identification of ADAM12 as a Novel Basigin Sheddase

Reidar Albrechtsen [1], Nicolai J. Wewer Albrechtsen [2], Sebastian Gnosa [1], Jeanette Schwarz [1,†], Lars Dyrskjøt [3] and Marie Kveiborg [1,*]

1 Biotech Research and Innovation Centre (BRIC), Faculty of Health and Medical Sciences, University of Copenhagen, 2200 Copenhagen, Denmark; reidar.albrechtsen@bric.ku.dk (R.A.); sebastian.gnosa@bric.ku.dk (S.G.); j.schwarz@ikmb.uni-kiel.de (J.S.)

2 Department of Biomedical Sciences and Department of Clinical Biochemistry, Rigshospitalet, Faculty of Health and Medical Sciences, University of Copenhagen, 2200 Copenhagen, Denmark; nicolai.albrechtsen@sund.ku.dk

3 Department of Molecular Medicine (MOMA), Aarhus University Hospital, 8200 Aarhus, Denmark; lars@clin.au.dk

* Correspondence: marie.kveiborg@bric.ku.dk

† Current affiliation: Institute of Clinical Molecular Biology, University Hospital Schleswig-Holstein and Institute of Clinical Chemistry, University Hospital Schleswig-Holstein, 24105 Kiel, Germany.

Received: 6 March 2019; Accepted: 17 April 2019; Published: 22 April 2019

Abstract: The transmembrane glycoprotein basigin, a member of the immunoglobulin superfamily, stimulates matrix metalloproteinase (MMP)-mediated extracellular matrix (ECM) degradation and thereby drives cancer cell invasion. Basigin is proteolytically shed from the cell surface and high concentrations of soluble basigin in the blood dictates poor prognosis in cancer patients. A positive correlation between basigin and a disintegrin and metalloproteinase (ADAM)-12 in serum from prostate cancer patients has been reported. Yet, the functional relevance of this correlation is unknown. Here, we show that ADAM12 interacts with basigin and cleaves it in the juxtamembrane region. Specifically, overexpression of ADAM12 increases ectodomain shedding of an alkaline phosphatase-tagged basigin reporter protein from the cell surface. Moreover, CRISPR/Cas9-mediated knockout of ADAM12 in human HeLa carcinoma cells results in reduced shedding of the basigin reporter, which can be rescued by ADAM12 re-expression. We detected endogenous basigin fragments, corresponding to the expected size of the ADAM12-generated ectodomain, in conditioned media from ADAM12 expressing cancer cell-lines, as well as serum samples from a healthy pregnant donor and five bladder cancer patients, known to contain high ADAM12 levels. Supporting the cancer relevance of our findings, we identified several cancer-associated mutations in the basigin membrane proximal region. Subsequent in vitro expression showed that some of these mutants are more prone to ADAM12-mediated shedding and that the shed ectodomain can enhance gelatin degradation by cancer cells. In conclusion, we identified ADAM12 as a novel basigin sheddase with a potential implication in cancer.

Keywords: a disintegrin and metalloproteinase; EMMPRIN; CD147; ectodomain shedding

---

1. Introduction

Basigin (BSG), also named CD147/EMMPRIN, a member of the immunoglobulin family of transmembrane proteins, exerts a wide range of both physiological and pathological functions [1–4]. Of particular note, BSG regulates the expression and cell surface localization of monocarboxylate transporters-1 (MCT1) and MCT4, allowing the efflux of lactate produced by aerobic glycolysis [5]. Moreover, it induces matrix metalloproteinase (MMP)-mediated extracellular matrix (ECM) degradation, thereby driving tumor invasion and metastasis [6–8].

BSG is upregulated in several types of tumors, and soluble BSG concentrations in blood or urine from cancer patients correlate with disease stage and poor prognosis [9,10]. Furthermore, BSG associates with chemotherapy response and survival in bladder cancer [11]. Full-length BSG can be released from tumor cells, macrophages, and stromal cells through micro-vesicle or exosome secretion [3]. However, BSG can also be released into the circulation as a soluble form through proteolytic shedding of its extracellular part (ectodomain). For example, Tang et al. showed a MMP-dependent generation of soluble BSG, lacking the C-terminal part [12], and Egawa et al. demonstrated that MMP14 sheds a 22-kDa N-terminal fragment of BSG into the media of tumor cells [13]. Also, it was recently shown that cholesterol-depletion induces ectodomain shedding of BSG by a disintegrin and metalloproteinase (ADAM)-10 in tumor cells [14].

Like BSG, several of the ADAMs play important roles in cancer. ADAMs are frequently upregulated in human tumors, with high levels of expression indicating worse prognosis [15,16]. Here, ADAM-mediated shedding of membrane bound substrates, such as growth factors, cytokines and adhesion molecules is thought to promote tumor growth and/or progression, best demonstrated by the role of ADAM17 in epidermal growth factor receptor (EGFR) signaling [17]. Another example is ADAM12, which among several substrates is capable of shedding some EGFR ligands [18–20] and adhesion molecules like VE-cadherin [21]. Moreover, ADAM12 exerts pro-tumorigenic effects independent of its own protease function [22,23], partly by regulating the activity of MMP14 [24]. A recent study showed a significant correlation between serum levels of ADAM12 and BSG from prostate cancer patients. The two proteins were therefore suggested as biomarker candidates for early diagnosis of prostate cancer [25]. However, the functional relevance of the correlation between ADAM12 and BSG in cancer remains elusive.

Our present study demonstrates novel functional aspects of the interaction between ADAM12 and BSG. We show that ADAM12 binds BSG and cleaves it in the juxta membrane region, thereby releasing a soluble BSG fragment to the extracellular space. Importantly, assessment of publicly available data revealed a number of cancer-associated BSG mutations in the ADAM12 cleavage region and when tested experimentally, we found that some of the mutations alter the susceptibility to ADAM12-mediated cleavage.

2. Results

2.1. ADAM12 Interacts with BSG in Human Cells

To understand the functional correlation between ADAM12 and BSG, we first examined whether the two proteins interact. To this end, we used 293-VnR cells that express little ADAM12, but substantial amounts of endogenous BSG (Table S1). We transiently expressed ADAM12-Δcyt (hereafter named ADAM12), tagged at the truncated C-terminus with green fluorescence protein (GFP), immunoprecipitated it with an antibody against GFP, and tested it for interaction with endogenous BSG by Western blot. Using an antibody against the C-terminal part of BSG, we detected an approximately 55 kDa protein band in immunoprecipitates from ADAM12-GFP expressing cells, but not from control cells expressing GFP alone (Figure 1A). Performing the reverse co-immunoprecipitation experiment on lysates from cells co-transfected with ADAM12 and full-length BSG, precipitated BSG was capable of pulling down the 68 kDa mature ADAM12 protein, whereas we did not detect ADAM12 when we used control IgG for precipitation (Figure 1B). As the overexpressed ADAM12 protein lacks the cytoplasmic tail, we predict that the interaction site between ADAM12 and BSG resides in the extracellular part of ADAM12. Also, we found that only the mature form of ADAM12 and not the pro form of ADAM12 co-immunoprecipitates with BSG. Since the mature form of ADAM12 is generated by pro protein processing in the secretory pathway [26], this indicates that the interaction between ADAM12 and BSG takes place after transit through the Golgi, possibly at the cell surface.

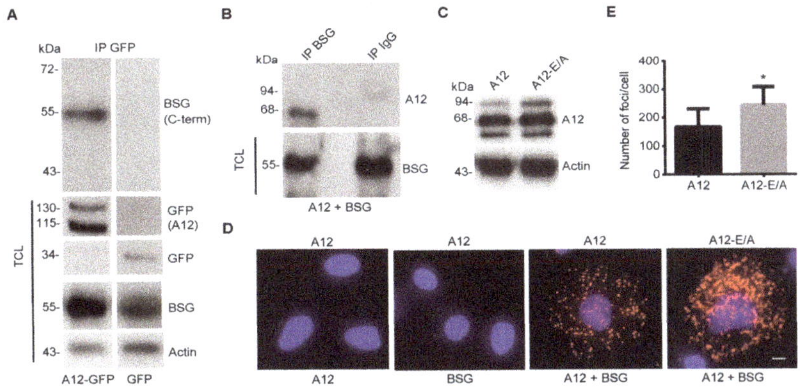

Figure 1. A disintegrin and metalloproteinase (ADAM)-12 interacts with basigin (BSG) in human cells. (**A**) 293-VnR cells were transfected with A12-GFP or GFP alone, immunoprecipitated with a polyclonal anti-GFP antibody and analyzed by Western blot, using an antibody against the C-terminal part of BSG. Total cell lysates (TCL) were tested for A12-GFP (pro and mature forms) and endogenous BSG protein expression. Actin was used as the loading control. (**B**) 293-VnR cells were co-transfected with A12 and full-length BSG, immunoprecipitated with an antibody against BSG and analyzed by Western blot, using a rabbit polyclonal antibody against A12. Immunoprecipitation with control rabbit IgG served as a negative control. (**C**) 293-VnR cells were transfected with A12 or the catalytically inactive A12-E/A mutant, together with full-length BSG and tested for A12 expression by Western blot, using actin as the loading control. (**D**) Cells from C were tested for A12-BSG co-localization by the proximity ligation assay (PLA) (red signal), using antibodies against the extracellular parts of A12 and BSG. Reactions using either of the two antibodies alone served as negative controls. Scale bar = 8 µm. (**E**) Number of red foci were automatically counted in 50 cells from three independent experiments, using MetaMorph microscopy analysis software. The average number of foci per cells ± SEM is shown in the graph. * $p < 0.05$, Student's t-test.

To examine whether indeed ADAM12 and BSG interact at the cell surface, we used the commercially available Duolink® kit, which is based on the in situ proximity ligation assay (PLA) [27]. We transfected 293-VnR cells with ADAM12 or the catalytically dead ADAM12-E/A mutant (Figure 1C) and stained for ADAM12 and/or BSG. As seen in Figure 1D and quantified in Figure 1E, over-expressed ADAM12 interacts with BSG and the interaction is significantly increased when we express the catalytically inactive ADAM12-E/A mutant.

2.2. ADAM12 Overexpression Increases Ectodomain Shedding of a Truncated BSG Reporter Substrate

To investigate a potential role of ADAM12 in shedding of BSG, we employed a cell-based shedding assay. The reporter substrate consisted of a truncated form of BSG, containing the extracellular membrane proximal region, the transmembrane domain and the entire cytoplasmic domain fused at the N-terminus to alkaline phosphatase (BSG-AP; Figure 2A). Transfection of 293-VnR cells with the BSG-AP construct, together with either ADAM12 or ADAM12-E/A showed a clear increase in the amount of shed BSG when we expressed the active protease (Figure 2B). Treating the cells with the broad-spectrum metalloproteinase inhibitor Batimastat, known to block ADAM12 activity [21], abrogated the ADAM12-mediated shedding (Figure 2B).

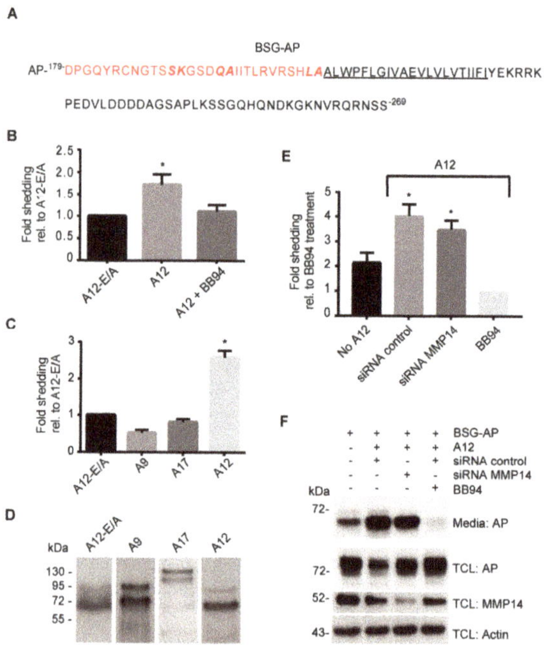

Figure 2. ADAM12 overexpression increases ectodomain shedding of a truncated BSG reporter substrate. (**A**) Amino acid sequence of the reporter substrate BSG-AP, consisting of the C-terminal part of BSG fused to alkaline phosphatase (AP). The truncated extracellular part of BSG is shown in red and the transmembrane domain is underlined. (**B**) Fold shedding in 293-VnR cells transfected with BSG-AP together with A12 or catalytically inactive A12-E/A, and treated with or without the metalloproteinase inhibitor Batimastat (BB94). Fold shedding is calculated as AP activity in the medium divided by the total AP activity in medium and cell lysate, and normalized to A12-E/A. (**C**) Fold shedding in 293-VnR cells transfected with BSG-AP and A12-E/A, A12, ADAM9 (A9) or ADAM17 (A17), calculated as in (**C**). (**D**) Western blot of total lysates from cells used in (**C**), showing comparable expression of the different ADAMs. (**E**) Fold shedding in 293-VnR cells transfected with BSG-AP alone (no A12) or with BSG-AP and A12 together in cells treated with control siRNA, siRNA against matrix metalloproteinase (MMP)-14, or the inhibitor BB94 as indicated. (**F**) Western blot of media and total cell lysates (TCL) from cells in (**E**), using actin as the loading control. For all graphs, values represent means ± SEM from three independent experiments. * $p < 0.05$, ANOVA.

Demonstrating the specificity of the proteolytic release, BSG-AP shedding was only seen when we overexpressed ADAM12, but not the related ADAM9 or ADAM17 (Figure 2C,D). Since ADAM12 regulates MMP14 [24], which has been previously implicated in the shedding of BSG [12], we tested whether MMP14 was required for the observed ADAM12-mediated shedding of BSG-AP. As expected, since the BSG-AP reporter substrate does not contain the previously reported MMP14 cleavage site [12], knocking down the expression of MMP14 by siRNA did not significantly reduce the increased BSG-AP shedding seen when overexpressing ADAM12 (Figure 2E). Complementing these shedding data, Western blot analysis of the conditioned media, using an antibody against alkaline phosphatase (AP), showed a band corresponding to the shed AP-fusion protein, which had approximately the same intensity when analyzing media from MMP14 and control siRNA transfected cells, whereas media from cells treated with Batimastat showed little to no shed protein (Figure 2F, top lane). Of note, the shedding assay showed a certain level of BSG-AP shedding in the absence of exogenous ADAM12 expression (Figure 2E), which we could block with the protease inhibitor Batimastat. This shedding activity may reflect the expression of other sheddases in 293-VnR cells.

2.3. CRISPR/Cas9-Mediated ADAM12 Knockout Reduces BSG Reporter Shedding

To corroborate the importance of ADAM12 for shedding of BSG, we knocked out ADAM12 expression in HeLa cells, known to express endogenous ADAM12 (Table S1), using CRISPR/Cas9 gene editing. Based on sequencing of established cell clones, we chose the knockout clone D7, which harbors deletions of 4 and 14 nucleotides, and the E7 clone that harbors 4 nucleotides deletion on both alleles. We used parental HeLa cells and the non-edited clone D4 as control cell lines. We validated the lack of ADAM12 expression in D7 and E7 cell lines by qPCR analysis (Figure 3A) and immunofluorescence staining (Figure 3B). When we tested clones D7 and E7 for BSG-AP shedding, we observed a decreased shedding for both ADAM12 knockout clones, as compared to parental HeLa cells or the D4 control cell line (Figure 3C). Importantly, re-expressing ADAM12 in both D7 and E7 cells partially rescued the lack of BSG-AP shedding (Figure 3C).

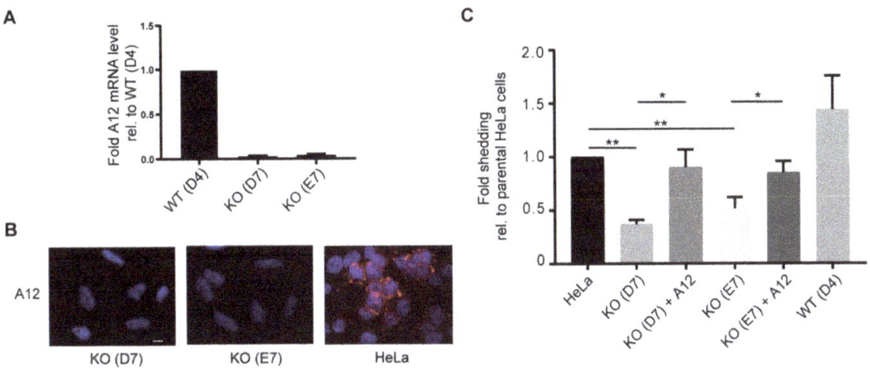

Figure 3. CRISPR/Cas9-mediated ADAM12 knockout reduces BSG reporter shedding. (**A**) qPCR analysis of A12 mRNA in CRISPR/Cas9 generated A12 knockout (clones D7 and E7) and non-edited wildtype (clone D4) HeLa cells. (**B**) Immunofluorescence staining of A12 in CRISPR/Cas9 generated A12 knockout HeLa clones (KO) D7 and E7, and parental HeLa cells. Nuclei were stained with 4′,6-diamidino-2-phenylindole (DAPI) and scale bar = 12 μm. (**C**) Fold shedding of BSG-AP transfected parental HeLa cells, CRISPR/Cas9-generated A12 knockout (KO) HeLa clones D7 or E7 with or without A12 re-expressed, and the non-edited wildtype (WT) clone D4, as indicated. Values represent means ± SEM from three independent experiments. * $p < 0.05$, ** $p < 0.005$, ANOVA.

2.4. ADAM12 Sheds Endogenous BSG

To assess ADAM12-mediated shedding of full-length endogenous BSG, we stably expressed ADAM12 in MCF7 human breast cancer cells, which express no detectable endogenous ADAM12 (Table S1). To detect the shed BSG ectodomain, we used an antibody recognizing a peptide (amino acids 70–206) located between the known MMP14 cleavage site and the region suggested to be cleaved by ADAM12 (Figure 4A). Immunofluorescence staining, using this antibody, detected a substantial amount of endogenous BSG at the cell surface of most MCF7 cells (Figure 4B). In contrast, the fluorescence signal was almost completely absent in MCF7 cells expressing ADAM12 (Figure 4B), indicating that the BSG ectodomain was shed from these cells. Also, using cell surface biotinylation and streptavidin pull-down of biotinylated proteins, we could detect more of the endogenous BSG ectodomain by Western blot in media from ADAM12 expressing MCF7 cells, as compared to wildtype MCF7 cells (Figure 4C). Supporting the notion that the soluble BSG protein is proteolytically shed from the cell surface rather than secreted (e.g., in exosomes), we did not detect any bands in conditioned media when blotting with an antibody against the intracellular C-terminal part of BSG (Figure 4C). In contrast, bands corresponding to full-length BSG and the cleaved C-terminal fragment were identified in total cell lysates when using this antibody (Figure 4C).

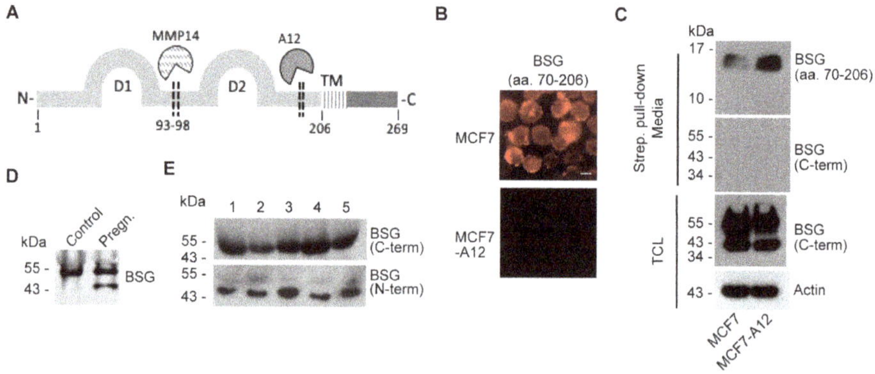

Figure 4. ADAM12 sheds endogenous BSG. (**A**) Schematic of the BSG structure, showing the two immunoglobulin domains (D1 and D2), the identified MMP14 and putative ADAM12 (A12) cleavage sites, and the transmembrane domain (TM). (**B**) Cytospin of wildtype or A12 expressing MCF7 cells were immuno-stained with the antibody PA5-29787, recognizing region aa. 70–206 of BSG. Scale bar = 8 µm. (**C**) Wildtype or A12 expressing MCF7 cells were surface biotinylated and incubated overnight in serum free media. Biotinylated cell surface proteins were pulled down with streptavidin–agarose from conditioned cell media or total cell lysates (TCL) and detected by Western blot, using antibodies against aa. 70–206 (PA5-29787) or the C-terminus of BSG. Actin was used as a loading control. (**D**) Western blot of BSG in serum from pregnant (pregn.) or non-pregnant (control) women. (**E**) Western blot of BSG in serum from five bladder cancer patients stained with antibodies recognizing either the C- or the N-terminus of BSG.

Human placenta produces high amounts of ADAM12, both the transmembrane ADAM12-L and the soluble ADAM12-S isoforms [28]. To assess the physiological relevance of ADAM12-mediated BSG shedding, we examined whether soluble BSG could be detected in serum from pregnant women. As shown by Western blot analysis, we detected a 55 kDa band, corresponding to full-length BSG, in pregnant or non-pregnant women, whereas we could only detect an approximately 48 kDa band in serum from pregnant women (Figure 4D). Similarly, we detected bands of a size corresponding to shed BSG in a Western blot of sera from five bladder cancer patients (Figure 4E), which we previously reported exhibit high levels of soluble ADAM12 in the circulation [29].

2.5. Cancer-Associated BSG Mutants Are Differentially Shed by ADAM12

When examining the extracellular membrane proximal region of BSG that is cleaved by ADAM12 (marked by red in the BSG-AP reporter substrate in Figure 2A), we noticed three amino acid pairs (SK, QA, LA), which were also found in the cleavage site of tumor necrosis factor (TNF)-α and constitutes the quenched fluorescence peptide substrate frequently used for measuring the catalytic activity of several ADAMs, including ADAM12 [30]. Searching the COSMIC (Catalogue Of Somatic Mutations In Cancer) database for mutations in the BSG gene associated with a broad range of cancers (https://cancer.sanger.ac.uk/cosmic), there were several somatic mutations in the part encoding this extracellular membrane proximal region (Figure 5A). To investigate the consequence of these mutations for BSG shedding, we introduced the identified individual single point mutations in the BSG-AP construct, co-transfected the constructs together with ADAM12, and performed shedding assays. Of the ten mutations tested, two of the cancer-associated mutations (T199A and A207T) were more efficiently shed compared to the control wildtype construct, whereas one mutation (A207V) resulted in significantly reduced shedding (Figure 5B). Notably, the BSG-AP mutants were expressed to approximately similar levels, but ran at slightly variable molecular weights, as shown by Western blot analysis (Figure 5C), indicating that some of the mutations changed the protein glycosylation pattern.

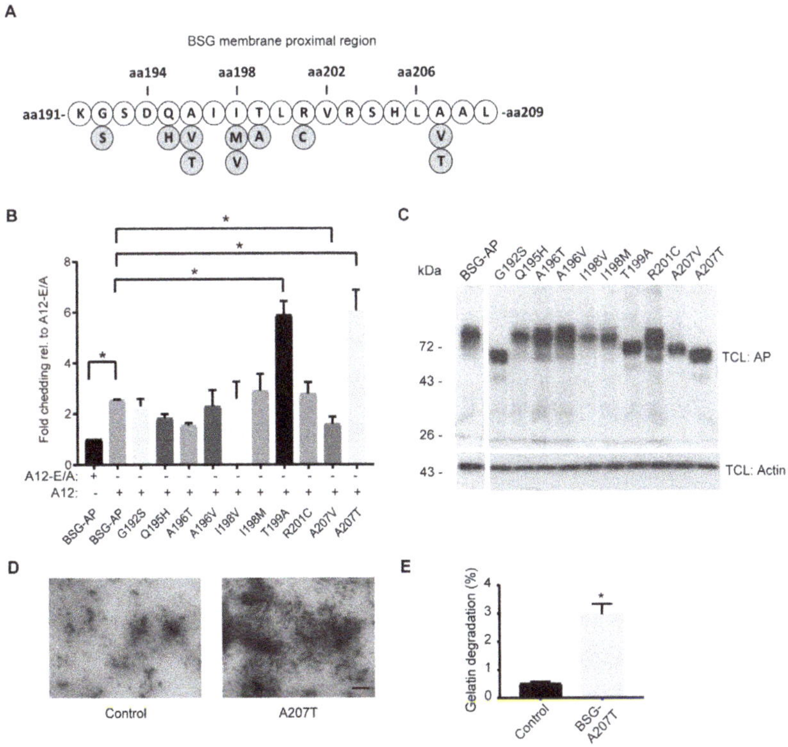

Figure 5. Cancer-associated BSG mutants are differentially shed by ADAM12. (**A**) Illustration of cancer-associated mutations (marked in grey) in the extracellular membrane proximal region (aa. 190–208) of BSG, identified in the Catalogue Of Somatic Mutations In Cancer (COSMIC) database (BSG, ENST00000333511; August 2017). (**B**) Fold shedding in 293-VnR cells transfected with A12 or the catalytically inactive A12-E/A mutant together with wildtype (BSG-AP), or various BSG-AP point mutations identified in (**A**). (**C**) Western blot of AP in total cell lysates (TCL) from cells in (**B**). Actin was used as loading control. (**D**) In situ gelatinase assay, using HT1080 cells grown on Oregon green-labeled gelatin and treated with conditioned media from untransfected 293-VnR cells (control) or 293-VnR cells transfected with ADAM12 together with a full-length BSG expression construct harboring the A207T mutation. Dark areas (without green fluorescence) represent gelatin degradation, scale bar = 10 µm. (**E**) Gelatin degradation was measured in µm^2 for 27 images, using MetaMorph software and the percentage degradation was calculated. For both graphs, values represent means ± SEM from three independent experiments. * $p < 0.05$, Student's *t*-test or ANOVA.

The biological role of BSG shedding is not fully understood; yet, several studies have shown a stimulatory effect of the MMP14-generated BSG ectodomain on MMP activation, particularly activation of the secreted MMP2 collagenase [2,3,8]. To test the effect of the ADAM12-generated ectodomain on collagen degradation, we performed an in situ gelatinase assay [24]. Specifically, we added conditioned media from control 293-VnR cells or cells expressing ADAM12, together with the shedding prone BSG-A207T mutant, onto HT1080 cells grown on Oregon green gelatin and incubated the cells for 24 h. Little gelatin degradation was observed when media from control cells was added to the HT1080 cells (Figure 5D); however, gelatin degradation was clearly enhanced when we added media from BSG-A207T expressing cells to the HT1080 cells (Figure 5D).

3. Discussion

The proteolytic shedding of BSG from the cell surface is well described [12–14]. However, our data are the first to demonstrate that, in addition to MMP14 and most recently ADAM10, also ADAM12 is capable of shedding BSG. By ADAM12 overexpression and siRNA- or CRISPR-Cas9-mediated ADAM12 depletion, we demonstrate that the transmembrane form of ADAM12 cleaves BSG in the extracellular membrane proximal region. In contrast, we observed no BSG shedding when overexpressing the shorter secreted ADAM12 isoform. While MMP14 has been shown to cleave BSG between the two immunoglobulin-like domains IgC2 (D1: aa. 25–101) and IgI (D2: aa. 106–200), generating a 22 kDa soluble fragment [13], we showed that ADAM12 generates an approximately 48 kDa ectodomain, containing both immunoglobulin-like domains (Figure 4A). The protein region harboring the ADAM12 cleavage site contains three amino acid pairs (SK, QA, LA), which were previously referred to by Moss et al. as the potential cleavage site found in a short peptide of TNFα, used to measure the catalytic activity of multiple MMPs and ADAMs [30]. While this suggests that other MMPs or ADAMs would likely also be able to cleave within this region, our data indicate that neither MMP14, nor ADAM9 or ADAM17 cleave at this site. As for ADAM10-mediated BSG shedding, the published study focused on the intracellular cytoplasmic fragment generated by intramembranous proteolysis following the initial extracellular cleavage event and did not mention the size of the shed ectodomain [14].

According to previous reports, the generated soluble BSG ectodomain acts to stimulate the activity of MMPs [12,13,31–33], for which reason BSG is also designated EMMPRIN—extracellular matrix metalloprotease inducer [6]. For example, it appears that soluble BSG, shed from the cell surface of cancer cells, acts in a paracrine manner to stimulate the expression of MMP2 in the surrounding stromal fibroblast [31]. Furthermore, it has been suggested that membrane bound BSG on fibroblasts functions as a receptor for the soluble BSG ectodomain, which upon binding gets internalized and activates the ERK1/2 signaling pathway, thereby inducing the expression of MMP1, MMP2, and MMP3 [34]. Based on our findings, the soluble BSG fragment generated by ADAM12 can increase gelatinase activity when added to cancer cells in vitro. In this context, it is worth mentioning that ADAM12 has been previously shown to enhance the gelatinase activity of MMP14, through a mechanism involving a protein complex composed of ADAM12, MMP14, and the adhesion receptor $\alpha_V\beta_3$ [24]. Thus, it appears that ADAM12 contributes to BSG shedding and MMP activation both directly and indirectly by enhancing MMP14 activity.

To assess the relevance of our in vitro findings, we examined serum samples from pregnant versus non-pregnant healthy women, as well as five bladder cancer patients by Western blot. Here we detected a protein band, most likely corresponding to full-length BSG released in exosomes. Moreover, we detected a smaller protein band of approximately 48 kDa in all cancer samples, as well as in pregnancy serum, but not in the non-pregnancy serum. Based on the band size and the fact that we only detected the band when we used an antibody against the N-terminal extracellular part, but not an antibody binding in the intracellular C-terminal part, this could potentially represent the ADAM12 generated BSG ectodomain. Supporting this idea, we previously reported that ADAM12 is highly upregulated in the same serum samples where we detected the 48 kDa band [29]. Thus, as previously suggested for ADAM12, it could be interesting to investigate if presence of this BSG fragment in plasma samples constitutes a potential prognostic marker.

In line with the shed BSG ectodomain being a potential biomarker in cancer, we found a number of cancer-associated mutations in the region harboring the ADAM12 cleavage site. When introducing the individual mutations in the exogenous BSG-AP substrate, we found two of the mutations (T199A and A207T) to be more susceptible to ADAM12-mediated shedding. Also, one of these BSG mutations (T199A) appeared to change the size of the shed BSG fragment, thereby potentially shifting the ADAM12 cleavage site. Indeed, some variability of the cleavage site has been reported for the shedding of BSG by MMP14 [13]. Looking into the structural information, BSG has been shown to function as a dimer [34], and the dimerization of the two BSG molecules were suggested to take place around amino acids 184–196 [35], which is exactly at the beginning of the region harboring the identified cancer mutations. While our data does not reveal any effects of BSG dimerization on shedding efficiency, our findings provide an example of how proteolytic events can be affected not only at the level of the protease, but also by altering the susceptibility of the substrate for being cleaved under particular pathological conditions.

In conclusion, the present study provides mechanistic insight on the pro-tumorigenic role of ADAM12 in cancer. ADAM12-mediated shedding of BSG could promote tumor progression by stimulating MMP activation and enabling cancer cell invasion.

4. Materials and Methods

4.1. Antibodies and Reagents

Antibodies against ADAM12 were as previously described [23,36,37]. The rabbit polyclonal antibodies N-19 (sc-9752) and H-200 (sc-13976) against BSG, and goat polyclonal antibody L-19 against PLAP (sc-9757) were from Santa Cruz Biotechnology (Dallas, TX, USA). Other antibodies used were rabbit polyclonal antibody PA5-29787 against BSG (Thermo Fischer Scientific, Waltham, MA, USA), mouse monoclonal antibodies against actin (MAB1501, Millipore Chemicon, Burlington, MA, USA) and FITC-conjugated anti-CD147 (555962, BD Biosciences, San Jose, CA, USA). Living Colors® GFP Monoclonal Antibody from Clontech was used for co-immunoprecipitation studies. Secondary antibodies used were horseradish peroxidase (HRP)-conjugated goat anti-mouse, goat anti-rabbit, and rabbit anti-goat immunoglobulins from DAKO A/S (Glostrup, Denmark). Alexa Fluor 488-conjugated rabbit anti-goat IgG and goat anti-mouse IgG, and Alexa Fluor 546-conjugated goat anti-mouse IgG $F(ab)_2$ fragment and goat anti-rabbit IgG $F(ab)_2$ fragment were from Invitrogen (Taastrup, Denmark). The metalloprotease inhibitors Batimastat, GM6001, and all other chemicals were from Merck KGaA (Darmstadt, Germany).

4.2. Plasmid

Mammalian expression constructs encoding full-length human ADAM12-L (ADAM12), human ADAM12-L fused to GFP (ADAM12-GFP), or human ADAM12-L lacking the cytoplasmic tail (ADAM12-Δcyt) were as previously described [24]. A point mutation in the catalytic site (E351Q) of ADAM12 (ADAM12-E/A) was introduced using the Phusion Site-directed mutagenesis kit (Thermo Fischer Scientific, Waltham, MA, USA). For retroviral transduction, cDNA encoding ADAM12 and ADAM12-E/A was cloned into the retroviral expression vector pRevTRE (Clontech, BD Biosciences). Full-length human BSG isoform 2 (OHu27639) cDNA was obtained from GeneScript (Piscataway, NJ, USA) and inserted into the pcDNA3.1 vector. The expression construct encoding alkaline phosphatase (AP)-tagged BSG (BSG-AP) was provided by Carl Blobel (New York, NY, USA), and have been described previously [21]. Using a PCR-based method [38], single point mutations were introduced into the BSG-AP construct or the full-length BSG construct as indicated. ADAM9 and ADAM17 expression constructs were from William English (Sheffield, UK), and Gillian Murphy (Cambridge, UK), respectively.

4.3. Cell Culture and Transfections

The HEK293 cell line stably expressing the vitronectin receptor $\alpha V\beta 3$ integrin, called 293-VnR, was previously described [39]. The cancer cell lines MCF-7, HeLa, and HT1080 were from ATCC (LGC Standards AB, Boras, Sweden) and cultured as previously described [36]. All cell lines were transiently transfected using X-tremeGENE9 Transfection Reagent (Roche Applied Science, Hvidovre, Denmark). MCF-7 cells stably expressing ADAM12-Δcyt in a tetracycline (tet)-Off system (MCF7-A12) were generated by retroviral transduction of ADAM12-Δcyt in the pRevTRE vector (Clontech, BD Sciences) as described previously [40]. The MCF7-A12 cell line was kept in growth medium supplied with 50 mg/mL hygromycin B (Roche Applied Science) and 100 mg/mL geneticin (Sigma-Aldrich, St. Louis, MO, USA). Small interfering RNAs (siRNAs) against MMP14 were obtained as siGENOME SMARTpool reagents from Thermo Scientific (Dharmacon, Waltham, MA, USA), and siRNA universal negative control was from Sigma-Aldrich. siRNA transfection was performed according to the manufacturer's instructions, using OPTI-MEM I and Lipofectamine^{TM} 2000 (Invitrogen).

4.4. CRISPR/Cas9 Gene Editing

HeLa ADAM12 knockout cells were generated using the CRISPR-Cas9 system. Guide RNAs (gRNA) were designed using the WTSI genome editing tool [41] and individually inserted in the vector pSpCas9(BB)-2A-GFP [42]. To determine the gRNA editing efficiency, HeLa cells were transfected with the pSpCas9(sgRNA)-2A-GFP vectors and verified by Indel Detection by Amplicon Analysis (IDAA). gRNA 5-TACCGTGTAATTTCGAGCGA-3, targeting exon 4, showed the highest editing efficiency and was subsequently transfected into HeLa cells. GFP positive cells were single cell sorted, expanded and tested by qPCR and Western blot for ADAM12 knockout. Additionally, we screened positive clones for biallelic frameshifts using Sanger sequencing. Two individual ADAM12 knockout clones, D7 and E7, as well as one non-edited clone D4, were used for further experiments.

4.5. Human Serum Samples

Control and pregnancy serum samples obtained from healthy individuals and serum samples from patients with bladder carcinomas were obtained as previously described [29]. Informed consent was obtained from all patients and the bladder cancer protocol was approved by The National Committee on Health Research Ethics (#1708266).

4.6. Shedding Assay

ADAM12-mediated shedding of AP-tagged substrates was determined as described previously [21]. The shedding activity was calculated as AP activity in conditioned medium divided by AP activity in the medium plus corresponding cell lysate after subtracting the background signal from non-transfected cells. The value from cells transfected with the AP construct alone or together with ADAM12-E/A was set to 1. Western blot analysis of total cell lysates was performed to ensure equal expression of the constructs.

4.7. Immunofluorescence Staining

Visualization of ADAM12 and BSG by immunofluorescence staining was performed using standard techniques as previously described [36]. Cell surface staining of BSG was done using a cytospin method, as described by Kawaguchi et al. [43]. In brief, cells were trypsinized and stained without permeabilization, fixed in 4% paraformaldehyde, and spun down in a cytospin centrifuge (Sandon, Thermo Fisher Scientific, Waltham, MA, USA). For staining of adherent cells, fixation by either cold methanol or 4% paraformaldehyde was used. BSG-stained cells were counted and the total grey value was measured at 546 nm using MetaMorph software and a multi wavelength cell-scoring program. For visualization of protein co-localizing at the cell surface, Duolink reagents from Olink (Uppsala, Sweden) were used to stain non-permeabilized cells, as recently described [26]. Fluorescence imaging was performed using an inverted Zeiss Axiovert 220 Apotome system equipped with a 63/1.4 Plan-Apochromat water immersion objective. The images were processed using the Axiovision program (Carl Zeiss, Oberkochen, Germany) and MetaMorph software. Average number of foci from 50 cells from ADAM12 or ADAM12-E/A transfected cells were compared.

4.8. In Situ Gelatinase Assay

The in situ gelatinase assay was performed as previously described [24]. In brief, cells were seeded on 3.5 cm dishes coated with gelatin (10 mg/mL) coupled to Oregon green 488 dye (G-13186) from Molecular Probes (Life Technologies, Taastrup, Denmark). Twenty-four hours after, gelatin degradation was quantified by measuring the area of black holes in the fluorescent gelatin relative to the total area, using MetaMorph software (MM45, Molecular Devices, San Jose, CA, USA).

4.9. Immunoprecipitation and Western Blot Analysis

Protein immunoprecipitation and Western blot analysis were performed using standard protocols, as previously described [21,24]. In brief, 293-VnR cells transfected with ADAM12-GFP or with basigin were extracted in RIBA buffer for 20 min, containing inhibitors as described [20,23]. The extracts were incubated with antibodies for 2 h at 4 °C with gentle agitation. Protein G-SepharoseTM 4 Fast Flow beads (GE Healthcare, Chicago, IL, USA) were added for an additional 1 h at 4 °C. Beads were gently washed three times in RIPA buffer (20 mM Tris-HCl (pH 7.5) 150 mM NaCl, 1 mM Na2EDTA, 1% Triton X-100). Bound proteins were eluted in 2× sample buffer, followed by Western blot analysis. Cell surface proteins were biotinylated using non-cleavable EZ-Link Sulfo-NHS-LC-Biotin, pulled down with streptavidin–agarose, and analyzed by Western blotting as previously described [44].

4.10. Quantitative PCR

Total RNA extraction was performed using GeneJet RNA Purification kit (Thermo Scientific), cDNA synthesis, and quantitative PCR (qPCR) and was carried out as described earlier [21]. They were performed with ADAM12 primers 5-CAGGCACAAAGTGTGCAGAT-3, 5-GCTTGTGCTTCCTCCAAAGC-3 and BSG primers 5-GACGTCCTGGATGATGACGA-3, 5-GAAGAGTTCCTCTGGCGGAC-3. The gene for ribosomal phosphoprotein (RPO) was used as a house keeping reference gene, and RPO primers 5-CAGCAGTTTCTCCAGAGC-3, 5-TTCATTGTGGGAGCAGAC-3, and data were analyzed using the $2(-\Delta\Delta C_T)$ method.

4.11. Statistical Analysis

Statistical analyses of all experiments were performed for three independent repeats using Student's t test for comparing two groups and ANOVA for multiple group comparisons; $p < 0.05$ was considered statistically significant.

Supplementary Materials: Supplementary materials can be found at http://www.mdpi.com/1422-0067/20/8/1957/s1.

Author Contributions: Conceptualization, R.A.; methodology, R.A., S.G. and J.S.; software, R.A.; validation, R.A.; formal analysis, R.A.; investigation, R.A.; resources, R.A. and L.D.; data curation, R.A.; writing—original draft preparation, R.A.; writing—review and editing, R.A., N.J.W.A., S.G., J.S., L.D. and M.K.; visualization, R.A., S.G., M.K.; supervision, M.K.; project administration, R.A.; funding acquisition, M.K.

Funding: This research was funded by the Danish Cancer Society (R146-A9211-16-S2).

Acknowledgments: The BSG-AP, ADAM9, and ADAM17 expression constructs were kindly provided by Carl Blobel (New York, USA), William English (Sheffield, UK), and Gillian Murphy (Cambridge, UK), respectively. We thank Ramya Kweder for technical contribution.

Conflicts of Interest: The authors declare no conflict of interest.

Abbreviations

ADAM	A disintegrin and metalloproteinase
AP	Alkaline phosphatase
BSG	Basigin
COSMIC	Catalogue of somatic mutations in cancer
ECM	Extracellular matrix
EGFR	Epidermal growth factor receptor
EMMPRIN	extracellular matrix metalloprotease inducer
HRP	Horseradish peroxidase
MCT	Monocarboxylate transporter
MMP	Matrix metalloproteinase
PLA	Proximity ligation assay
TNFα	Tumor necrosis factor alpha

References

1. Gabison, E.E.; Hoang-Xuan, T.; Mauviel, A.; Menashi, S. EMMPRIN/CD147, an MMP modulator in cancer, development and tissue repair. *Biochimie* **2005**, *87*, 361–368. [CrossRef]
2. Halestrap, A.P. The SLC16 gene family—Structure, role and regulation in health and disease. *Mol. Asp. Med.* **2013**, *34*, 337–349. [CrossRef]
3. Muramatsu, T. Basigin (CD147), a multifunctional transmembrane glycoprotein with various binding partners. *J. Biochem.* **2016**, *159*, 481–490. [CrossRef] [PubMed]
4. Perez-Escuredo, J.; Van Hee, V.F.; Sboarina, M.; Falces, J.; Payen, V.L.; Pellerin, L.; Sonveaux, P. Monocarboxylate transporters in the brain and in cancer. *Biochim. Biophys. Acta* **2016**, *1863*, 2481–2497. [CrossRef] [PubMed]
5. Kirk, P.; Wilson, M.C.; Heddle, C.; Brown, M.H.; Barclay, A.N.; Halestrap, A.P. CD147 is tightly associated with lactate transporters MCT1 and MCT4 and facilitates their cell surface expression. *EMBO J.* **2000**, *19*, 3896–3904. [CrossRef] [PubMed]
6. Biswas, C.; Zhang, Y.; DeCastro, R.; Guo, H.; Nakamura, T.; Kataoka, H.; Nabeshima, K. The human tumor cell-derived collagenase stimulatory factor (renamed EMMPRIN) is a member of the immunoglobulin superfamily. *Cancer Res.* **1995**, *55*, 434–439.
7. Gallagher, S.M.; Castorino, J.J.; Wang, D.; Philp, N.J. Monocarboxylate transporter 4 regulates maturation and trafficking of CD147 to the plasma membrane in the metastatic breast cancer cell line MDA-MB-231. *Cancer Res.* **2007**, *67*, 4182–4189. [CrossRef] [PubMed]
8. Wu, J.; Hao, Z.W.; Zhao, Y.X.; Yang, X.M.; Tang, H.; Zhang, X.; Song, F.; Sun, X.X.; Wang, B.; Nan, G.; et al. Full-length soluble CD147 promotes MMP-2 expression and is a potential serological marker in detection of hepatocellular carcinoma. *J. Transl. Med.* **2014**, *12*, 190. [CrossRef]
9. Huang, X.; Shen, W.; Xi, H.; Zhang, K.; Cui, J.; Wei, B.; Chen, L. Prognostic role of extracellular matrix metalloproteinase inducer/CD147 in gastrointestinal cancer: A meta-analysis of related studies. *Oncotarget* **2016**, *7*, 81003–81011. [CrossRef]
10. Bauman, T.M.; Ewald, J.A.; Huang, W.; Ricke, W.A. CD147 expression predicts biochemical recurrence after prostatectomy independent of histologic and pathologic features. *BMC Cancer* **2015**, *15*, 549. [CrossRef]
11. Als, A.B.; Dyrskjot, L.; von der Maase, H.; Koed, K.; Mansilla, F.; Toldbod, H.E.; Jensen, J.L.; Ulhoi, B.P.; Sengelov, L.; Jensen, K.M.; et al. Emmprin and survivin predict response and survival following cisplatin-containing chemotherapy in patients with advanced bladder cancer. *Clin. Cancer Res.* **2007**, *13*, 4407–4414. [CrossRef]
12. Tang, Y.; Kesavan, P.; Nakada, M.T.; Yan, L. Tumor-stroma interaction: Positive feedback regulation of extracellular matrix metalloproteinase inducer (EMMPRIN) expression and matrix metalloproteinase-dependent generation of soluble EMMPRIN. *Mol. Cancer Res.* **2004**, *2*, 73–80.
13. Egawa, N.; Koshikawa, N.; Tomari, T.; Nabeshima, K.; Isobe, T.; Seiki, M. Membrane type 1 matrix metalloproteinase (MT1-MMP/MMP-14) cleaves and releases a 22-kDa extracellular matrix metalloproteinase inducer (EMMPRIN) fragment from tumor cells. *J. Biol. Chem.* **2006**, *281*, 37576–37585. [CrossRef]

14. Wu, B.; Cui, J.; Yang, X.M.; Liu, Z.Y.; Song, F.; Li, L.; Jiang, J.L.; Chen, Z.N. Cytoplasmic fragment of CD147 generated by regulated intramembrane proteolysis contributes to HCC by promoting autophagy. *Cell Death Dis.* **2017**, *8*, e2925. [CrossRef]
15. Murphy, G. The ADAMs: Signalling scissors in the tumour microenvironment. *Nat. Rev. Cancer* **2008**, *8*, 929–941. [CrossRef]
16. Edwards, D.R.; Handsley, M.M.; Pennington, C.J. The ADAM metalloproteinases. *Mol. Asp. Med.* **2008**, *29*, 258–289. [CrossRef] [PubMed]
17. Blobel, C.P. ADAMs: Key components in EGFR signalling and development. *Nat. Rev. Mol. Cell Biol.* **2005**, *6*, 32–43. [CrossRef] [PubMed]
18. Horiuchi, K.; Le Gall, S.; Schulte, M.; Yamaguchi, T.; Reiss, K.; Murphy, G.; Toyama, Y.; Hartmann, D.; Saftig, P.; Blobel, C.P. Substrate selectivity of epidermal growth factor-receptor ligand sheddases and their regulation by phorbol esters and calcium influx. *Mol. Biol. Cell* **2007**, *18*, 176–188. [CrossRef] [PubMed]
19. Higashiyama, S.; Nanba, D. ADAM-mediated ectodomain shedding of HB-EGF in receptor cross-talk. *Biochim. Biophys. Acta* **2005**, *1751*, 110–117. [CrossRef]
20. Miller, M.A.; Moss, M.L.; Powell, G.; Petrovich, R.; Edwards, L.; Meyer, A.S.; Griffith, L.G.; Lauffenburger, D.A. Targeting autocrine HB-EGF signaling with specific ADAM12 inhibition using recombinant ADAM12 prodomain. *Sci. Rep.* **2015**, *5*, 15150. [CrossRef]
21. Frohlich, C.; Klitgaard, M.; Noer, J.B.; Kotzsch, A.; Nehammer, C.; Kronqvist, P.; Berthelsen, J.; Blobel, C.; Kveiborg, M.; Albrechtsen, R.; et al. ADAM12 is expressed in the tumour vasculature and mediates ectodomain shedding of several membrane-anchored endothelial proteins. *Biochem. J.* **2013**, *452*, 97–109. [CrossRef]
22. Frohlich, C.; Nehammer, C.; Albrechtsen, R.; Kronqvist, P.; Kveiborg, M.; Sehara-Fujisawa, A.; Mercurio, A.M.; Wewer, U.M. ADAM12 produced by tumor cells rather than stromal cells accelerates breast tumor progression. *Mol. Cancer Res.* **2011**, *9*, 1449–1461. [CrossRef] [PubMed]
23. Kveiborg, M.; Frohlich, C.; Albrechtsen, R.; Tischler, V.; Dietrich, N.; Holck, P.; Kronqvist, P.; Rank, F.; Mercurio, A.M.; Wewer, U.M. A role for ADAM12 in breast tumor progression and stromal cell apoptosis. *Cancer Res.* **2005**, *65*, 4754–4761. [CrossRef] [PubMed]
24. Albrechtsen, R.; Kveiborg, M.; Stautz, D.; Vikesa, J.; Noer, J.B.; Kotzsh, A.; Nielsen, F.C.; Wewer, U.M.; Frohlich, C. ADAM12 redistributes and activates MMP-14, resulting in gelatin degradation, reduced apoptosis and increased tumor growth. *J. Cell Sci.* **2013**, *126*, 4707–4720. [CrossRef] [PubMed]
25. Bilgin Dogru, E.; Dizdar, Y.; Aksit, E.; Ural, F.; Sanli, O.; Yasasever, V. EMMPRIN and ADAM12 in prostate cancer: Preliminary results of a prospective study. *Tumour. Biol.* **2014**, *35*, 11647–11653. [CrossRef] [PubMed]
26. Kveiborg, M.; Albrechtsen, R.; Couchman, J.R.; Wewer, U.M. Cellular roles of ADAM12 in health and disease. *Int. J. Biochem. Cell Biol.* **2008**, *40*, 1685–1702. [CrossRef] [PubMed]
27. Bagchi, S.; Fredriksson, R.; Wallen-Mackenzie, A. In Situ Proximity Ligation Assay (PLA). *Methods Mol. Biol.* **2015**, *1318*, 149–159. [CrossRef]
28. Kokozidou, M.; Drewlo, S.; Bartz, C.; Raven, G.; Brandenburg, L.O.; Wruck, C.J.; Pufe, T. Complex patterns of ADAM12 mRNA and protein splice variants in the human placenta. *Ann. Anat.* **2011**, *193*, 142–148. [CrossRef]
29. Frohlich, C.; Albrechtsen, R.; Dyrskjot, L.; Rudkjaer, L.; Orntoft, T.F.; Wewer, U.M. Molecular profiling of ADAM12 in human bladder cancer. *Clin. Cancer Res.* **2006**, *12*, 7359–7368. [CrossRef]
30. Moss, M.L.; Rasmussen, F.H. Fluorescent substrates for the proteinases ADAM17, ADAM10, ADAM8, and ADAM12 useful for high-throughput inhibitor screening. *Anal. Biochem.* **2007**, *366*, 144–148. [CrossRef]
31. Taylor, P.M.; Woodfield, R.J.; Hodgkin, M.N.; Pettitt, T.R.; Martin, A.; Kerr, D.J.; Wakelam, M.J. Breast cancer cell-derived EMMPRIN stimulates fibroblast MMP2 release through a phospholipase A(2) and 5-lipoxygenase catalyzed pathway. *Oncogene* **2002**, *21*, 5765–5772. [CrossRef]
32. Knutti, N.; Kuepper, M.; Friedrich, K. Soluble extracellular matrix metalloproteinase inducer (EMMPRIN, EMN) regulates cancer-related cellular functions by homotypic interactions with surface CD147. *FEBS J.* **2015**, *282*, 4187–4200. [CrossRef]
33. Hatanaka, M.; Higashi, Y.; Fukushige, T.; Baba, N.; Kawai, K.; Hashiguchi, T.; Su, J.; Zeng, W.; Chen, X.; Kanekura, T. Cleaved CD147 shed from the surface of malignant melanoma cells activates MMP2 produced by fibroblasts. *Anticancer Res.* **2014**, *34*, 7091–7096.

34. Belton, R.J., Jr.; Chen, L.; Mesquita, F.S.; Nowak, R.A. Basigin-2 is a cell surface receptor for soluble basigin ligand. *J. Biol. Chem.* **2008**, *283*, 17805–17814. [CrossRef]
35. Yu, X.L.; Hu, T.; Du, J.M.; Ding, J.P.; Yang, X.M.; Zhang, J.; Yang, B.; Shen, X.; Zhang, Z.; Zhong, W.D.; et al. Crystal structure of HAb18G/CD147: Implications for immunoglobulin superfamily homophilic adhesion. *J. Biol. Chem.* **2008**, *283*, 18056–18065. [CrossRef]
36. Albrechtsen, R.; Stautz, D.; Sanjay, A.; Kveiborg, M.; Wewer, U.M. Extracellular engagement of ADAM12 induces clusters of invadopodia with localized ectodomain shedding activity. *Exp. Cell Res.* **2011**, *317*, 195–209. [CrossRef]
37. Iba, K.; Albrechtsen, R.; Gilpin, B.J.; Loechel, F.; Wewer, U.M. Cysteine-rich domain of human ADAM 12 (meltrin alpha) supports tumor cell adhesion. *Am. J. Pathol.* **1999**, *154*, 1489–1501. [CrossRef]
38. Hansson, M.D.; Rzeznicka, K.; Rosenback, M.; Hansson, M.; Sirijovski, N. PCR-mediated deletion of plasmid DNA. *Anal. Biochem.* **2008**, *375*, 373–375. [CrossRef]
39. Sanjay, A.; Houghton, A.; Neff, L.; DiDomenico, E.; Bardelay, C.; Antoine, E.; Levy, J.; Gailit, J.; Bowtell, D.; Horne, W.C.; et al. Cbl associates with Pyk2 and Src to regulate Src kinase activity, alpha(v)beta(3) integrin-mediated signaling, cell adhesion, and osteoclast motility. *J. Cell Biol.* **2001**, *152*, 181–195. [CrossRef]
40. Ronnov-Jessen, L.; Villadsen, R.; Edwards, J.C.; Petersen, O.W. Differential expression of a chloride intracellular channel gene, CLIC4, in transforming growth factor-beta1-mediated conversion of fibroblasts to myofibroblasts. *Am. J. Pathol.* **2002**, *161*, 471–480. [CrossRef]
41. Hodgkins, A.; Farne, A.; Perera, S.; Grego, T.; Parry-Smith, D.J.; Skarnes, W.C.; Iyer, V. WGE: A CRISPR database for genome engineering. *Bioinformatics* **2015**, *31*, 3078–3080. [CrossRef]
42. Ran, F.A.; Hsu, P.D.; Wright, J.; Agarwala, V.; Scott, D.A.; Zhang, F. Genome engineering using the CRISPR-Cas9 system. *Nat. Protoc.* **2013**, *8*, 2281–2308. [CrossRef]
43. Kawaguchi, N.; Sundberg, C.; Kveiborg, M.; Moghadaszadeh, B.; Asmar, M.; Dietrich, N.; Thodeti, C.K.; Nielsen, F.C.; Moller, P.; Mercurio, A.M.; et al. ADAM12 induces actin cytoskeleton and extracellular matrix reorganization during early adipocyte differentiation by regulating beta1 integrin function. *J. Cell Sci.* **2003**, *116*, 3893–3904. [CrossRef] [PubMed]
44. Stautz, D.; Leyme, A.; Grandal, M.V.; Albrechtsen, R.; van Deurs, B.; Wewer, U.; Kveiborg, M. Cell-surface metalloprotease ADAM12 is internalized by a clathrin- and Grb2-dependent mechanism. *Traffic* **2012**, *13*, 1532–1546. [CrossRef] [PubMed]

International Journal of
Molecular Sciences

MDPI

Article

Matrix Metalloproteinase Response of Dendritic Cell, Gingival Epithelial Keratinocyte, and T-Cell Transwell Co-Cultures Treated with *Porphyromonas gingivalis* Hemagglutinin-B

Amber M. Bates [1,*], Carol L. Fischer [2], Vrushali P. Abhyankar [3], Georgia K. Johnson [4], Janet M. Guthmiller [5], Ann Progulske-Fox [6] and Kim A. Brogden [1,4]

1 Iowa Institute for Oral Health Research, College of Dentistry, The University of Iowa, Iowa City, IA 52242, USA; kim-brogden@uiowa.edu
2 Department of Biology, Waldorf University, Forest City, IA 50436, USA; carol.fischer@waldorf.edu
3 Department of Periodontology, College of Dentistry, University of Tennessee Health Science Center, Memphis, TN 38103, USA; vabhyank@uthsc.edu
4 Department of Periodontics, College of Dentistry, The University of Iowa, Iowa City, IA 52242, USA; georgia-johnson@uiowa.edu
5 College of Dentistry, University of Nebraska Medical Center, Lincoln, NE 68583, USA; janet.guthmiller@unmc.edu
6 Center for Molecular Microbiology and Department of Oral Biology, University of Florida, Gainesville, FL 32603, USA; APFOX@dental.ufl.edu
* Correspondence: ambates@wisc.edu; Tel.: +608-563-5316

Received: 20 November 2018; Accepted: 4 December 2018; Published: 7 December 2018

Abstract: Matrix metalloproteinases (MMPs) are enzymes involved in periodontal tissue destruction. Hemagglutinin B (HagB) from the periodontal pathogen *Porphyromonas gingivalis* induces an elevated MMP response in dendritic cells, but responses from cultures of single-cell types do not reflect the local tissue environment. The objective of this study was to measure HagB-induced MMP responses in a transwell co-culture system containing dendritic cells, gingival epithelial (GE) keratinocytes, and CD4+ T-cells. Transwell co-cultures were assembled and treated with or without HagB. Immunoassays were used to determine production of MMP1, MMP7, MMP9, and MMP12 in response to HagB up to 64 h. Control responses were subtracted from HagB-induced responses. A two-way fixed effect ANOVA was fit to log-transformed concentrations and pairwise group comparisons were conducted ($p < 0.05$). At 64 h, dendritic cells produced elevated MMP1 and MMP9 responses, which were attenuated in the 3-cell co-culture ($p < 0.05$). There were also significant differences in MMP7 and MMP12 production between single-cell cultures and co-cultures. These results support the need to use multiple cell types in culture models to evaluate a more representative response to proinflammatory agonists. This three-cell transwell co-culture model may help us better understand the inflammatory process in periodontal disease and test novel therapeutic approaches.

Keywords: hemagglutinin-B; transwell co-cultures; matrix metalloproteinases

1. Introduction

Periodontal disease, a chronic inflammatory disease that causes tissue damage in the oral cavity, affects approximately 46% of U.S. adults [1]. Periodontal disease results from exposure of gingival tissue to microorganisms and microbial products in adhered polymicrobial biofilms and the subsequent host inflammatory and immune responses they induce [2]. Biofilms contain a number of early and late colonizers, including *Porphyromonas gingivalis*. The prolonged inflammatory response to this biofilm results in elevated matrix metalloproteinase (MMP) production, which in turn leads to degradation of collagen fibers and subsequent periodontal attachment and bone loss, and ultimately tooth loss if left untreated.

The periodontal inflammatory response is a complex process involving communication between cells and strict regulation of cell activities necessary for host defense [3]. A network of cytokines and chemokines directs cell migration, survival, and immune cell function. MMPs are enzymes that play a critical role in the inflammatory response either by modifying proteins post-translationally to promote rapid delivery to other cells or by inactivating these proteins to initiate or terminate the immune process [4]. These functions allow MMPs to direct systematic inflammation and regulate cytokine biosynthesis through activation of signal transduction pathways. In periodontal disease, there is an elevated MMP response associated with the inflammatory response.

Matrix metalloproteinases are major enzymes implicated in extracellular matrix degradation and tissue remodeling [5]. MMPs are part of a large family of calcium-dependent, zinc-containing endopeptidases categorized into six groups based on their respective substrates [6,7]. Healthy adults express low levels of MMPs compared to the elevated levels expressed in adults with inflammatory diseases [6]. MMPs play an important role in the physiological response during wound healing, tissue repair, morphogenesis, tooth eruption, cell communication, and remodeling after injury, and they are also implicated in diseases such as rheumatoid arthritis, atherosclerosis, cancer, and periodontal disease. With respect to periodontal disease, biofilm induced inflammation amplifies MMP expression leading to increased tissue damage.

Recently, we reported that dendritic cells produce an enhanced cytokine and MMP response when treated with Hemagglutinin B (HagB), a virulence factor of the periodontal pathogen *P. gingivalis* [8]. Because cells in single-cell culture systems cannot receive signals from surrounding cells as they would in the natural host environment, the results from single-cell culture studies are not representative of the host response. Thus, studies encompassing multiple cell types are needed to better represent the *in vivo* host response. In this study, we utilized a three-cell transwell co-culture, composed of dendritic cells, gingival epithelial (GE) keratinocytes, and CD4+ T-cells, three cell types that play an important role in the innate immune response in periodontal disease. CD4+ helper T-cells were included in the transwell co-culture because they are the most dominant type of T-cell in the gingiva and have been suggested to play roles in homeostasis and immunopathology [9,10]. Dendritic cells in the periodontal tissues detect bacteria and initiate an immune response by migrating to the lymph nodes, leading to activation of other cell types and their recruitment to these tissues [11].

The transwell co-culture allows for the exchange of extracellular biomarkers from one cell type to stimulate co-cultured cells in an allogeneic system. In this study, we explored the use of immune cells and epithelial cells from different hosts in a three-cell transwell co-culture where the cells do not have physical contact with one another. Our study compared the *P. gingivalis* HagB-induced MMP response in a uniquely stacked, three-cell transwell co-culture system containing dendritic cells, GE keratinocytes, and T-cells to the MMP response from single-cell cultures.

2. Results

2.1. Experimental Set-Up of the Three-Cell Transwell Co-Culture.

The production of MMPs in response to HagB was assessed in a modified three-cell transwell co-culture (Figure 1). The production of MMPs was first examined in samples at 0, 2, 4, 8, 16, and 32 h after HagB addition or HagB diluent ($n = 3$, Figure 2a, Figure 3a, Figure 4a, Figure 5a, and Supplemental Table S1) and then more closely examined at 64 h ($n = 9$, Figure 2b, Figure 3b, Figure 4b, Figure 5b, and Supplemental Table S2). Note thatFigure 2a, Figure 3a, Figure 4a, and Figure 5a are from one experiment with three replicate reads at each time point ($n = 3$) from 0 to 32 h, while Figure 2b, Figure 3b, Figure 4b, and Figure 5b are based on three experiments with three replicate reads each ($n = 9$) at 64 h. These values were determined by subtracting the response to the HagB diluent from the HagB-induced response. This not only eliminates the response of any reaction to the diluent, but also eliminates the cell responses induced by the other cell types in the co-cultures and any baseline MMP production from the cells when no agonist is present. In most cases, MMP production was first observed from 8 to 16 h.

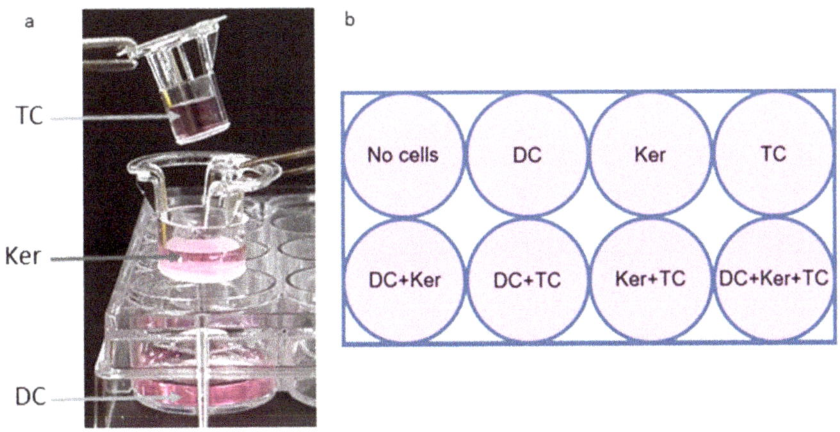

Figure 1. (**a**) Set-up of the three-cell transwell co-culture. The top insert contained CD4+ T-cells (TC), the middle insert contained GE keratinocytes (Ker) and the bottom well contained dendritic cells (DC). Cells were suspended in LGM-3. The inserts stack into each other. A porous membrane is found on the bottom of each insert, allowing small molecules, such as cytokines and MMPs, to travel between the cell layers while cells remain in their designated tiers. This plate was inserted in the incubator on a shaker plate to facilitate diffusion between the inserts. (**b**) Template used to establish the one-cell, two-cell, and three-cell co-cultures. Two plates were constructed; one was treated with HagB (test) and the other with diluent (control).

2.2. *Single-Cell Culture Responses to HagB*

The transwell co-cultures were set up with only one cell type. Each cell type was in its respective tier: Dendritic cells (bottom), GE keratinocytes (middle), and T-cells (top). By 64 h, HagB stimulated single cultures of dendritic cells, GE keratinocytes, and T-cells produced varying concentrations of MMP1, MMP7, MMP9, and MMP12 (Figure 2b, Figure 3b, Figure 4b, and Figure 5b). Of the single-cell cultures at 64 h, dendritic cells produced significantly higher amounts of MMP1 and MMP9 in response to HagB ($p < 0.05$), while GE keratinocytes and T-cells produced little to no MMP1 and MMP9. This trend of MMP1 and MMP9 production was observed throughout the 64 h, starting at 8 h (Figure 2a and Figure 4a). There was little to no production of MMP7 and MMP12 in the single-cell cultures, as there was no significant difference ($p > 0.05$) from the no cell control at 64 h (Figure 3b and Figure 5b). However, from the detection of MMP12 from 0 to 32 h, dendritic cells were the highest single-cell culture to produce MMP12 in response to HagB. In general, GE keratinocytes and T-cells single-cell cultures produced little to no MMP response to HagB of the four MMPs examined.

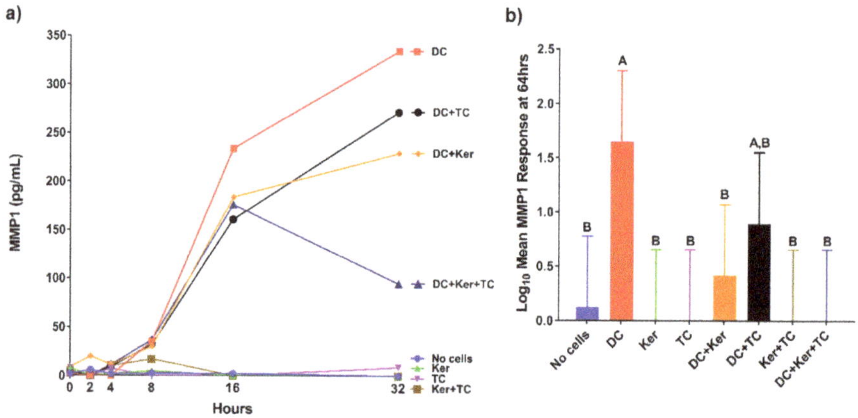

Figure 2. (**a**) Matrix metalloproteinase 1 (MMP1) responses of single cell and co-cultures when treated with hemagglutinin B (HagB) ($n = 3$). HagB was added at Time 0 to single-cell cultures and co-cultures of T-cells (TC), GE keratinocytes (Ker), and dendritic cells (DC). MMP1 concentrations were determined at 0, 2, 4, 8, 16, and 32 h. The MMP1 responses from the HagB diluent were subtracted from the HagB-induced MMP1 responses. (**b**) MMP1 responses to HagB at 64 h ($n = 9$). The MMP1 responses from the HagB diluent were subtracted from the HagB-induced MMP1 responses. A two-way fixed effect ANOVA was fit to log-transformed concentrations of MMP1. Pairwise group comparisons were conducted using the post-hoc Tukey's honest significance test. A 0.05 level was used determine statistically significant differences between groups. Shared letters above bars indicate the groups are not statistically different from one another ($p > 0.05$), while differing letters indicate significantly different mean values ($p < 0.05$).

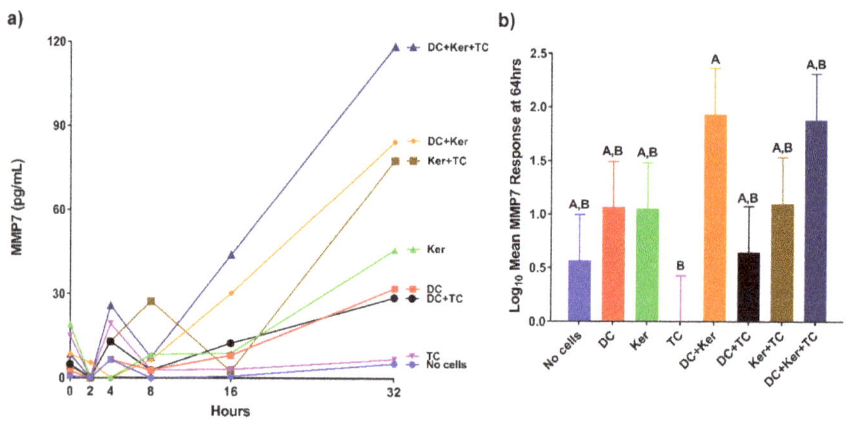

Figure 3. (**a**) MMP7 responses of single cell and co-cultures when treated with HagB ($n = 3$). HagB was added at Time 0 to single-cell cultures and co-cultures of T-cells (TC), GE keratinocytes (Ker), and dendritic cells (DC). MMP7 concentrations were determined at 0, 2, 4, 8, 16, and 32 h. The MMP7 responses from the HagB diluent were subtracted from the HagB-induced MMP7 responses. (**b**) MMP7 responses to HagB at 64 h ($n = 9$). The MMP7 responses from the HagB diluent were subtracted from the HagB-induced MMP7 responses. A two-way fixed effect ANOVA was fit to log-transformed concentrations of MMP7. Pairwise group comparisons were conducted using the post-hoc Tukey's honest significance test. A 0.05 level was used determine statistically significant differences between groups. Shared letters above bars indicate the groups are not statistically different from one another ($p > 0.05$), while differing letters indicate significantly different mean values ($p < 0.05$).

2.3. Multiple Cell Types in the Three-cell Transwell Co-Culture System

As noted above, dendritic cells produced the highest amount of MMP1 in single-cell cultures, but when dendritic cells were cultured with GE keratinocytes or GE keratinocytes + T-cells, production of MMP1 was attenuated ($p < 0.05$). At 64 h, the dendritic cell MMP1 response was attenuated in dendritic cells + GE keratinocytes ($p < 0.05$) and dendritic cells + GE keratinocytes + T-cells transwell co-cultures ($p < 0.05$) (Figure 2b).

Dendritic cells were also the main drivers of the MMP9 response to HagB, as all cultures containing dendritic cells had the greatest response (Figure 4). However, at 64 h the three-cell co-culture values were significantly less than the dendritic cells alone, dendritic cells + GE keratinocytes, and dendritic cells + T-cells cultures, suggesting the presence of both T-cells and GE keratinocytes caused a decrease in the production of MMP9 by dendritic cells in response to HagB.

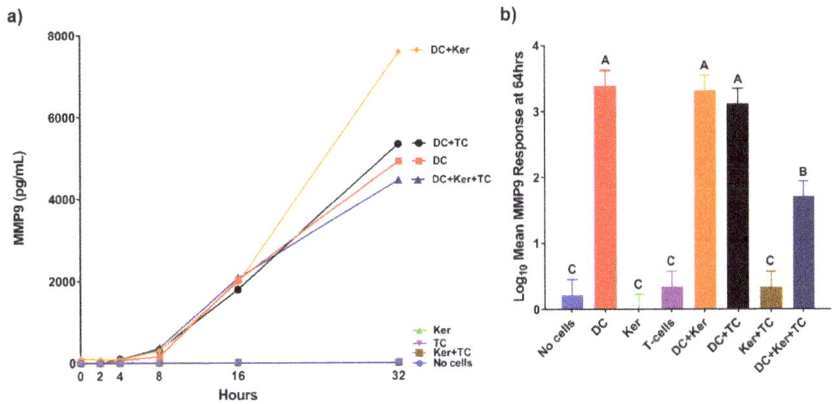

Figure 4. (**a**) MMP9 responses of single cell and co-cultures when treated with HagB ($n = 3$). HagB was added at Time 0 to single-cell cultures and co-cultures of T-cells (TC), GE keratinocytes (Ker), and dendritic cells (DC). MMP9 concentrations were determined at 0, 2, 4, 8, 16, and 32 h. The MMP9 responses from the HagB diluent were subtracted from the HagB-induced MMP9 responses. (**b**) MMP9 responses to HagB at 64 h ($n = 9$). The MMP9 responses from the HagB diluent were subtracted from the HagB-induced MMP9 responses. A two-way fixed effect ANOVA was fit to log-transformed concentrations of MMP9. Pairwise group comparisons were conducted using the post-hoc Tukey's honest significance test. A 0.05 level was used determine statistically significant differences between groups. Shared letters above bars indicate the groups are not statistically different from one another ($p > 0.05$), while differing letters indicate significantly different mean values ($p < 0.05$).

Like the single-cell cultures above, there was little to no production of MMP7 and MMP12 in the transwell co-cultures in response to HagB, as there was no significant difference from the no cell control at 64 h (Figure 3b and Figure 5b). However, from the detection of MMP7 and MMP12 from 0 to 32 h, it seems dendritic cell cultures may have been the main producer of MMP12 as it was produced in all cultures containing dendritic cells (Figure 3a and Figure 5a). Dendritic cells + GE keratinocytes and dendritic cells + GE keratinocytes + T-cells produced the greatest amount of MMP7 (Figure 3). MMP7 and MMP12 responses were suggestively enhanced in the dendritic cells + GE keratinocytes + T-cells three-cell transwell co-culture relative to the single-cell cultures, although that enhancement was not significant (Figures 3 and 5).

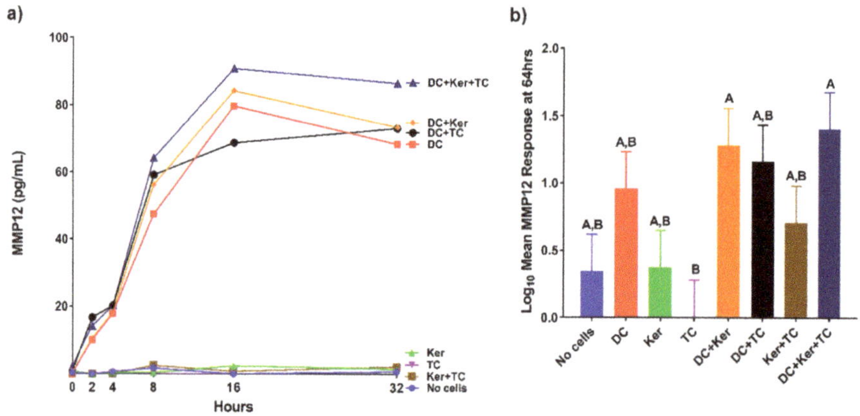

Figure 5. (**a**) MMP12 responses of single cell and co-cultures when treated with HagB (n = 3). HagB was added at Time 0 to single-cell cultures and co-cultures of T-cells (TC), GE keratinocytes (Ker), and dendritic cells (DC). MMP12 concentrations were determined at 0, 2, 4, 8, 16, and 32 h. The MMP12 responses from the HagB diluent were subtracted from the HagB-induced MMP12 responses. (**b**) MMP12 responses to HagB at 64 h (n = 9). The MMP12 responses from the HagB diluent were subtracted from the HagB-induced MMP12 responses. A two-way fixed effect ANOVA was fit to log-transformed concentrations of MMP12. Pairwise group comparisons were conducted using the post-hoc Tukey's honest significance test. A 0.05 level was used determine statistically significant differences between groups. Shared letters above bars indicate the groups are not statistically different from one another ($p > 0.05$), while differing letters indicate significantly different mean values ($p < 0.05$).

3. Discussion

There has been little research on characterizing MMP responses of cells grown in three-cell transwell co-culture systems using primary GE keratinocytes from the oral cavity. We used the three-cell transwell co-cultures to elucidate cell-to-cell communication that regulates MMP production induced by proinflammatory agonists such as HagB. Co-cultures produced significantly different MMP responses than single-cell cultures, and cytokine and chemokine responses also differed [12].

In this model, dendritic cells drove MMP production, but the expression of MMPs changed in the presence of GE keratinocytes and T-cells. Using transwell co-cultures, we demonstrated how the MMP responses of three cell types differed in a co-culture environment when exposed to the proinflammatory agonist HagB. This experimental approach highlights the contribution and influence multiple cell types in co-culture can have on the MMP response, which plays a role in the pathogenesis of periodontal tissue destruction. Studying the MMP response in this model provides insight into intra- and inter-cellular regulation of MMPs and the role of multiple cell types in the overall MMP inflammatory response.

Knowing more about the MMP response of the immune system can provide us with a better understanding of regulation of MMP expression. For example, MMP9 cleaves and activates interleukin-8 (IL8), a cytokine that enhances neutrophil recruitment to the site of infection [13]. MMP9 has a major role in extracellular matrix degradation and tissue remodeling [14]. HagB induced a high MMP9 response from dendritic cells in this study. MMP9 is upregulated during wound healing, development, and in inflammatory pathologies. Furthermore, in inflammatory diseases such as arthritis, cardiovascular disease, cancer, and diabetes, MMP9 stimulates the immune response initiating pathogenesis and increasing disease progression. TNFα causes upregulation of MMP9 expression [14]. MMP9 can also recruit eosinophils and T-cells to the site of infection. In this study, the dendritic cell MMP9 response was significantly decreased in the presence of T-cells and GE keratinocytes when all three cell types were cultured together, but not when dendritic cells were cultured with GE keratinocytes or T-cells (Figure 4b). The presence of both GE keratinocytes and T-cells caused an attenuation of the dendritic cell response, suggesting a synergistic inhibition effect of MMP9 expression to prevent further immune cell recruitment. One possibility is that T-cells signal the GE keratinocytes to produce tissue inhibitors of metalloproteinases, which in turn leads to a reduction in MMP9 production, but this has yet to be shown.

Similarly, dendritic cells produced more collagenase MMP1 than other cell types, but this response was significantly attenuated when all three cell types were cultured together (Figure 2b). MMP1 is elevated during tissue remodeling, wound healing, and repair [15]. Therefore, we expected MMP1 expression to be enhanced by the presence of an agonist like HagB, which occurred in the dendritic cell single-cell culture. When cultured with GE keratinocytes, the dendritic cells' response is significantly attenuated. This suggests GE keratinocytes may inhibit MMP1 expression in dendritic cells. One possibility is that the GE keratinocytes are inhibiting one of the cytokine inducers of MMP1 expression such as growth factors and interleukins [15].

Co-culture models differ from other systems because of their unique ability to focus on the cellular communication between different cell types over time. Standard *in vitro* culture systems are limited by the inability to represent all of the immune mediators found in the host environment. In particular, a more complex multi-dimensional system that can more closely mimic the *in vivo* inflammatory response is required to generate a comprehensive understanding of the process underlying periodontal disease [16]. There are numerous co-culture models, each with their own intrinsic advantages and disadvantages, which have been extensively described in the current literature [17–23]. These models generally consist of autologous cells cultured together (e.g., organoids, spheroids, etc.), or allogeneic cells cultured together but physically separated (e.g., transwell co-cultures, etc.). They can be constructed as 2D or 3D systems and may involve different cell types, different types of membranes, and different sources of collagen or matrix.

The three-cell transwell co-culture system developed for this study is not without its own advantages and disadvantages. We used a heterotypic, transwell co-culture system that does not allow contact among the different cell types by separation using membranes as barriers between the cells. By removing cell–cell physical interactions, we can focus on the effects of cell signals elicited by the agonist or from other cells in the co-culture. Preventing cell-to-cell contact allows us to use cells from different individuals that would otherwise be incompatible. One limitation of this study is that the cells are not necessarily in the same ratios as they would be *in vivo*. With this in mind, we chose to use equal concentrations of each cell type because it is easily repeatable and can be plated in the transwell plates at a reasonable density.

In conclusion, a number of factors could be causing the attenuation of dendritic cell expression of MMP9 and MMP1 in the three-cell co-culture. This study provided insight into MMP regulation in the host environment and the effects of a proinflammatory agonist, HagB. Learning more about these signaling pathways and MMP involvement may lead to identification of possible targets for therapeutic approaches. Further analysis will look for connections between the MMP responses over time and the associated cytokine responses, which will provide more information on the inflammatory response.

In silico models using bioinformatics are being developed to help us study and analyze inflammatory responses in environments like the periodontal tissues where multiple factors are involved [24]. Previously, we used a computational model to examine the HagB-induced MMP response [10]. Similarly, the results from this study will be used to modify this computational model so that it can be used to search drug databases for potential MMP inhibitors. These drugs can then be tested in the three-cell transwell co-culture to verify their effectiveness in suppressing MMP and inflammatory cytokine expression. This will help us identify and test possible therapeutic approaches for suppressing the inflammatory response in diseases such as periodontal disease.

4. Materials and Methods

4.1. Preparation of HagB

Hemagglutinin B (HagB) from *P. gingivalis (Pg strain 381)* was cloned and His-tagged into the vector pQE31 and expressed in *Escherichia coli* (M15 pREP4 pQE-31-TX1, QIAGEN Inc., Valencia, CA, USA), which was obtained and prepared similarly to what was previously described [25]. Briefly, recombinant *E. coli* was grown at 37 °C in selective Luria-base (LB) broth with ampicillin and kanamycin A. After 4 h, 1 mM Isopropyl B-D-1-thiogalactopyranoside (IPTG) per liter of culture was added to induce HagB expression. After an additional 4 h of incubation, 10.0 mL of 1.54 M sodium azide (NaN_3) per liter of culture was added to arrest growth. The culture was pelleted and washed in 10.0 mL of 140 mM phosphate buffered saline (PBS). The PBS was removed, and the cells were resuspended in 10 mL of Buffer A (6.0 M Guanidine-HCl, 0.1 M NaH_2PO_4, 0.01 M Tris acid, pH 8.0) to lyse the *E. coli*. The cells were pelleted and the supernatant containing the HagB was collected.

To purify the HagB, affinity chromatography was performed using the BioRad Biologic low-pressure pump and a column containing Profinity IMAC Ni-charged resin (Bio-Rad Laboratories, Inc., Hercules, CA, USA), which binds His-tagged proteins. After the column was equilibrated with Buffer A, the supernatant containing HagB was passed through the column. This elution was collected and discarded since the proteins remained in the column. The affinity-attached HagB was washed with 8.0 M Urea, 0.1 M NaH_2PO_4, 0.01 M Tris acid, pH 8.0, followed by a second wash with 8.0 M Urea, 0.1 M NaH_2PO_4, 0.01 M Tris acid, pH 6.3. To refold the HagB, a buffer containing 6.0 M Urea, 0.5 M NaCl, 0.01 M Tris acid, 20.0% glycerol, pH 7.4 was applied to the column. The next refolding buffer applied was the same buffer but with 3.0 M Urea followed by buffer without Urea. The last buffer (0.25 M Imidazole, 0.5 M NaCl, 0.01 M Tris acid, 20.0% glycerol, pH 7.4) added to the column was a refolding buffer with imidazole to elute the HagB using a fraction collector. The protein concentration of each fraction was obtained using a NanoDrop (NanoDrop Technologies, Inc., Wilmington, DE, USA), and fractions containing the most protein were pooled.

This protein sample was applied to a desalting column (Bio-Rad Laboratories, Inc., Hercules, CA, USA) to remove imidazole and for further purification using the BioRad Biologic low-pressure pump. The column was equilibrated with a refolding buffer of 0.5 M NaCl, 0.01 M Tris acid, 20.0% glycerol, pH 7.4. The pooled fraction sample was run through the column followed by the refolding buffer without urea. The protein concentration was quantified using a NanoDrop, and fractions containing the most protein were pooled together.

4.2. Analysis of HagB

To determine the protein concentration of the sample more precisely, a BCA protein concentration assay was utilized, and found to be approximately 0.445 mg/mL (Pierce bicinchoninic acid (BCA) Protein Assay Kit, ThermoFisher Scientific, Rockford, IL, USA). The final concentration of LPS was 2.44 ng LPS/1.0 μg HagB (QCL-1000, Chromogenic Limulus Amebocyte Lysate Assay, Lonza, Inc. Walkersville, MD, USA). An SDS-Page of the sample was performed to verify purification of HagB using a NuPAGE Novex 4% to 12% Bis-Tris protein gel. This produced a solid band at about 46 kDa. A MALDI-TOF mass spectrometry analysis confirmed the presence of HagB in this band with the greatest number of matches and sequences of all proteins found. HagB was the only protein found from *P. gingivalis*.

4.3. Dendritic cells, GE keratinocytes, and T-cells

Dendritic cells, GE keratinocytes, and T-cells were chosen for this experiment due to their presence in the periodontium and their innate immune functions. Human monocyte-derived immature myeloid dendritic cells (AllCells, Alameda, CA, USA) were thawed, washed with LGM-3 (Lymphocyte Growth Media-3, Lonza, Walkersville, MD, USA), counted, pelleted by centrifugation (400× *g*) at 4 °C for 10 min, and then resuspended in LGM-3 to 1×10^5 viable cells/mL. CD4+ T-cells (StemCell Technologies, Inc.,

Vancouver, BC, Canada) were activated using a human T-cell Activation/Expansion kit (Miltenyi Biotec, San Diego, CA, USA) in TexMACS medium (Miltenyi Biotec) for 3 days at 37 °C with 5% CO_2. The T-cells were then collected, pooled, and a viable cell count was obtained. The T-cells were resuspended to 1×10^5 viable activated cells/mL in LGM-3. Primary first passage GE keratinocytes were used, which were previously prepared and stored in liquid nitrogen [26]. These cells were isolated from healthy, non-smoking individuals undergoing crown lengthening or canine exposure procedures. In compliance with a protocol approved by the University of Iowa Institutional Review Board for the Use of Human Subjects in Research (199811030, approval date 11 June 2005), informed consent was previously obtained from these patients.

All three cell types were centrifuged at 400× *g* at 4 °C for 10 min and resuspended in LGM-3 to 1×10^5 viable cells/mL. LGM-3 was used because all three cell types can survive and remain active in this serum-free media.

4.4. Establishment of Co-Culture

The three-cell transwell co-culture system was made using transwell 12-well polystyrene plates (3104, Corning, Inc., Corning, NY, USA) and inserts from the transwell 24-well polystyrene plates (3413, Corning, Inc., Corning, NY, USA). 200 µL of 1×10^5 cells/mL of dendritic cells were added to the bottom well of the transwell 12-well plate along with 800 µL of LGM-3. 200 µL of 1×10^5 cells/mL of GE keratinocytes were added to the middle 12 mm insert along with 300 µL of LGM-3. 200 µL of 1×10^5 cells/mL of activated T-cells were added to the top 6.5 mm insert. These inserts have a permeable membrane with a pore size of 0.4 um to allow media to be shared between the three cell types. The cell layers are organized in this manner due to cell properties. For example, dendritic cells are placed in the bottom well so they can spread out and adhere to the plate. GE Keratinocytes also somewhat adhere so they were given the middle tier, while the T-cells do not adhere and therefore do not require as much surface area on the plate. The plates were placed on a plate shaker at low speed in the incubator at 37 °C with 5% CO_2 (Figure 1a). This created a diffusible 3-tier platform. The total volume was 1700 µL/well.

To compare the responses between different cell types, single-cell responses, and the response of all three cell types together, combinations of cell types and single cells were set up in the same manner (Figure 1b). In total, there were three replicates of each well. LGM-3 was added in place of cell suspensions not present in a well. The HagB diluent (refolding buffer: 0 M Urea, 0.5 M NaCl, 0.01 M Tris acid, 20% glycerol, pH 7.4) was used as the treatment control, which allows any responses from the buffer to be subtracted out. The plates were incubated for 2 h at 37 °C to allow the cells to settle and adjust.

HagB was then added at a concentration of 10.0 µg/mL. Preliminary tests were used to determine the appropriate concentration of HagB to use by testing varying concentrations and their respective dendritic cell response. An equal amount of buffer was added to control wells. Plates were placed on a titer plate shaker at a low speed for 60 s to diffuse the treatment. 200 µL of supernatant was collected from each well as a time zero sample, and 200 µL of LGM-3 was added back to each well. Samples were centrifuged for five minutes at 0.8× *g* to pellet any cells or debris, and supernatants were stored at −80 °C. The plates were placed on the shaker and incubated at 37 °C, 5% CO_2 for 64 h. Samples were collected in this same manner at 2, 4, 8, 16, and 32 h. At 64 h, 2.0 mL centrifugal filter tubes were used to pool the media, one tier at a time.

4.5. Determination of MMP Concentrations

The concentrations of MMP1, MMP7, MMP9, and MMP12 were first evaluated to examine production over 32 h. This was done using multiplex immunoassays (Human Magnetic Luminex Screening Assay, R & D Systems, Minneapolis, MN, USA) read on the Luminex^{100} (Luminex 100 IS Instrument, Luminex, Austin, TX, USA) (n = 3) (Figure 2a, Figure 3a, Figure 4a, and Figure 5a). After the experiment was repeated twice more, MMP concentrations at 64 h were further evaluated and compared ($n = 9$) using multiplex immunoassays (Milliplex MAP Human MMP Magnetic Bead Panels, EMD Millipore, Billerica, MA, USA) read on the Luminex^{100} (Figure 2b, Figure 3b, Figure 4b, and Figure 5b). These immunoassay kits use antibody-coated magnetic beads to bind desired analytes and use a standard curve of known concentrations to determine unknown concentrations as previously described [27].

4.6. Statistical Analysis

MMP concentrations were interpolated from curves constructed from the MMP standards and their respective median fluorescence intensity (MFI) readings on the Luminex^{100} and readout files (Milliplex Analyst v5.1 software, EMD Millipore, Billerica, MA, USA). First, the 0 to 32 h results from the immunoassay (Human Magnetic Luminex Screening Assay, R & D Systems, Minneapolis, MN, USA) were analyzed ($n = 3$). The responses of the control plate (HagB diluent) were subtracted from the responses of the test plate (HagB) to assess only the effect of the presence of HagB. A two-way fixed effect ANOVA was fit to log-transformed concentrations of the MMPs (Analysis performed using JMP, Version 10.0, SAS, Cary, NC, USA). Pairwise group comparisons were conducted using the post-hoc Tukey's honest significance test. A significance level of 0.05 was used determine statistically significant differences between groups.

The co-culture experiment was then repeated three times, and three repeats from each experiment at 64 h were included in the immunoassay ($n = 9$). Responses from the control plate (HagB diluent) were subtracted from the responses from the test plate (HagB) to assess only the effect of HagB. Once again, results were analyzed using a two-way fixed effect ANOVA fit to log-transformed data and pairwise group comparisons were made.

Supplementary Materials: Supplementary materials can be found at http://www.mdpi.com/1422-0067/19/12/3923/s1.

Author Contributions: Conceptualization, A.M.B., C.L.F., V.P.A. and K.A.B.; Data curation, A.M.B. and K.A.B.; Formal analysis, A.M.B. and K.A.B.; Funding acquisition, K.A.B.; Investigation, A.M.B.; Methodology, A.M.B., C.L.F. and K.A.B.; Project administration, K.A.B.; Resources, G.K.J., J.M.G., A.P.-F. and K.A.B.; Supervision, C.L.F. and K.A.B.; Validation, A.M.B. and K.A.B.; Visualization, A.M.B.; Writing–original draft, A.M.B.; Writing–review and editing, A.M.B., C.L.F., V.P.A., G.K.J., J.M.G., A.P.-F. and K.A.B.

Funding: This research was supported by a grant from the National Institute of Dental and Craniofacial Research (NIDCR) of the National Institutes of Health (R01 DE014390) and a training grant from the National Institute of Dental and Craniofacial Research (NIDCR) of the National Institutes of Health (T90 DE023520).

Conflicts of Interest: The authors declare no conflicts of interest.

References

1. Eke, P.I.; Dye, B.A.; Wei, L.; Slade, G.D.; Thornton-Evans, G.O.; Borgnakke, W.S.; Taylor, G.W.; Page, R.C.; Beck, J.D.; Genco, R.J. Update on Prevalence of Periodontitis in Adults in the United States: NHANES 2009 to 2012. *J. Periodontol.* **2015**, *86*, 611–622. [CrossRef] [PubMed]
2. Giannobile, W.V. Host-response therapeutics for periodontal diseases. *J. Periodontol.* **2008**, *79*, 1592–1600. [CrossRef] [PubMed]
3. Kinane, D.F.; Preshaw, P.M.; Loos, B.G. Periodontology Working Group 2 of Seventh European Workshop on. Host-response: Understanding the cellular and molecular mechanisms of host-microbial interactions–consensus of the Seventh European Workshop on Periodontology. *J. Clin. Periodontol.* **2011**, *38*, 44–48. [CrossRef] [PubMed]
4. Khokha, R.; Murthy, A.; Weiss, A. Metalloproteinases and their natural inhibitors in inflammation and immunity. *Nat. Rev. Immunol.* **2013**, *13*, 649–695. [CrossRef] [PubMed]
5. Nawaz, M.; Shah, N.; Zanetti, B.R.; Maugeri, M.; Silvestre, R.N.; Fatima, F.; Neder, L.; Valadi, H. Extracellular Vesicles and Matrix Remodeling Enzymes: The Emerging Roles in Extracellular Matrix Remodeling, Progression of Diseases and Tissue Repair. *Cells* **2018**, *7*, 167. [CrossRef] [PubMed]

6. Sapna, G.; Gokul, S.; Bagri-Manjrekar, K. Matrix metalloproteinases and periodontal diseases. *Oral Dis.* **2014**, *20*, 538–550. [CrossRef]
7. Nagase, H.; Visse, R.; Murphy, G. Structure and function of matrix metalloproteinases and TIMPs. *Cardiovasc Res.* **2006**, *69*, 562–573. [CrossRef] [PubMed]
8. Harvey, L.E.; Kohlgraf, K.G.; Mehalick, L.A.; Raina, M.; Recker, E.N.; Radhakrishnan, S.; Prasad, S.A.; Vidva, R.; Progulske-Fox, A.; Cavanaugh, J.E.; et al. Defensin DEFB103 bidirectionally regulates chemokine and cytokine responses to a pro-inflammatory stimulus. *Sci. Rep.* **2013**, *3*, 1232. [CrossRef]
9. Dutzan, N.; Konkel, J.E.; Greenwell-Wild, T.; Moutsopoulos, N.M. Characterization of the human immune cell network at the gingival barrier. *Mucosal Immunol.* **2016**, *9*, 1163–1172. [CrossRef]
10. Raina, M.; Bates, A.M.; Fischer, C.L.; Progulske-Fox, A.; Abbasi, T.; Vali, S.; Brogden, K.A. Human beta defensin 3 alters matrix metalloproteinase production in human dendritic cells exposed to *Porphyromonas gingivalis* hemagglutinin B. *J. Periodontol.* **2018**, *89*, 361–369. [CrossRef]
11. Song, L.; Dong, G.; Guo, L.; Graves, D.T. The function of dendritic cells in modulating the host response. *Mol. Oral Microbiol.* **2018**, *33*, 13–21. [CrossRef] [PubMed]
12. Abhyankar, V.P.; Bates, A.M.; Fischer, C.L.; Johnson, G.K.; Guthmiller, J.M.; Progulske-Fox, A.; Brogden, K.A. Chemokine and cytokine responses of dendritic cell, gingival epithelial keratinocyte, and T-cell 3D transwell co-cultures treated with *Porphyromonas gingivalis* hemagglutinin B. *Data Brief* **2018**. Manuscript submitted.
13. Van den Steen, P.E.; Proost, P.; Wuyts, A.; Van Damme, J.; Opdenakker, G. Neutrophil gelatinase B potentiates interleukin-8 tenfold by aminoterminal processing, whereas it degrades CTAP-III, PF-4, and GRO-alpha and leaves RANTES and MCP-2 intact. *Blood* **2000**, *96*, 2673–2681. [PubMed]
14. Yabluchanskiy, A.; Ma, Y.; Iyer, R.P.; Hall, M.E.; Lindsey, M.L. Matrix metalloproteinase-9: Many shades of function in cardiovascular disease. *Physiology* **2013**, *28*, 391–403. [CrossRef] [PubMed]
15. Pardo, A.; Selman, M. *MMP-1:* The elder of the family. *Int. J. Biochem. Cell Biol.* **2005**, *37*, 283–288. [CrossRef] [PubMed]
16. Kabadi, P.K.; Vantangoli, M.M.; Rodd, A.L.; Leary, E.; Madnick, S.J.; Morgan, J.R.; Kane, A.; Boekelheide, K. Into the depths: Techniques for in vitro three-dimensional microtissue visualization. *Biotechniques* **2015**, *59*, 279–286. [CrossRef] [PubMed]
17. Golinski, P.A.; Groger, S.; Herrmann, J.M.; Bernd, A.; Meyle, J. Oral mucosa model based on a collagen-elastin matrix. *J. Periodontal Res.* **2011**, *46*, 704–711. [CrossRef] [PubMed]
18. Augustine, T.N.; Dix-Peek, T.; Duarte, R.; Candy, G.P. Establishment of a heterotypic 3D culture system to evaluate the interaction of TREG lymphocytes and NK cells with breast cancer. *J. Immunol. Methods* **2015**, *426*, 1–13. [CrossRef]
19. Broutier, L.; Mastrogiovanni, G.; Verstegen, M.M.; Francies, H.E.; Gavarro, L.M.; Bradshaw, C.R.; Allen, G.E.; Arnes-Benito, R.; Sidorova, O.; Gaspersz, M.P.; et al. Human primary liver cancer-derived organoid cultures for disease modeling and drug screening. *Nat. Med.* **2017**, *23*, 1424–1435. [CrossRef]
20. Bao, K.; Papadimitropoulos, A.; Akgul, B.; Belibasakis, G.N.; Bostanci, N. Establishment of an oral infection model resembling the periodontal pocket in a perfusion bioreactor system. *Virulence* **2015**, *6*, 265–273. [CrossRef]
21. Sobue, T.; Bertolini, M.; Thompson, A.; Peterson, D.E.; Diaz, P.I.; Dongari-Bagtzoglou, A. Chemotherapy-induced oral mucositis and associated infections in a novel organotypic model. *Mol. Oral Microbiol.* **2018**, *33*, 212–223. [CrossRef] [PubMed]
22. Gursoy, U.K.; Gursoy, M.; Kononen, E.; Sintim, H.O.; Uitto, V.J.; Syrjanen, S. Construction and characterization of a multilayered gingival keratinocyte culture model: The TURK-U model. *Cytotechnology* **2016**, *68*, 2345–2354. [CrossRef] [PubMed]
23. Griffith, L.G.; Swartz, M.A. Capturing complex 3D tissue physiology in vitro. *Nat. Rev. Mol. Cell Biol.* **2006**, *7*, 211–224. [CrossRef] [PubMed]
24. Meyle, J.; Dommisch, H.; Groeger, S.; Giacaman, R.A.; Costalonga, M.; Herzberg, M. The innate host response in caries and periodontitis. *J. Clin. Periodontol.* **2017**, *44*, 1215–1225. [CrossRef] [PubMed]
25. Borgwardt, D.S.; Martin, A.D.; Van Hemert, J.R.; Yang, J.; Fischer, C.L.; Recker, E.N.; Nair, P.R.; Vidva, R.; Chandrashekaraiah, A.; Progulske-Fox, A.; et al. Brogden. Histatin 5 binds to *Porphyromonas gingivalis* hemagglutinin B (HagB) and alters HagB-induced chemokine responses. *Sci. Rep.* **2014**, *4*, 3904. [CrossRef] [PubMed]

26. Joly, S.; Organ, C.C.; Johnson, G.K.; McCray, P.B., Jr.; Guthmiller, J.M. Correlation between beta-defensin expression and induction profiles in gingival keratinocytes. *Mol. Immunol.* **2005**, *42*, 1073–1084. [CrossRef] [PubMed]
27. Thunell, D.H.; Tymkiw, K.D.; Johnson, G.K.; Joly, S.; Burnell, K.K.; Cavanaugh, J.E.; Brogden, K.A.; Guthmiller, J.M. A multiplex immunoassay demonstrates reductions in gingival crevicular fluid cytokines following initial periodontal therapy. *J. Periodontal Res.* **2010**, *45*, 148–152. [CrossRef]

International Journal of
Molecular Sciences

MDPI

Article

Metalloproteinases TACE and MMP-9 Differentially Regulate Death Factors on Adult and Neonatal Monocytes After Infection with *Escherichia coli*

Stephan Dreschers [1,†], Christopher Platen [1,†], Andreas Ludwig [2], Christian Gille [3], Natascha Köstlin [2] and Thorsten W. Orlikowsky [1,*]

1 Department of Neonatology, University Children's Hospital, Aachen 52074, Germany; sdreschers@ukaachen.de (S.D.); cplaten@ukaachen.de (C.P.)
2 Department of Pharmacology and Toxicology, University Hospital, Aachen 52074, Germany; aludwig@ukaachen.de (A.L.); natascha.koestlin@med.uni-tuebingen.de (N.K.)
3 Department of Neonatology, University Children's Hospital, Tuebingen 72074, Germany; christian.gille@med.uni-tuebingen.de
* Correspondence: torlikowsky@ukaachen.de; Tel.: +0049-241-8089214; Fax: 0049-241-802437
† These authors contributed equally to this work.

Received: 28 December 2018; Accepted: 28 February 2019; Published: 20 March 2019

Abstract: Background: Cleaving ligands and receptors of the tumor necrosis factor (TNF) superfamily can critically regulate the induction of apoptosis. Matrix metalloproteinases (MMPs) such as MMP-9 and tumor necrosis factor-α-converting enzyme (TACE) have been shown to cleave CD95-Ligand (CD95L) and TNF/(TNF receptor-1) TNFR1 which induce phagocytosis induced cell death (PICD) in adult monocytes. This process is reduced in neonatal monocytes. **Methods:** Here we tested in vitro, whether *Escherichia coli* infection mounts for activation of MMP-9 and TACE in monocytes and whether this process regulates PICD. **Results:** The surface expression of TACE was most prominent on infected adult monocytes. In contrast, surface presentation of MMP-9 was highest on infected neonatal monocytes. Selective blocking of MMP-9 decreased CD95L secretion, while inhibition of TACE left CD95L secretion unaltered. Blocking of MMP-9 increased surface CD95L (memCD95L) expression on infected neonatal monocytes to levels comparable to infected adult monocytes. Moreover, MMP-9 inhibition raised PICD of infected neonatal monocytes to levels observed for infected adult monocytes. In contrast, TACE inhibition decreased PICD in infected monocytes. Addition of extracellular TNF effectively induced memCD95L presentation and PICD of adult monocytes and less of neonatal monocytes. **Conclusion:** MMP-9 activity is crucial for downregulating cell-contact dependent PICD in *E. coli* infected neonatal monocytes. By this mechanism, MMP-9 could contribute to reducing sustained inflammation in neonates.

Keywords: matrix-metalloproteinase; monocytes; inflammation; phagocytosis; apoptosis

1. Introduction

Perinatal mortality and morbidity is often caused by preterm delivery. Although modern intensive care has increased the number of surviving preterm infants, they may suffer from complications and lifelong handicaps. Preterm delivery can be triggered by intrauterine infections and numerous environmental and genetic factors, which give raise to an inappropriate cytokine response [1]. Granulocytes, monocytes and monocyte-derived macrophages build up the first line of defence against infections. Especially, macrophages secrete pro-inflammatory cytokines that trigger the antibacterial response. This promotes recruitment of phagocytes to the site of infection and subsequent phagocytosis and killing of the infectious pathogens. After the clearance of bacteria, the pro-inflammatory response is shifted into an anti-inflammatory situation. In this process, monocytes releasing anti-inflammatory mediators become more prominent while phagocytosing monocytes are removed via apoptosis [2]. This phagocytosis induced cell death (PICD) extenuates the pro-inflammatory response of monocytes after the acute phase of infection. Monocytes of neonates exhibit reduced PICD, resulting in a prolonged pro-inflammatory phase compared to adults [2]. There is increasing evidence that the inadequate termination of inflammation in prematurity ("sustained inflammation") is of major impact on tissue and organ damage later in life [3].

We have previously shown that monocytes derived from neonatal cord blood (CBMO) undergo less apoptosis following infection with *E. coli* or group B-streptococci (GBS), while monocytes from adult donors (PBMO) show a strong apoptotic response to infection [4,5]. Moreover, infected PBMO were able to induce apoptosis of monocytes, which were not infected (bystander-apoptosis), thereby actively terminating inflammatory immune response [6]. The mechanisms by which cell death was induced were activation of internal and external apoptosis pathways mediated through CD95L/CD95 (receptor) [7], and apoptosis induction by tumor-necrosis-factor-alpha (TNF) via tumor-necrosis-factor-receptor-1 (TNFR1) followed by caspase-cleaving and consecutive cell death [8].

Previous work revealed that CBMO show less CD95L-expression, TNFR1-internalization and TNF secretion compared to PBMO. This difference could be the result of a distinct posttranslational regulation of these pro-apoptotic factors in adults and neonates. One possible regulatory mechanism is the limited proteolysis of surface. This process, designated as shedding, can critically reduce ligands/receptors on the cell surface and thereby down-regulate the signal transmission [9,10]. The responsible enzymes are termed ectodomain sheddases and often represent zinc containing metalloproteinases.

The most relevant metalloproteinase for shedding of TNF and TNFR-1 is the tumor necrosis factor-converting enzyme (TACE/ADAM17). TACE belongs to the ADAM family proteases that share the characteristics of type-1-transmembrane proteins containing a catalytically active metalloproteinase, a disintegrin, an EGF and a transmembrane domain. Many other substrates for TACE have been identified including the IL-6 receptor [10], CD62L [11,12] and growth factors such as AREG and EGF [13].

Cleavage of CD95L can be mediated by the matrix metalloproteinase-9 (MMP-9, Gelatinase-B). This protease is well known to cleave extracellular matrix proteins such as collagen IV and V. De novo synthesized MMP-9 (pro-MMP-9) is activated by a two-step auto-proteolysis [14,15]. Moreover, MMP-9 activity can be regulated by TNF [16] and lipopolysaccharide (LPS), therefore playing an important role in endotoxin tolerance [17].

To better understand differences in PICD between neonatal and adult monocytes we here study the regulation of MMP-9 and TACE expression in CBMO and PBMO after *E. coli* infection. By blocking experiments with distinct metalloproteinase inhibitors we then investigate the role of MMP-9 and TACE for surface-expression and secretion of death ligands in monocytes from adults and neonates. Next, we perform blocking experiments to obtain information on the involvement of metalloproteinase activities for PICD of CBMO and PBMO. Finally, we compare the relevance of cell contact formation for bystander apoptosis induced in CBMO and PBMO target cells. Our results suggest that MMP9 and TACE are differentially regulated and have opposite functions in PICD. MMP9 expression is comparatively high in CBMO and seems in part responsible for shedding of CD95L and the low rate of PICD in these cells.

2. Results

2.1. Plasma-Membrane Expression of Metalloproteinases is Induced by E. coli Infection

In previous studies we observed that phagocytosing PBMO and CBMO express similar levels cell associated TNF whereas CBMO secrete much lower levels of soluble TNF than PBMO 24 h and 48 h post-infection (p.i.), (Supplemental Figure S1). These findings led us to the question whether ectodomain sheddases responsible for release of TNF and other death ligands (e.g., CD95L) could be differentially regulated in PBMO and CBMO. We here focus on the two metalloproteinase TACE and MMP-9 known as major shedding enzymes for membrane expressed TNF (memTNF) and memCD95L, respectively. First, we studied the surface expression level of TACE and MMP9 by flow cytometry before and after infection with *E. coli*. Infection caused a significant up-regulation of TACE on PBMO, whereas the expression level remained low in CBMO (Figure 1A). In contrast to TACE, the expression level of MMP-9 was significantly higher in CBMO compared to PBMO (Figure 1B).

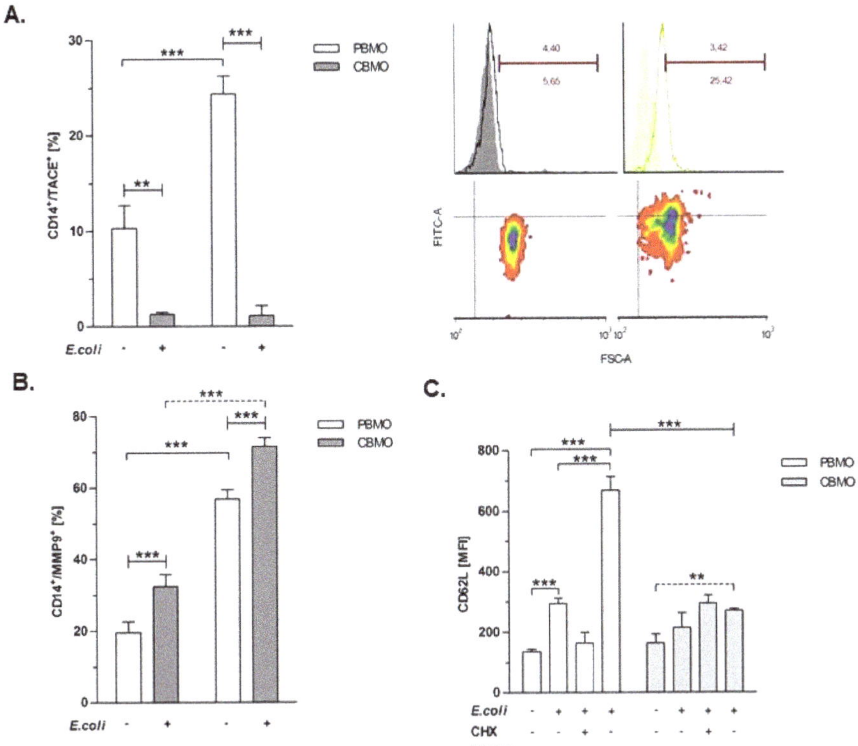

Figure 1. TACE and MMP-9 expression is induced four hours p.i. by infection with *E. coli*. TACE expressing monocytes were assessed before and after infection with *E. coli*. (**A**) Histogram plots to the right compare isotype controls (filled) and anti-TACE stained PBMO which were non-infected or *E. coli* infected. Density plots (below) detail the distribution of *E. coli* infected PBMO regarding TACE surface expression (compare to the non-infected PBMO to the left). *E. coli* MMP-9 expressing monocytes were assessed before and after infection with *E. coli*. (**B**) ($n = 5$, *** $p < 0.001$, forked bars represent Student's *t*-test, blunt-ended bars represent ANOVA). Blocking of TACE activity increases its substrate CD62L. (**C**) Monocytes were infected with GFP-*E. coli* for 4 h with or without indicated inhibitors and surface expression of CD62L was determined. (**A–C**); ($n = 5$, ** $p < 0.01$, *** $p < 0.001$, forked bars represent Student's *t*-test, blunt-ended bars represent ANOVA).

To check whether expression profiles were correlated to enzyme activities and to exclude the possibility that infection caused alterations in the presence of MMP-9 and TACE only coincidentally, we performed a shedding assay for CD62L as an indicator for TACE-activity (Figure 1C). CD62L plasma membrane expression was induced after *E. coli* infection in PBMO but less in CBMO. Administration of the TACE inhibitor TAPI-0 considerably increased CD62L expression on infected PBMO and less on CBMO, indicating that the expression level of TACE is reflected by the shedding of one of the major substrates. By contrast, CHX treatment, which is known to inhibit MMP-9 activity [18] had no effect on CD62L expression (Figure 1C). These data show that under our experimental conditions PBMO exhibit higher TACE expression and activity while CBMO exhibit higher expression of MMP9. Furthermore, the results show effectiveness of the used inhibitors.

2.2. Metalloproteinase-Inhibition Down-Regulates the Expression of CD95L and TNF

Next we checked whether TACE and MMP-9 activities on *E. coli* infected monocytes could affect the expression of the two reported substrates CD95L and TNF, respectively. To this end, we added the metalloproteinase inhibitors TAPI-0 and CHX to infected monocytes (Figure 2). Addition of these inhibitors had no effect on the phagocytosis index under the chosen conditions (Supplemental Figure S2).

PBMO showed reduced levels of memCD95L after treatment with both inhibitors and an increase of memCD95L after *E. coli* infection. CBMO showed no difference regarding memCD95L levels. The memTNF levels increased significantly after *E. coli* infection in PBMO and CBMO (Supplemental Figure S3). Inhibition of MMP-9 by CHX was associated with a two-fold increase of memCD95L on PBMO and CBMO, whereas TACE inhibitor TAPI-0 had no effect (Figure 2A, compare columns 1, 2 and 3). The results are in line with the CD62L shedding assay, which shows a clear specificity of CHX for MMP-9 (Figure 1B). Infection with *E. coli* increased the number of memCD95L-positive PBMO eight-fold. This effect was much lower (four-fold) in CBMO (Figure 2A, compare columns 1 and 4). CHX but not TAPI-0 increased the number of memCD95L expressing infected PBMO. CHX also increased the number of memCD95L expressing CBMO to levels that were observed for infected but not CHX treated PBMO (Figure 2A, compare columns 4 and 11).

CHX had no effect on the expression of memTNF. However, treatment with TAPI-0 clearly increased the number of memTNF expressing PBMO and CBMO (Figure 2B, compare columns 1 and 3). Infection with *E. coli* amplified the expression level of memTNF three-fold on PBMO and two-fold on CBMO (Figure 2B, $p < 0.05$ vs. non-treated, non-infected monocytes). After blockage of TACE, nearly all PBMO expressed memTNF, while only 60% of CBMO did (Figure 2B, compare columns 4 and 6).

The results mirrored the effects of CHX and TAPI-0 on memCD95L and memTNF: While increasing memCD95L on infected monocytes (Figure 2A), CHX significantly diminished the concentration of secreted CD95L (Figure 2C). Similar results were obtained for TAPI-0 with respect to secreted TNF (Figure 2D).

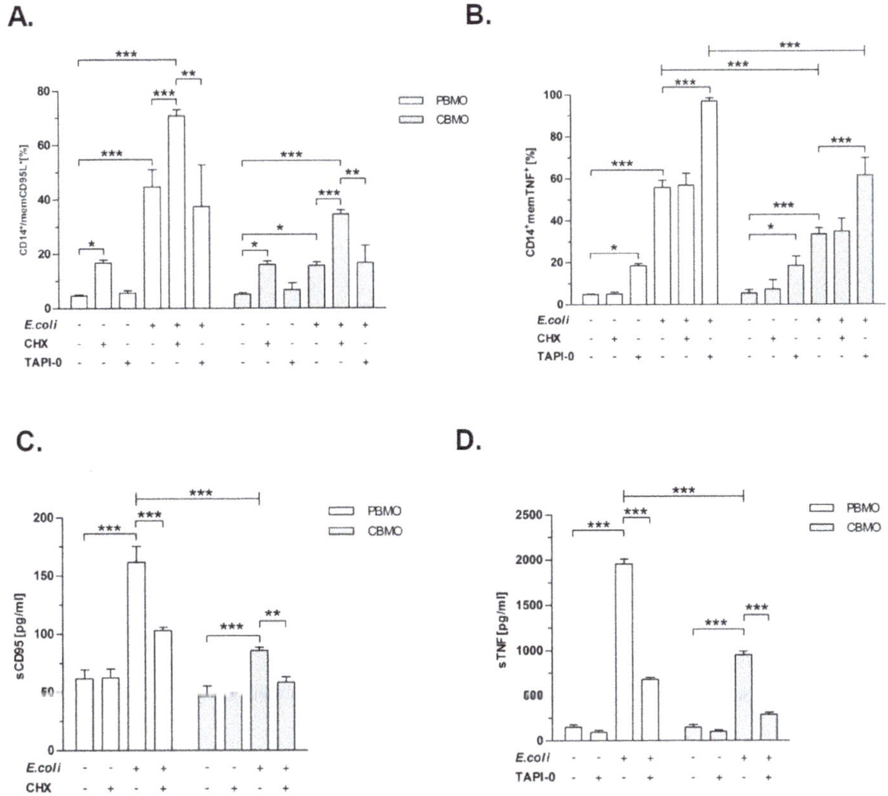

Figure 2. Blockage of metalloproteinases modulated expression of memCD95L and memTNF. PBMO and CBMO were infected and/or treated with the indicated metalloproteinase inhibitors. Expression of memCD95L (**A**) and memTNF (**B**) were assessed on CD14-positive gated monocytes. The same groups were analyzed for secreted (s) CD95 (**C**) and TNF (**D**) 24 h p.i. via ELISA ($n = 5$, * $p < 0.05$, ** $p < 0.01$, *** $p < 0.001$, forked bars represent Student's t-test, blunt-ended bars represent ANOVA).

2.3. Inhibition of TACE Reduced PICD, but Inhibition of MMP-9 Restored PICD in CBMO

Since our experiments indicated the involvement of metalloproteinases in the regulation of death receptors and ligands we addressed the question as to whether inhibition of metalloproteinases affects the induction of apoptosis in uninfected and infected monocytes (Figure 3). Treatment with CHX doubled apoptosis rates in uninfected PBMO and CBMO (compare columns 1 and 2). TAPI-0 did not show this effect (compare columns 1 and 3). In concordance with our previous results, PICD of CBMO was diminished compared to PBMO (column 4). Treatment with CHX increased the PICD in infected PBMO significantly by about 10%. In infected CBMO, the effect of CHX on PICD was much higher and doubled PICD rates on *E. coli.* infected monocytes. PICD rates of CHX treated infected CBMO were two times higher than infected PBMO (Figure 3, right panel) In contrast, TAPI-0 addition reduced PICD in PBMO by about 50% ($p < 0.05$ vs. infected PBMO). In CBMO PICD was found to be unaltered (colums 6).

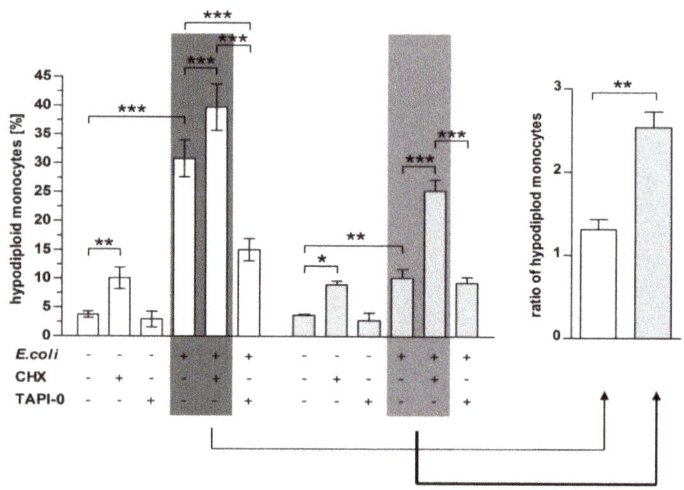

Figure 3. Blockage of MMP-9 increased PICD of CBMO. PBMO and CBMO were infected and/or treated with the indicated metalloproteinase inhibitors. Apoptosis was determined by Nicoletti assay 24 h p.i. The right panel gives the ratio of the values highlighted in dark-grey and grey, respectively ($n = 5$, * $p < 0.05$, ** $p < 0.01$, *** $p < 0.001$, forked bars represent Student's *t*-test, blunt-ended bars represent ANOVA).

2.4. TNF and E. coli Infection Triggers Expression of memCD95L

We had observed that infected PBMO and CBMO release considerable amounts of soluble TNF in a manner that most likely involves TACE (Figure 2D). Moreover, this inhibition was associated with a reduction of CD95L expression (Figure 2A). This may suggest that a cross-talk between TNF production and memCD95L expression exists. To address this possibility, monocytes were either incubated with soluble recombinant TNF and/or infected with *E. coli* and memCD95L expression was analyzed. Both addition of TNF and *E. coli* infection strongly up-regulated memCD95L on PBMO and to a lesser degree in CBMO (Figure 4, compare groups 2 and 3). As an inhibitor of TNF, we applied the TNF inhibitor etanercept to one group (Figure 4A, columns 2 and 4) resulting in considerable reduction of CD95L expression in PBMO, while the effect was not significant in CBMO. Furthermore, pre-treatment with etanercept reduced apoptosis rates significantly in TNF-incubated and *E. coli*-infected PBMO, but had no effect on CBMO (Figure 4A). Comparing the mean intensity values (MFI) of memCD95L (Figure 4B) reflected the specific effect of TNF, since addition of etanercept reduced the concentration of memCD95L on both PBMO and CBMO after TNF treatment and infection, respectively.

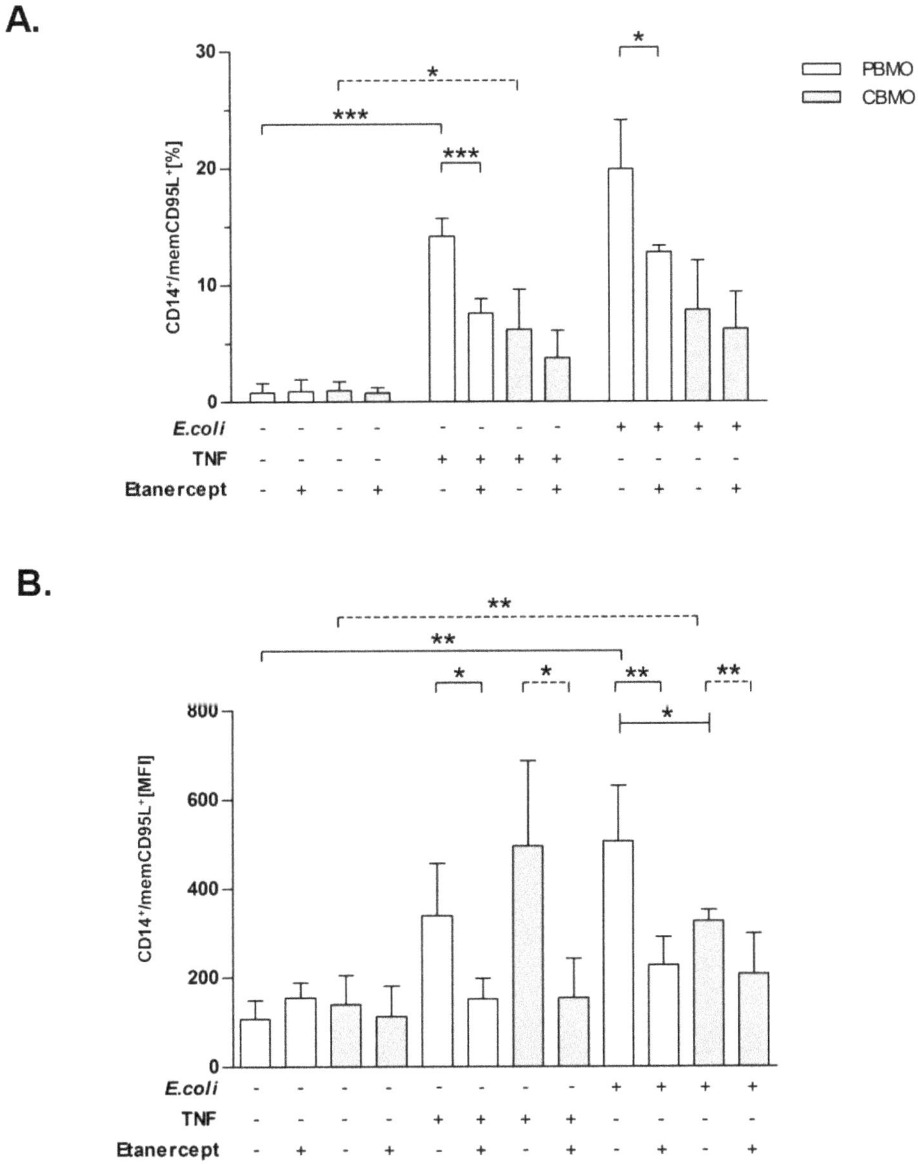

Figure 4. TNF triggers expression of CD95L. PBMO and CBMO were stimulated with TNF and/or infected with *E. coli* for four hours and analyzed for CD95L expression. Some groups were pre-treated with etanercept. The number of PBMO and CBMO expressing memCD95L were determined as the percentage (**A**) and the mean concentration given as the mean intensity (MFI), (**B**); for A and B, $n = 5$, * $p < 0.05$, ** $p < 0.01$, *** $p < 0.001$, forked bars represent Student's t-test, blunt-ended bars represent ANOVA.

2.5. Apoptosis of Bystander Monocytes is Reduced Without Cellular Contact

The surface expression levels of TNF and CD95L are separately regulated by TACE and MMP-9 leading to the release of soluble mediators. This process may affect bystander apoptosis in which cells could be directed to undergo apoptosis by either cell-to-cell contact or soluble mediators. In fact, bystander-apoptosis in PBMO required intercellular contact with phagocytosing monocytes, as we have previously shown in a transwell chamber setup [8]. To compare this aspect in CBMO and PBMO, we infected monocytes with *E. coli*-GFP in one (upper) transwell chamber and co-cultivated non-infected monocytes of the same donor in the other (lower) chamber (Figure 5A, sketch). In the upper chamber apoptosis occurred in PBMO in a significantly higher percentage compared to CBMO, as was expected. In the lower chamber, apoptosis of PBMO was reduced by 70% compared to PBMO in the upper chamber (Figure 5A). In CBMO, contact-independent ("trans") apoptosis was nearly absent.

We then questioned whether the expression of death receptors or shedding enzymes on cells that had either bound or ingested *E. coli* would be different from that on those cells, that had no contact to bacteria. As previously reported [6], we distinguished monocytes that had either bound or ingested *E. coli* (GFP-positive, GFP^+) from those that had no contact with bacteria (GFP-negative, GFP^-). CD95L was significantly higher up-regulated on the surface of GFP^+ PBMO as compared to GFP^- PBMO (Figure 5B). For CBMO, the upregulation of CD95L on GFP^+ cells was less compared to on GFP^+ PBMO ($p < 0.005$). Moreover, for on GFP- cells much less upregulation of CD95L was seen on CBMO than on PBMO.

Since CD95L expression can be regulated via TACE through TNF signalling and via MMP9 by cleavage of CD95L we analyzed TACE and MMP-9 expression on GFP^+ and GFP^- PBMO and CBMO, respectively. GFP^+ PBMO expressed more TACE and MMP-9 compared to GFP^- PBMO (Figure 5C, left panel, $p < 0.001$). MMP-9 was also upregulated on GFP^+ CBMO compared to GFP^- CBMO (Figure 5C, right panel). By contrast TACE expression was very low and comparable on GFP^+ and GFP^- CBMO (Figure 5C). These results suggest that CBMO are less effective in inducing bystander apoptosis than PBMO. This correlates with less upregulation of CD95L on CBMO compared to PBMO target cells, which is again associated with a low expression of TACE but still high expression of MMP-9 on CBMO compared to PBMO target cells. This correlation further supports our notion that activity of MMP-9 rather than of TACE contributes to reduced levels of PICD that are typically observed in neonate monocytes.

Additionally, the percentage of memCD95L-, memTNF-positive PBMO as well as the density of memTNF on PBMO was higher than on CBMO. In contrast, with less memCD95L-positive CBMO, their mean density was higher than on PBMO (Figure 5D).

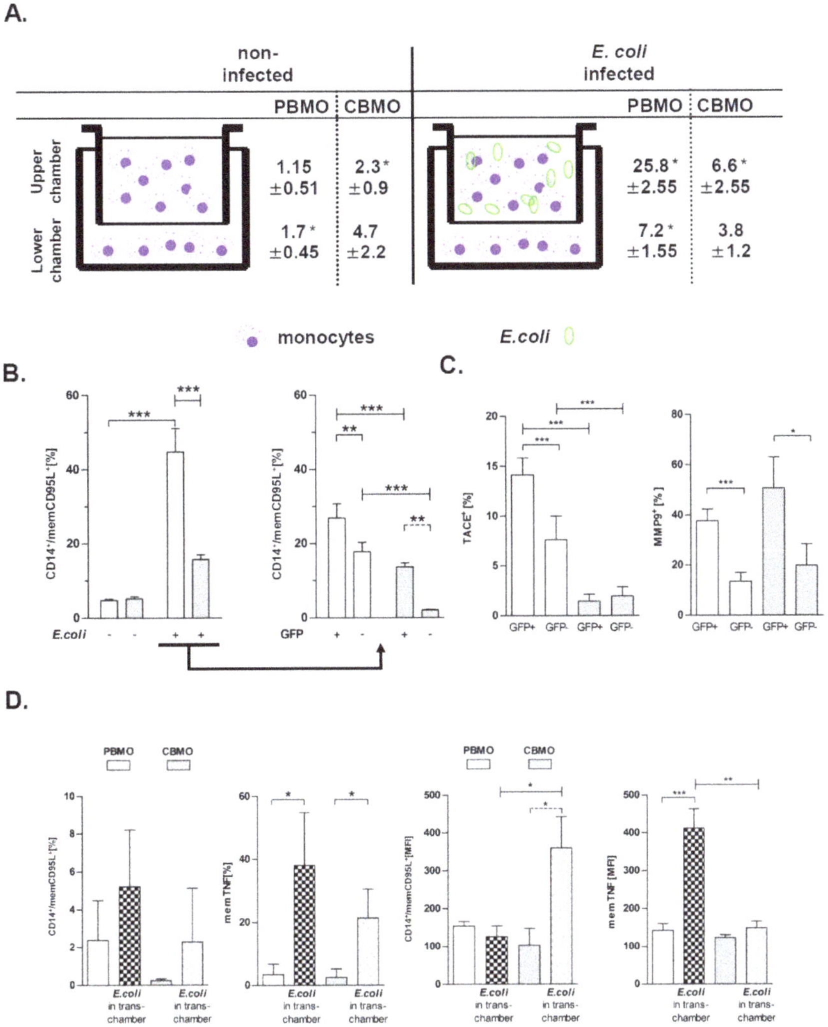

Figure 5. Monocyte PICD is enhanced by cell–cell contact. Monocytes were infected with GFP-*E. coli* for one hour, washed and re-incubated for another 24 h in transwell chambers (experimental setup sketch); non-infected cells served as controls. Monocytes from the same donor were co-cultivated in compartments, separated by teflon membranes. Apoptosis was detected by hypodiploid DNA-content (**A**, $n = 6$ for PBMO and $n = 4$ for CBMO; * $p < 0.05$). PBMO and CBMO were sub-gated (**B**) in phagocyting/binding (GFP^+) and non-phagocyting (GFP^-) monocytes and analyzed for memCD95L expression. Subgated PBMO and CBMO were further analyzed for TACE (**C**, left panel) and MMP9 (**C**, right panel) expression. (**D**) The expression of memCD95L and memTNF of monocytes in the lower ("trans") chamber was assessed. The percentage of surface expressing monocytes (panels to the left) and the mean values (MFI, panels to the right) were determined ($n = 5$, * $p < 0.05$, ** $p < 0.01$, *** $p < 0.001$, forked bars represent Student's *t*-test, blunt-ended bars represent ANOVA).

3. Discussion

The present study compared the induction of PICD in infected and non-infected monocytes (bystander cell death) from cord blood and adult blood, utilizing an in vitro *E. coli* infection model with respect to a distinct regulation pattern of the metalloproteinases TACE and MMP-9 (Figure 1). MMP expression was distinctly regulated by infection in PBMO and CBMO. Whereas infected CBMO exhibited less TACE expression and activity, they reacted with stronger expression of MMP-9 (Figure 1) as compared to PBMO. Depending on the expression pattern of MMP-9 and TACE, death ligands CD95L and TNF were either presented on the surface or shedded by distinct monocyte populations and thereby functionally controlled in regard to PICD (Figure 2, Supplemental Figure S3). We speculate that this regulation may be phagocytosis dependent (see Figure 5B), since the percentage of memCD95L appears higher on GFP+ phagocytosing monocytes (Figure 5B) than on non-phagocytosing (GFP-) monocytes. This observation was one of the reasons for monitoring the shedding of memCD95L and TNF via ELISA (Figure 2C,D). We could show, that PICD in CBMO was diminished (Figure 3), confirming earlier results [4]. TNF secretion triggered expression of memCD95L on PBMO but critically less on CBMO (Figure 4) which may explain that induction of PICD is predominantly contact-dependent (Figure 5). The data suggest that phagocytosis regulates expression of MMP-9 but not TACE in *E. coli* infected CBMO (Figure 5). To our knowledge, data on the expression and activity of metallomatrixproteinases in CBMO have been scarcely published.

The results of the actual study indicate an interlocked action of TNF and memCD95L. They suggest a functional model, where infection with *E. coli* activates TNF and consecutive TACE secretion in PBMO. TACE cleaves memTNF allowing predominantly ligation to TNFR1, since soluble TNF preferentially binds to TNFR1 [19,20]. The TNFR1/TNF ligand receptor complex is internalized to induce apoptosis [8,20]. This is in line with the findings that TAPI-0 decreases concentrations of soluble TNF resulting in reduced apoptosis (Figure 2B,D and Figure 3) However, our experiments provide evidence that soluble TNF also induces memCD95L (Figure 4). Pro-apoptotic signaling of CD95L/CD95R (receptor) can be distinguished from death receptor signaling by acting in a cell–cell contact dependent manner [21]. Increasing the amount of memCD95L by the MMP-9 inhibitor CHX is accompanied by an increase in apoptosis rates (Figure 2A,C and Figure 3).

This model is further supported by the low abundance of secreted TNF, TNFR1 [6] and CD95L in CBMO resulting in reduced PICD [4]. It explains also the significantly reduced trans-activation of PICD in PBMO and CBMO (Figure 5A). Phagocytosing monocytes express more TNF (Supplemental Figure S1A) which may cause a higher level of memCD95L in GFP^+ monocytes (Figure 5B). Previous data have shown that blockage of TNF almost abolished PICD in PBMO [6]. The fact that CBMO secrete significantly less TNF than PBMO (Supplemental Figure S1B), but TNF stimulation increases the memCD95L-expression (Figure 4) may explain the observation of a reduced PICD initiation in cord blood (Figure 3). However, PBMO and CBMO showed a different reaction to factors which were secreted by phagocytosing mates in the transwell experiments (Figure 5D), pointing to a fundamental different meaning of cell-contact independent signaling in neonatal monocytes. Future experiments will elucidate these differences.

Various microbial, fungal and viral agents such as *Streptococcus pneumoniae* [22], zymosan [23] and HIV [24] were reported to initialize bystander apoptosis in monocytes. For resolving inflammation, bystander apoptosis may in the same way be essential [25]. Disbalancing this system may result in overwhelming infection or prolonged inflammation, thus the reduced bystander apoptosis of CBMO may be critical for the resulting immune reaction.

The mechanism of infection induced MMP-9/TACE expression should be clarified by future experiments. It was recently published, that soluble CD95L as well as memTNF activate the NFκ-B pathway which in turn targets metalloproteinases such as MMP-9 [26]. TACE, which also was examined in this study, is activated by infection-induced p38 MAP kinase and reactive oxygen species (ROS) [27].

MMP-9 and TACE are integrative members of a feedback control system which can skew the signal transduction of TNF-α and CD95L from pro-apoptotic to pro-inflammatory by balancing soluble and plasma-membrane bound forms and vice versa [28,29]. Reduced TACE activity was shown to promote endotoxin tolerance [30] without engagement of TLR4 [31], which circumvents an overwhelming inflammation upon infection.

The damage of organ tissues can be curtailed by controlling the expression of MMP-9. Enhanced expression of MMP-9 was correlated to COPD [32,33]. MMP-9 activation plays also a role in amniotic membrane rupture during labor [34] and enhanced MMP levels were associated with sepsis [35].

In this study we attributed a co-expression of MMP-9 and TACE with memCD95L and memTNF to the induction of apoptosis and proved this by simultaneous exposure to inhibitors. As a limitation it has to be mentioned that substrate specificity of the two enzymes is still controversial. The cleavage of memTNF by TACE is well documented [36,37] but the cleavage of the receptor TNFR1 in vivo has been discussed [38]. Furthermore it was reported that pathogenic *E. coli* synthesize proteins with metalloproteinase like properties [39,40] which could contribute to the shedding of TNF and CD95L. The results of our study are based on relatively small sample sizes and did not follow up on kinetics of cord and peripheral blood of newborns longitudinally. The neonatal situation several days postnatal is therefore not properly reflected by this study.

PICD and bystander apoptosis represent important tools to shape the immune response upon infection. Our data suggest phagocytosis-related cell death of non-phagocytosing monocytes, designated as monocytic bystander kill after infection with *E. coli*. It provides evidence that if monocytes are "at the wrong place at the wrong time", i.e., near or in contact to phagocytosing monocytes, they may be killed by fratricide. To what extent this may influence the course of neonatal sepsis, remains to be elucidated in vivo.

4. Materials and Methods

4.1. Patients

The study protocol was approved by the Ethics Committees of Aachen University Hospital (Permission No: EK150/09, 6 October 2009, signed by Profs G. Schmalzing and U. Buell, respectively). All adult participants involved gave written consent to use their blood samples. All term neonates were delivered spontaneously and did not exhibit signs of infection, as defined by clinical status, white blood cell count and C-reactive protein. Mothers with amnion infection and prolonged labour (>12 h) were excluded. Umbilical cord blood was placed in heparin-coated tubes (4 IE/mL blood), immediately following cord ligation as described before [7].

4.2. Bacteria

E. coli-GFP

E. coli DH5α, an encapsulated K12 laboratory strain, carrying the green fluorescent protein (*gfp*)-mut2 gene (*E. coli*-GFP) was a generous gift from Prof. Dr. Dehio (University of Basel, Switzerland) and was used for phagocytosis as previously described. Bacteria were freshly grown in Lennox-L-Broth-medium (Invitrogen) until early logarithmic growth, resuspended in phosphate-buffered-saline (PBS) and used immediately. Infection was performed at a multiplicity of infection (MOI) of 25 which was achieved by dilution with PBS. The phagocytosis assays were performed as described [7]. The phagocytosis index ($CD14^+GFP^+$ monocytes: $CD14^+$ monocytes) was analyzed by flow cytometry. In some indicated experiments, *E. coli*-GFP was replaced by *E. coli*-EOS-FP (Supplemental Figure S1).

4.3. Reagents

Antibodies to CD14 (clone MEM18), MMP-9 (clone 56129), TACE (clone FAB9301P), CD62L (L-selectin; clone LT-TD180) and Ig-matched controls (IgG1, IgG2b) were purchased from R&D Systems and Immunotools (Abingdon, UK and Friesoythe, Germany), respectively. FITC-labeled CD95L antibody (clone SB93a) was purchased from SouthernBiotech (Birmingham, USA) and corresponding mouse IgG2b kappa isotype control (clone eBMG2b) was obtained from eBioscience (Waltham, USA). The secondary anti-mouse-PB antibody (F'ab fragment) was purchased from

Invitrogen (Darmstadt, Germany). Propidium iodide (PI), isopropyl-β-D-thiogalactopyranoside (IPTG) and antibiotics were purchased from Sigma (Munich, Germany). For the blocking of TNF an anti-TNF antibody (a chimeric molecule combining the ligand-binding domain of the TNF-receptor 2 and the Fc-domain of human IgG1 (Etanercept, Pfizer–Wyeth, Hamburg, Germany), final concentration 1 μg/mL) was administered 1 h prior to infection. TACE inhibitor N-(R)-(2-(Hydroxyaminocarbonyl)Methyl)-4-Methylpentanoyl-L-Naphthylalanyl-L-Alanine amide (TAPI-0; Merck Millipore, Darmstadt, Germany) was added to a final concentration of 100 nm/L 30 min prior to infection. PBMO and CBMO were pre-incubated with MMP-9 inhibitor CHX (chlorhexidine; Santa Cruz Biotechnologies, Heidelberg, Germany) to a final concentration of 20 μg/mL 1 h prior to the infection. This concentration was chosen since it had been previously shown to be effective in MMP-9 inhibition [18].

TNF was purchased from eBiosciences (eBiosciences-Natutec, Frankfurt, Germany), aliquoted freshly after dilution in PBS and used in apoptosis induction assays in final concentrations of 5 ng/mL.

4.4. Mononuclear Cell Cultures

Peripheral blood cells from adults and cord blood mononuclear cells (PBMC and CBMC) were isolated by density gradient centrifugation on Ficoll cushions (Amersham, Freiburg, Germany) as described previously [7]. Washed cells were resuspended in VLE RPMI-1640 (Biochrom, Berlin, Germany). For analysis of post-phagocytic reactions, cells were counted in an ultraplane Neubauer hemocytometer, placed at 2×10^6 cells/ml in flat bottom 24 well cell culture plates (Costar, Bodenheim, Germany), containing 10% heat-inactivated fetal calf serum (FCS, Biochrom) and were incubated at 37 °C.

4.5. Flow Cytometry

A daily calibrated FACS-Canto flow cytometer (Becton Dickinson, Mountain View, CA, USA) was used to perform phenotypic analysis. To prevent nonspecific binding, cells were incubated with 10% fetal calf serum on ice for 10 min before staining with appropriate fluorophore coupled secondary antibodies, or isotype-specific immunoglobulin-labelled monoclonal antibodies for 20 min over ice in the dark. Monocytes were gated by forward (FSC), side scatter (SSC), and CD14 expression. For intracellular cytokine staining, monocytes were fixed in 2% paraformlaldehyde/PBS for 30 min at room temperature (RT) and washed three times with PBS. Afterwards monocytes were permeabilized utilizing a permeabilization buffer (purchased from Thermo Fisher Scientific, Hennef, Germany) according to the manufacturers' recommendations. Data was analyzed using the FCS Express V4.0 research Edition software (DeNovo Software, Glendale, California, USA).

4.6. Detection of Hypodiploid Nuclei

DNA fragmentation was assessed according to Nicoletti and previously described [4]. In brief, washed cells were slowly resuspended in 2 mL of −20 °C ethanol 70% with continuous vortexing and stored for four hours at −20 °C. Cells were washed twice, resuspended in 50 μL PBS containing 13 units RNAse (DNAse free; Sigma, Taufkirchen, Germany) and incubated for 15 min at 37 °C. 180 μL of PI (70 μg/mL) was added, incubated for 20 min and analysis was performed immediately. Alternatively, mononuclear cells were stained with CD14 antibody for 15 min at RT to identify monocytes. A fixation with paraformaldehyde (2% *v/v* in PBS) for 2 h at RT replaced the ethanol fixation. Afterwards, cells were permeabilized by incubation in PBS-T (PBS, Triton X-100 0.1% *v/v*) for 20 min at RT, washed twice in PBS, resuspended in PBS-PI (PBS, 70 μg/mL PI and 13 units RNAse) and incubated for 10 min at RT before analysis by flow cytometry. Cell-doublets were discriminated by assessment of PI-width/PI-area.

4.7. Transwell Experiments

Transwell plates (pore diameter 0.4 μm, purchased from Corning, NY, USA) were used. Cells in the upper chamber were untreated or infected as described above. Cells separated by the Teflon membrane had no contact to bacteria as assessed by plating on appropriate growth-medium and FACS analysis.

4.8. ELISA

The TNF enzyme-linked immunosorbent assay (ELISA) was purchased from eBiosciences (eBiosciences-Natutec, Frankfurt, Germany) and the CD95L ELISA from Hölzel Diagnostika (Hölzel Diagnostika Handels GmbH, Cologne, Germany). Both were used according to the manufacturer's recommendations. The read-out was executed in a spectra max 340PC ELISA reader (Molecular Devices, Sunnyvale, CA, USA) with a sensitivity from 4–500 pg/mL.

4.9. Statistical Analysis

Results are expressed as mean +/− standard deviation. Error bars represent standard deviations. Values of $p < 0.05$ were considered as significant. Analyses were done with statistical software (performing two-way ANOVA adjusted according to Bonferroni-Holm for multiple group comparisons as provided by GraphPad Software Statistical Package, La Jolla, CA 92037 USA).

5. Conclusions

Matrix metalloproteases regulate the expression of death ligands and their receptors. On neonatal monocytes CD95L density is reduced upon infection due to lower activity of MMP-9. Enhancing MMP-9 expression could be a target to restore PICD in neonatal monocytes.

Supplementary Materials: Supplementary materials can be found at http://www.mdpi.com/1422-0067/20/6/1399/s1.

Author Contributions: S.D., conception, acquisition of data, analysis, interpretation of data; C.P., conception, acquisition of data, analysis, interpretation of data; A.L., drafting the article, critical revise; N.K., interpretation of data, drafting the article, critical revise; C.G., drafting the article, critical revise; T.O., conception, acquisition of data, analysis, interpretation of data.

Funding: This research received no external funding.

Acknowledgments: Thanks to the team of midwives (Roja Fakhrabadi) and the Department of Obstetrics and Gynaecology at the University Hospital of Aachen (E.Stickeler), for coordinating the cord blood sampling.

References

1. Cockle, J.V.; Gopichandran, N.; Walker, J.J.; Levene, M.I.; Orsi, N.M. Matrix metalloproteinases and their tissue inhibitors in preterm perinatal complications. *Reprod. Sci.* **2007**, *14*, 629–645. [CrossRef] [PubMed]
2. Gille, C.; Steffen, F.; Lauber, K.; Keppeler, H.; Leiber, A.; Spring, B.; Poets, C.F.; Orlikowsky, T.W. Clearance of apoptotic neutrophils is diminished in cord blood monocytes and does not lead to reduced IL-8 production. *Pediatr. Res.* **2009**, *66*, 507–512. [CrossRef] [PubMed]
3. Dammann, O.; Leviton, A. Intermittent or sustained systemic inflammation and the preterm brain. *Pediatr. Res.* **2014**, *75*, 376–380. [CrossRef] [PubMed]
4. Gille, C.; Leiber, A.; Spring, B.; Kempf, V.A.; Loeffler, J.; Poets, C.F.; Orlikowsky, T.W. Diminished phagocytosis-induced cell death (PICD) in neonatal monocytes upon infection with *Escherichia coli*. *Pediatr. Res.* **2008**, *63*, 33–38. [CrossRef]
5. Gille, C.; Dreschers, S.; Leiber, A.; Lepiorz, F.; Krusch, M.; Grosse-Opphoff, J.; Spring, B.; Haas, M.; Urschitz, M.; Poets, C.F. The CD95/CD95L pathway is involved in phagocytosis-induced cell death of monocytes and may account for sustained inflammation in neonates. *Pediatr. Res.* **2013**, *73*, 402–408. [CrossRef]

6. Dreschers, S.; Gille, C.; Haas, M.; Grosse-Ophoff, J.; Schneider, M.; Leiber, A.; Bühring, H.-J.; Orlikowsky, T.W. Infection-induced bystander-apoptosis of monocytes is TNF-alpha-mediated. *PLoS ONE* **2013**, *8*, e53589. [CrossRef]
7. Gille, C.; Leiber, A.; Mundle, I.; Spring, B.; Abele, H.; Spellerberg, B.; Hartmann, H.; Poets, C.F.; Orlikowsky, T.W. Phagocytosis and postphagocytic reaction of cord blood and adult blood monocyte after infection with green fluorescent protein-labeled *Escherichia coli* and group B Streptococci. *Cytom. B Clin. Cytom.* **2009**, *76*, 271–284. [CrossRef]
8. Dreschers, S.; Gille, C.; Haas, M.; Seubert, F.; Platen, C.; Orlikowsky, T.W. Reduced internalization of TNF-a/TNFR1 down-regulates caspase dependent phagocytosis induced cell death (PICD) in neonatal monocytes. *PLoS ONE* **2017**, *12*, e0182415. [CrossRef]
9. Vandooren, J.; Van den Steen, P.E.; Opdenakker, G. Biochemistry and molecular biology of gelatinase B or matrix metalloproteinase-9 (MMP-9): The next decade. *Crit. Rev. Biochem. Mol. Biol.* **2013**, *48*, 222–272. [CrossRef]
10. Levine, S.J. Molecular mechanisms of soluble cytokine receptor generation. *J. Biol. Chem.* **2008**, *283*, 14177–14181. [CrossRef]
11. Scheller, J.; Chalaris, A.; Garbers, C.; Rose-John, S. ADAM17: A molecular switch to control inflammation and tissue regeneration. *Trends Immunol.* **2011**, *32*, 380–387. [CrossRef] [PubMed]
12. Li, Y.; Brazzell, J.; Herrera, A.; Walcheck, B. ADAM17 deficiency by mature neutrophils has differential effects on L-selectin shedding. *Blood* **2006**, *108*, 2275–2279. [CrossRef] [PubMed]
13. Murthy, A.; Defamie, V.; Smookler, D.S.; Di Grappa, M.A.; Horiuchi, K.; Federici, M.; Sibilia, M.; Blobel, C.P.; Khokha, R. Ectodomain shedding of EGFR ligands and TNFR1 dictates hepatocyte apoptosis during fulminant hepatitis in mice. *J. Clin. Investig.* **2010**, *120*, 2731–2744. [CrossRef] [PubMed]
14. Scannevin, R.H.; Alexander, R.; Haarlander, T.M.; Burke, S.L.; Singer, M.; Huo, C.; Zhang, Y.-M.; Maguire, D.; Spurlino, J.; Deckman, I.; et al. Discovery of a highly selective chemical inhibitor of matrix metalloproteinase-9 (MMP-9) that allosterically inhibits zymogen activation. *J. Biol. Chem.* **2017**, *292*, 17963–17974. [CrossRef]
15. Kiaei, M.; Kipiani, K.; Calingasan, N.Y.; Wille, E.; Chen, J.; Heissig, B.; Rafii, S.; Lorenzl, S.; Beal, M.F. Matrix metalloproteinase-9 regulates TNF-alpha and FasL expression in neuronal, glial cells and its absence extends life in a transgenic mouse model of amyotrophic lateral sclerosis. *Exp. Neurol.* **2007**, *205*, 74–81. [CrossRef] [PubMed]
16. Vaday, G.G.; Hershkoviz, R.; Rahat, M.A.; Lahat, N.; Cahalon, L.; Lider, O. Fibronectin-bound TNF-alpha stimulates monocyte matrix metalloproteinase-9 expression and regulates chemotaxis. *J. Leukoc. Biol.* **2000**, *68*, 737–747. [PubMed]
17. Martínez, E.; Fernández-Ruíz, I.; Fresno, C.; Soares-Schanoski, A.; Jurado, T.; Siliceo, M.; Toledano, V.; Fernández-Palomares, R.; García-Rio, F.; et al. Role of MMPs in orchestrating inflammatory response in human monocytes via a TREM-1-PI3K-NF-kappaB pathway. *J. Leukoc. Biol.* **2012**, *91*, 933–945.
18. Gendron, R.; Grenier, D.; Sorsa, T.; Mayrand, D. Inhibition of the activities of matrix metalloproteinases 2, 8, and 9 by chlorhexidine. *Clin. Diagn. Lab. Immunol.* **1999**, *6*, 437–439. [PubMed]
19. Eissner, G.; Kolch, W.; Scheurich, P. Ligands working as receptors: Reverse signaling by members of the TNF superfamily enhance the plasticity of the immune system. *Cytokine Growth Factor Rev.* **2004**, *15*, 353–366. [CrossRef]
20. Schutze, S.; Schneider-Brachert, W. Impact of TNF-R1 and CD95 internalization on apoptotic and antiapoptotic signaling. *Results Probl. Cell Differ.* **2009**, *49*, 63–85.
21. Suda, T.; Hashimoto, H.; Tanaka, M.; Ochi, T.; Nagata, S. Membrane Fas ligand kills human peripheral blood T lymphocytes, and soluble Fas ligand blocks the killing. *J. Exp. Med.* **1997**, *186*, 2045–2050. [CrossRef] [PubMed]
22. Dockrell, D.H.; Lee, M.; Lynch, D.H.; Read, R.C. Immune-mediated phagocytosis and killing of Streptococcus pneumoniae are associated with direct and bystander macrophage apoptosis. *J. Infect. Dis.* **2001**, *184*, 713–722. [CrossRef] [PubMed]
23. Brown, S.B.; Savill, J. Phagocytosis triggers macrophage release of Fas ligand and induces apoptosis of bystander leukocytes. *J. Immunol.* **1999**, *162*, 480–485. [PubMed]

24. Holm, G.H.; Zhang, C.; Gorry, P.R.; Peden, K.; Schols, D.; De Clercq, E.; Gabuzda, D. Apoptosis of bystander T cells induced by human immunodeficiency virus type 1 with increased envelope/receptor affinity and coreceptor binding site exposure. *J. Virol.* **2004**, *78*, 4541–4551. [CrossRef] [PubMed]
25. Medzhitov, R.; Janeway, C., Jr. Innate immunity. *N. Engl. J. Med.* **2000**, *343*, 338–344. [CrossRef] [PubMed]
26. Cai, Z.; Yang, F.; Yu, L.; Yu, Z.; Jiang, L.; Wang, Q.; Yang, Y.; Wang, L.; Cao, X.; Wang, J. Activated T cell exosomes promote tumor invasion via Fas signaling pathway. *J. Immunol.* **2012**, *188*, 5954–5961. [CrossRef]
27. Scott, A.J.; O'Dea, K.P.; O'Callaghan, D.; Williams, L.; Dokpesi, J.O.; Tatton, L.; Handy, J.M.; Hogg, P.J.; Takata, M. Reactive oxygen species and p38 mitogen-activated protein kinase mediate tumor necrosis factor alpha-converting enzyme (TACE/ADAM-17) activation in primary human monocytes. *J. Biol. Chem.* **2011**, *286*, 35466–35476. [CrossRef]
28. Black, R.A.; Rauch, C.T.; Kozlosky, C.J.; Peschon, J.J.; Slack, J.L.; Wolfson, M.F.; Castner, B.J.; Stocking, K.L.; Reddy, P.; Srinivasan, S.; et al. A metalloproteinase disintegrin that releases tumour-necrosis factor-alpha from cells. *Nature* **1997**, *385*, 729–733. [CrossRef]
29. Walcheck, B.; Herrera, A.H.; St Hill, C.; Mattila, P.E.; Whitney, A.R.; Deleo, F.R. ADAM17 activity during human neutrophil activation and apoptosis. *Eur. J. Immunol.* **2006**, *36*, 968–976. [CrossRef]
30. Kirchner, S.; Boldt, S.; Kolch, W.; Haffner, S.; Kazak, S.; Janosch, P.; Holler, E.; Andreesen, R.; Eissner, G. LPS resistance in monocytic cells caused by reverse signaling through transmembrane TNF (mTNF) is mediated by the MAPK/ERK pathway. *J. Leukoc. Biol.* **2004**, *75*, 324–331. [CrossRef]
31. Muthukuru, M.; Cutler, C.W. Resistance of MMP9 and TIMP1 to endotoxin tolerance. *Pathog. Dis.* **2015**, *73*, ftu003. [CrossRef]
32. Papakonstantinou, E.; Karakiulakis, G.; Batzios, S.; Savic, S.; Roth, M.; Tamm, M.; Stolz, D. Acute exacerbations of COPD are associated with significant activation of matrix metalloproteinase 9 irrespectively of airway obstruction, emphysema and infection. *Respir. Res.* **2015**, *16*, 78. [CrossRef] [PubMed]
33. Sakthivel, P.; Breithaupt, A.; Gereke, M.; Copland, D.A.; Schulz, C.; Gruber, A.D.; Dick, A.D.; Schreiber, J.; Bruder, D. Soluble CD200 Correlates with Interleukin-6 Levels in Sera of COPD Patients: Potential Implication of the CD200/CD200R Axis in the Disease Course. *Lung* **2017**, *195*, 59–68. [CrossRef]
34. Sukhikh, G.T.; Kan, N.E.; Tyutyunnik, V.L.; Sannikova, M.V.; Dubova, E.A.; Pavlov, K.A.; Amiraslanov, E.Y.; Dolgushina, N.V. The role of extracellular inducer of matrix metalloproteinases in premature rupture of membranes. *J. Matern.-Fetal Neonatal Med.* **2016**, *29*, 656–659. [CrossRef] [PubMed]
35. Alqahtani, M.F.; Smith, C.M.; Weiss, S.L.; Dawson, S.; Ralay Ranaivo, H.; Wainwright, M.S. Evaluation of New Diagnostic Biomarkers in Pediatric Sepsis: Matrix Metalloproteinase-9, Tissue Inhibitor of Metalloproteinase-1, Mid-Regional Pro-Atrial Natriuretic Peptide, and Adipocyte Fatty-Acid Binding Protein. *PLoS ONE* **2016**, *11*, e0153645. [CrossRef]
36. Milla, M.E.; Gonzales, P.E.; Leonard, J.D. The TACE zymogen: Re-examining the role of the cysteine switch. *Cell Biochem. Biophys.* **2006**, *44*, 342–348. [CrossRef]
37. Lee, J.H.; Choi, Y.J.; Heo, S.H.; Lee, J.M.; Cho, J.Y. Tumor necrosis factor-alpha converting enzyme (TACE) increases RANKL expression in osteoblasts and serves as a potential biomarker of periodontitis. *BMB Rep.* **2011**, *44*, 473–477. [CrossRef] [PubMed]
38. Mohan, M.J.; Seaton, T.; Mitchell, J.; Howe, A.; Blackburn, K.; Burkhart, W.; Moyer, M.; Patel, I.; Waitt, G.M.; Becherer, J.D.; et al. The tumor necrosis factor-alpha converting enzyme (TACE): A unique metalloproteinase with highly defined substrate selectivity. *Biochemistry* **2002**, *41*, 9462–9469. [CrossRef] [PubMed]
39. Tapader, R.; Bose, D.; Pal, A. YghJ, the secreted metalloprotease of pathogenic *E. coli* induces hemorrhagic fluid accumulation in mouse ileal loop. *Microb. Pathog.* **2017**, *105*, 96–99. [CrossRef] [PubMed]
40. Stinemetz, E.K.; Gao, P.; Pinkston, K.L.; Montealegre, M.C.; Murray, B.E.; Harvey, B.R. Processing of the major autolysin of E. faecalis, AtlA, by the zinc-metalloprotease, GelE, impacts AtlA septal localization and cell separation. *PLoS ONE* **2017**, *12*, e0186706. [CrossRef]

International Journal of
Molecular Sciences

MDPI

Article

Isolation of a Novel Metalloproteinase from Agkistrodon Venom and Its Antithrombotic Activity Analysis

Jin Huang [1,†], Hui Fan [1,†], Xiaojian Yin [1,2,*] and Fang Huang [1,2,*]

1 State Key Laboratory of Natural Medicines, China Pharmaceutical University, Nanjing 210009, China
2 Chinese Medicine College, China Pharmaceutical University, Nanjing 210009, China
* Correspondence: xiaojian@cpu.edu.cn (X.Y.); huangfang@cpu.edu.cn (F.H.); Tel.: +86-25-86185559 (X.Y.); +86-181-5162-0528 (F.H.); Fax: +86-25-83271379 (X.Y.)
† These authors contributed equally to this work.

Received: 1 July 2019; Accepted: 18 August 2019; Published: 21 August 2019

Abstract: Snake venom contains large amounts of active proteins and peptides. In this study, a novel snake protein, metalloproteinase SP, was successfully isolated from the venom of *Agkistrodon acutus* by multi-gel chromatography. The isolated protein exhibits anti-platelet aggregation activity. Animal experiments showed that it exhibited defibration, anticoagulation, and antithrombotic effects and contributes to improved blood rheology and antiplatelet aggregation. In vivo experiments demonstrated that it prolonged clotting time, partial thromboplastin time, prothrombin time, thrombin time, fibrinogen time and reduced fibrinogen content of mice. Also, metalloproteinase SP inhibited carrageenan-induced tail thrombosis, ADP-induced acute pulmonary embolism, and ADP, Arachidonic acid (AA), or collagen-induced platelet aggregation. In vitro experiments showed that the protein cleaved the α, β, and γ chains of fibrinogen. Metabolomic analysis upon metalloproteinase SP treatment revealed that 14 metabolites, which are mainly involved in phenylalanine, tyrosine, and tryptophan biosynthesis, responded to metalloproteinase SP treatment. In summary, the isolated snake venom protein inhibits formation of acute pulmonary embolism probably through regulating and restoring perturbed energy, lipid, and amino acid metabolism.

Keywords: Agkistrodon venom; metalloproteinase; fibrinogen; antithrombotic; metabolomics

1. Introduction

Thrombotic disease is one with a high morbidity and mortality rate in the world, accounting for approximately 40% of deaths yearly [1]. Thrombotic diseases include pulmonary embolism (PE), deep vein thrombosis (DVT), myocardial infarction, coronary atherosclerosis, ischemic stroke, and so on [2,3]. The formation of thrombus is closely related to platelet aggregation, blood coagulation, and fibrin network formation. At present, antithrombotic drugs are categorized into three groups: anti-platelet aggregation drugs, anticoagulants, and thrombolytic drugs.

Evidence from recent research indicates that snake venom contains many active protein or peptides with defibration, anticoagulation, antiplatelet aggregation, and antithrombotic functions [4]. The anti-thrombotic action of snake venom protein is due to its ability to cleave the fibrinogen, reduce the content of fibrinogen, activate fibrinolytic enzyme, inhibit the activation of FXa, FIIa, thrombin, and other coagulation factors [5–7]. A large number of snake venom protease components have been isolated and some snake venom preparations are widely used in clinical practice.

Deinagkistrodon is one of the most virulent snake species in China that mostly lives in China's coastal and southwestern regions [8,9]. Snake venom contains a variety of active ingredients related to blood coagulation, such as phospholipase A2, serine protease, thrombin, metalloproteinase, and thrombin-like protein. Several new proteases and peptides have been isolated from the venom of Agkistrodon, such as Agacutas, Pt-A (Glu-Asn-Trp), Pt-B (Glu-Gln-Trp), ACH-11 [10–12]. These proteases affect many coagulation factors and cascades in the hemostatic system, such as cleavage of the alpha or beta chain of fibrinogen, inhibition of FXa, and activation of plasmin [13]. Most of the studied venomous snake venom polypeptides belong to serine proteases rather than metalloproteinases [14]. Metalloproteinases constitute an important components of the Agkistrodon genus. Metalloproteinase is an essential anticoagulant factor and exhibits fibrinogenolytic activity. However, there are few studies on the isolation and characterization of metalloproteinases from the venom of snakes belonging to the Agkistrodon genus. There is also paucity of information on the pharmacological effects of the isolated metalloproteinases. Metabolomics is an emerging discipline that compares the changes in body metabolites under different physiological and pathological conditions. In recent years, metabolomics has been widely used in drug research. It can be applied to study the mechanism of drug action by examining the changes of metabolic components in the body using serum, urine, and tissue fluid. It has been widely used in the study of monomeric compounds and traditional Chinese medicinal compounds. This study, to our best knowledge, is the first to report the anti-thrombotic effect mechanism of snake venom monomer using metabolomics

This work aimed to isolate and purify a novel metalloproteinase SP, determine its chemical properties and amino acid sequence, and investigate its pharmacological effects via in vivo studies. In order to study the mechanism of action of the novel snake venom protein, we investigated its antithrombotic effect using fibrinogen solubilization assay combined with UPLC-Q/TOF-MS-based non-targeted plasma metabolomics.

2. Results

2.1. Isolation of Metalloproteinase SP from Agkistrodon Venom

Based on the molecular weight differences of the snake venom proteins, an efficient separation and purification was achieved by the functional group ion exchange. The metalloproteinase SP snake venom monomer was then isolated and purified by sequential preparation, including ultracentrifugation, molecular exclusion chromatography on Sephadex TM G-75 (Figure 1A), anion exchange chromatography on DEAE-SphadexA-50 (Figure 1B), Sephadex TM G-75 (Figure 1C), and ultrafiltration.

2.2. Protein Identification

Based on LC-MS/MS analysis and database search analysis, it was found that three specific amino acid sequence fragments including SFGEWR, STEFQR, ENPPCILNKP were identified to belong to metalloproteinase SP (Figure 1D and Figure S1). Metalloproteinase SP consists of 202 amino acids with a molecular weight of 22.945 KDa and an isoelectric point of 5.78.

2.3. Anti-Thrombotic Activity

2.3.1. Anti-Coagulant Activity

Compared to the control group, the coagulation time after metalloproteinase SP treatment was significantly prolonged (*** $p < 0.001$, Figure 2A,B). Anticoagulant activity of metalloproteinase SP was evaluated by the classical coagulation assays. The parameters including activated partial thromboplastin time (APTT), thrombin time (TT), prothrombin time (PT), and Fibrinogen time (FT) were checked (Figure 2C–F). Metalloproteinase SP protein also significantly prolonged the clotting time, which inevitably influenced the changes in blood components and affect blood rheology (Tables 1 and 2).

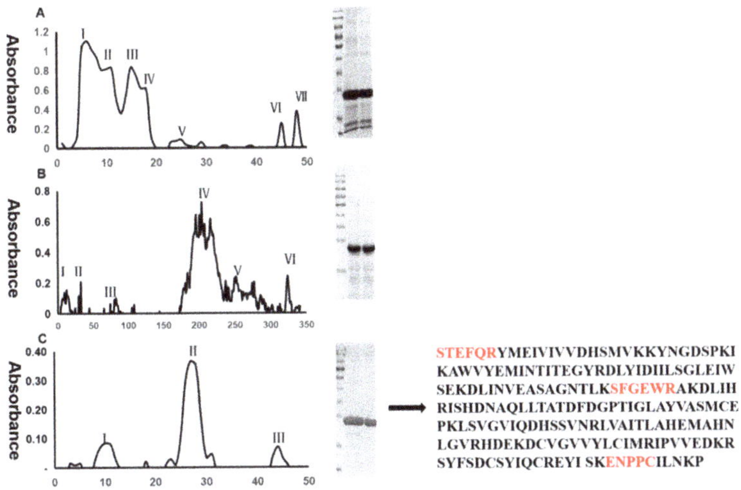

Figure 1. Isolation of metalloproteinase SP from Agkistrodon venom. (**A**) Snake venom (1000 mg) was subjected to Sephadex TM G-75 chromatography (1 × 100 cm) by eluting with 5 mM Tris-HCl (pH 7.4). The fraction III contains the target molecular weight protein. (**B**) Fraction III was further subjected to separation on DEAE-SphadexA-50 column (2 × 150 cm), eluted with 5 mM Tris-HCl (pH 7.4) and a segmented concentration gradient of 0.2 M NaCl. The fraction I contain the target molecular weight protein. (**C**) Fraction I was then separated on Sephadex TM G-75 column (1 × 50 cm), with 5 mM Tris-HCl (pH 7.4). Fraction II of this separation contains the target molecular weight protein. M, marker of protein molecular weight. (**D**) The peptide sequence of targeted protein. Red colors indicated unique peptides identified by LC-MS/MS, matching to targeted protein.

Table 1. Hemodynamic examination of arterial blood in clopidogrel and metalloproteinase SP treated mice.

	Red Blood Cell Count (10^{12}/L)	Hematocrit (%)	Blood Platelet Count (10^9/L)	Plateletcrit (%)	Mean Platelet Volume (fL)	Platelet Volume Distribution Width (%)	Platelet Larger Cell Ratio (%)
Blank	5.50 ± 0.29	29.4 ± 1.81	812 ± 36.7	0.440 ± 0.03	5.45 ± 0.16	4.15 ± 0.20	5.93 ± 0.93
Clopidogrel sulfate	5.61 ± 0.22	29.2 ± 0.81	783 ± 27.5	0.430 ± 0.02	5.47 ± 0.41	4.22 ± 0.22	4.42 ± 0.50 ***
Metalloproteinase SP	5.47 ± 0.20	28.9 ± 0.96	777 ± 26.5 *	0.410 ± 0.02 **	5.42 ± 0.27	4.21 ± 0.30	6.45 ± 0.26

In the clopidogrel sulfate treated group, the platelet-larger cell ratio was decreased compared with the blank group. In the metalloproteinase SP group, the platelet count and plateletcrit were decreased compared to the blank group. All data were expressed as mean ± SD. * $p < 0.05$, ** $p < 0.01$, and *** $p < 0.001$ compared with the blank control.

Table 2. Hemodynamic examination of venous blood in clopidogrel and metalloproteinase SP treated mice.

	Red Blood Cell Count (10^{12}/L)	Hematocrit (%)	Blood Platelet Count (10^9/L)	Plateletcrit (%)	Mean Platelet Volume (fL)	Platelet Volume Distribution Width (%)	Platelet-Larger Cell Ratio (%)
Blank	6.20 ± 0.57	31.4 ± 2.51	701 ± 86.3	0.391 ± 0.04	5.46 ± 0.19	4.37 ± 0.20	6.21 ± 0.79
Clopidogrel sulfate	5.37 ± 0.31 **	27.9 ± 1.66 **	733 ± 99.5	0.382 ± 0.06	5.29 ± 0.21	4.45 ± 0.25	4.60 ± 0.94 **
Metalloproteinase SP	5.63 ± 0.24 *	30.1 ± 2.56	613 ± 61.6 *	0.341 ± 0.02 **	5.41 ± 0.29	4.36 ± 0.60	5.39 ± 1.21

In the clopidogrel sulfate treated group, red blood cell count, hematocrit, and platelet-larger ratio cell were decreased compared with the blank group. In the metalloproteinase SP treated group, the red blood cell count, the number of platelets and plateletcrit were decreased. All data were expressed as mean ± SD. * $p < 0.05$ and ** $p < 0.01$ compared with the blank group.

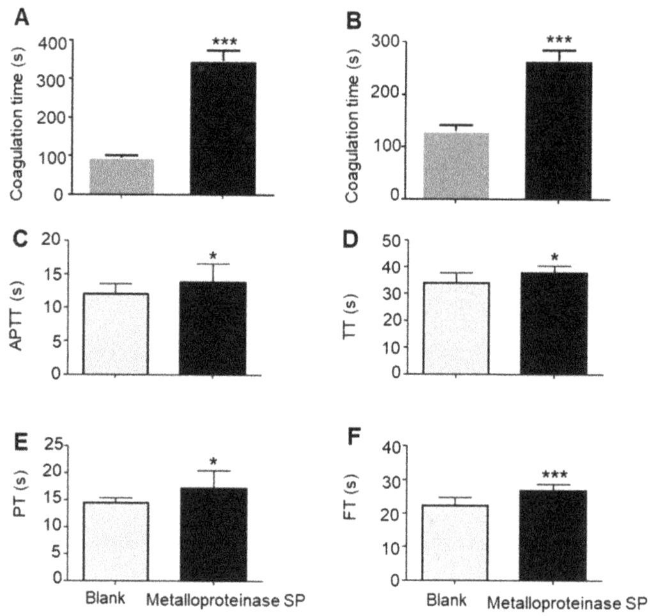

Figure 2. Anticoagulant activity of metalloproteinase SP was evaluated by coagulation time. (**A**) Blood coagulation time was determined by using the capillary technique. (**B**) Blood coagulation time was determined by the slide method. (**C**) Activated partial thromboplastin time (APTT). (**D**) Thrombin time (TT). (**E**) Prothrombin time (PT). (**F**) Fibrinogen time (FT). All data were expressed as mean ± SD, $n = 10$, * $p < 0.05$ and *** $p < 0.01$, compared with the blank group.

2.3.2. Anti-Thrombotic Activity

Compared with the blank group, the red blood cell and platelet content in the model group were significantly changed (** $p < 0.01$), indicating the successful establishment of model (Table 3). Metalloproteinase SP significantly inhibited acute pulmonary embolism formation induced by ADP (Figure 3A,B). It also inhibited Carrageenan-induced tail thrombus (Figure 3C and Figure S2). CCK-8 assay indicated that metalloproteinase SP did not have cytotoxicity within concentration of 0.5 ug/mL (Figure S3). Then, metalloproteinase SP activity was evaluated by the classical coagulation assays, thus, activated partial thromboplastin time (APTT), thrombin time (TT), prothrombin time (PT), and fibrinogen time (FT) (Figure 4A–D).

Table 3. Hemodynamic examination of venous blood to check clopidogrel and metalloproteinase SP effects on acute pulmonary embolism mice.

	Red Blood Cell Count (10^{12}/L)	Hematocrit (%)	Blood Platelet Count (10^9/L)	Plateletcrit (%)	Mean Platelet Volume (fL)	Platelet Volume Distribution Width (%)
Blank	8.01 ± 0.33	37.4 ± 2.36	865 ± 51.6	0.400 ± 0.04	4.16 ± 1.83	4.15 ± 0.21
Control	7.54 ± 0.40	34.6 ± 1.64	720 ± 73.5	0.350 ± 0.04	4.44 ± 0.65	3.99 ± 0.34
Clopidogrel sulfate	6.93 ± 0.16 **	31.3 ± 1.83 **	704 ± 56.9	0.320 ± 0.01	4.57 ± 0.167	3.91 ± 0.36
Metalloproteinase SP	6.92 ± 0.19 **	32.0 ± 1.16 **	640 ± 64.3 *	0.300 ± 0.03 *	4.70 ± 0.20	3.84 ± 0.28

In the clopidogrel sulfate treated group, red blood cell count and hematocrit were decreased compared with the blank group. In the metalloproteinase SP treated group, the red blood cell level, hematocrit, platelet count, and plateletcrit were decreased compared with blank. All data were expressed as mean ± SD. * $p < 0.05$ and ** $p < 0.01$ compared with the blank control group.

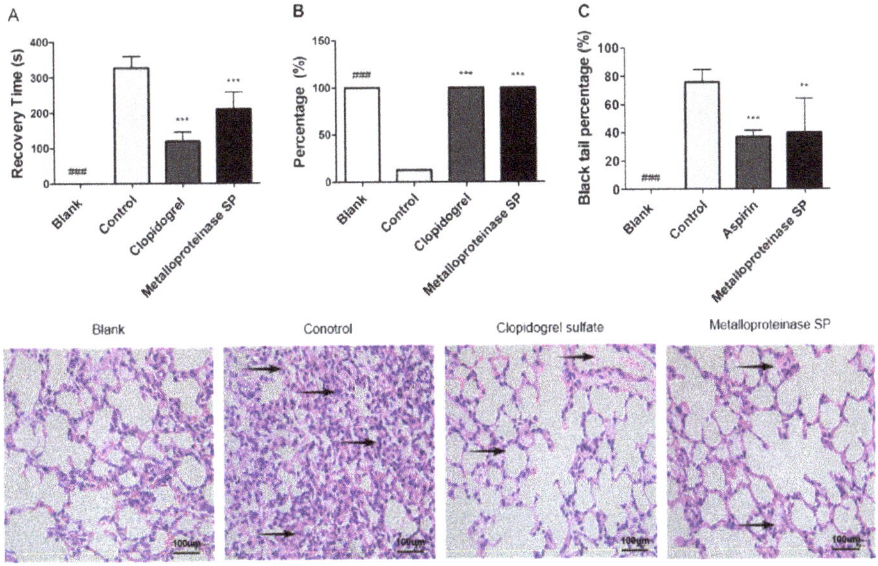

Figure 3. Effects of metalloproteinase SP on thrombosis. (**A**) Metalloproteinase SP inhibited the tail thrombus induced by the Carrageenan. (**B**,**C**) Metalloproteinase SP inhibited ADP-induced acute pulmonary thrombosis in mice. ADP (250 mg/kg) was intravenously injected to induce acute pulmonary thrombosis. The time that the mice righting reflex recovery was recorded as the recovery time. The hematoxylin and eosin stained section of the lung tissue shows the one for the control mouse to be dominated by thrombi. Effect of metalloproteinase SP on the ADP-induced formation of acute pulmonary thromboembolism in mice. Compared with the control group, the thrombus of the metalloproteinase SP and the clopidogrel group exhibited significant difference. All data were expressed as mean ± SD, compared with the control group, ** $p < 0.01$, *** $p < 0.001$ and compared with the blank group ### $p < 0.001$.

2.3.3. Antiplatelet Aggregation Activity

In vitro, metalloproteinase SP showed powerful antiplatelet activity by inhibiting the stimulatory effect of ADP, AA, and collagen (Figure 5A–C). In vivo, it exhibited a dose-dependent inhibition of the ADP-induced aggregation of PRP (Figure 5D–F). However, the inhibitory anti-platelet aggregation activity in vitro was more pronounced than in vivo.

2.4. Fibrinogenolytic Activity

In vivo, metalloproteinase SP protein cleaved α, β, and γ chains of fibrinogen, and decomposed the fibrinogen chain sequence in the order, $\alpha > \beta > \gamma$. In addition to degrading the α, β, and γ chains of fibrinogen, metalloproteinase SP also degraded the primary and secondary decomposition products to produce stable small molecular peptides (Figure 6A). Fibrinogen was also used to evaluate the effects of different inhibitors on the activity of metalloproteinase SP. The specific serine protease inhibitors such as benzamidine, butyl boronic acid, trasylol, trypsin, and PMSF were not able to significantly inhibit its enzymatic activity. However, EDTA significantly inhibited its enzymatic activity to a statistically significant level (Figure 6B). Regarding ideal temperature conditions, the enzyme showed high activity at 30–60 °C (Figure 6C). Metalloproteinase SP showed higher enzymatic activity at neutral pH values, with optimum catalysis at pH 6–7 (Figure 6D).

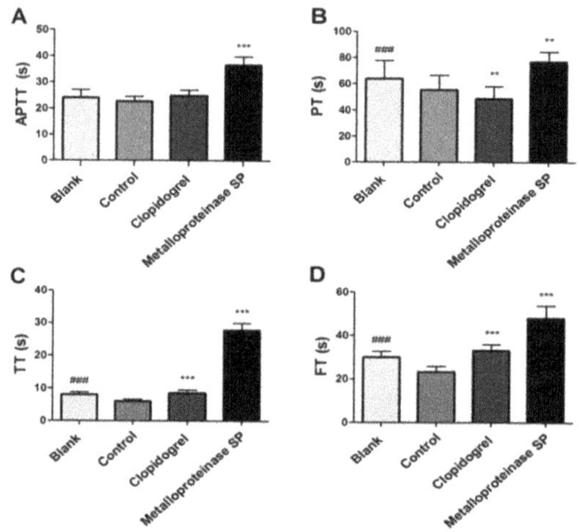

Figure 4. Effects of metalloproteinase SP on acute pulmonary embolism coagulation. Compared with blank group, the activated partial thromboplastin time (APTT), thrombin time (TT), prothrombin time (PT), and fibrinogen time (FT) of the control group exhibited significant difference (*** $p < 0.001$). Compared with control group, the metalloproteinase SP group could prolong the APTT, PT, TT, FT. (**A**) Activated partial thromboplastin time (APTT). (**B**) Thrombin time (TT). (**C**) Prothrombin time (PT). (**D**) Fibrinogen time (FT). All data were expressed as mean ± SD, compared with the control group, ** $p < 0.01$ ***, $p < 0.001$, and compared with the blank group ### $p < 0.01$.

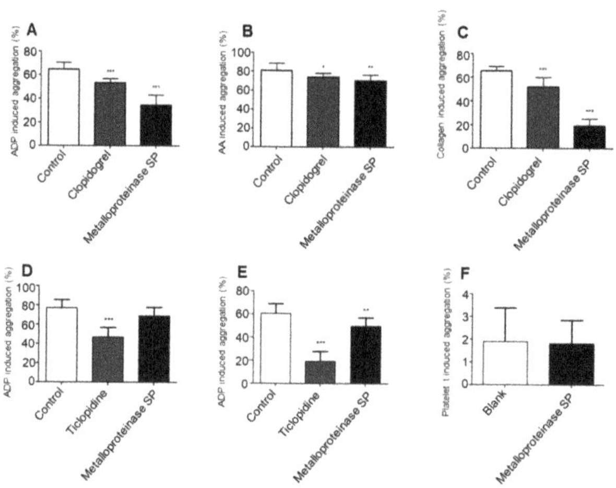

Figure 5. The in vitro and in vivo anti-platelet aggregation activity of metalloproteinase SP. (**A**) Compared with the blank group, the clopidogrel and the metalloproteinase SP groups specifically inhibited ADP-induced platelet aggregation. Metalloproteinase SP inhibited ADP-induced platelet aggregation with an inhibition rate of 44.28%. (**B**) Metalloproteinase SP inhibited AA-induced platelet aggregation with an inhibition rate of 13.19%. (**C**) Metalloproteinase SP inhibited collagen-induced platelet aggregation with an inhibition rate of 71.09%. The anti-platelet aggregation activity of metalloproteinase SP in vitro. (**D**,**E**) Metalloproteinase SP exhibited a dose-dependent inhibition of the ADP-induced aggregation of PRP. Compared with blank group, 0.5 μg of metalloproteinase SP significantly inhibited ADP-induced platelet aggregation with an inhibition rate of 18.27%. (**F**) Metalloproteinase SP not exhibited the promoted platelet aggregation activity. All data were expressed as mean ± SD, compared with the blank control group, * $p < 0.05$, ** $p < 0.01$, *** $p < 0.01$.

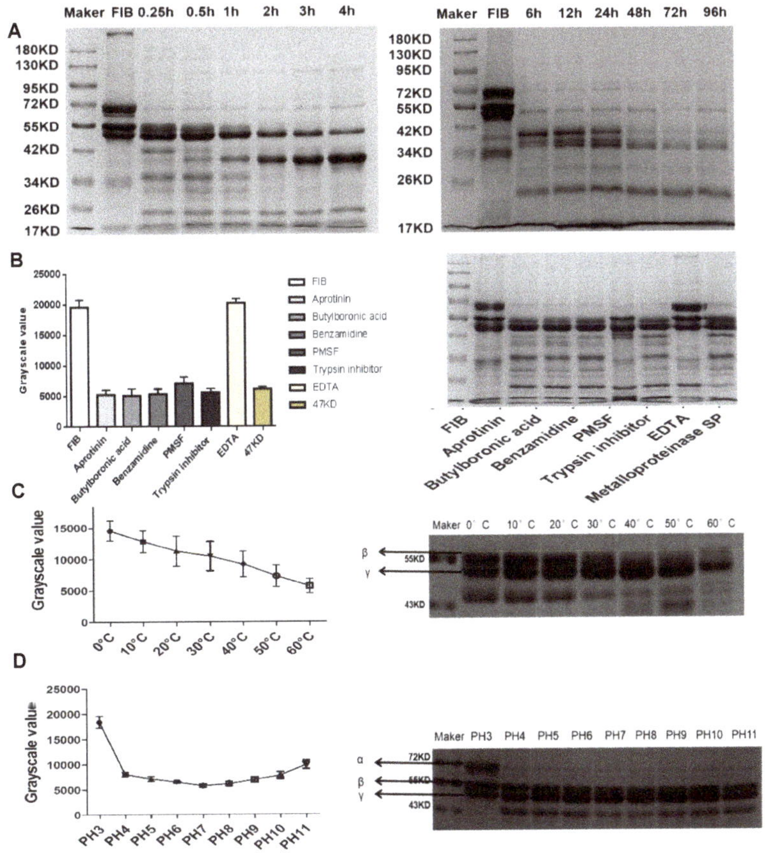

Figure 6. (**A**) Time-dependent effect of metalloproteinase SP on fibrinogen. Fibrinogenolytic activity was evaluated on 12% SDS-PAGE after incubation of metalloproteinase SP (5 μg) with human fibrinogen at 37 °C for different time periods. (**B**) Effects of inhibitors on metalloproteinase SP. Metalloproteinase SP was pre-incubated at 37 °C for 30 min at the presence of benzamidine (5 mM), ethylenediaminetetraacetic acid (EDTA, 5 mM), butyl boronic acid (20 mM), Trasylol (400 IU), trypsin inhibitor (20 mM), and PMSF (5 mM), respectively, prior to the addition of fibrinogen. (**C**) Temperature dependent effect of metalloproteinase SP on fibrinogen. Effect of temperature on fibrinogenolytic activity was evaluated on 12% SDS-PAGE after incubation of metalloproteinase SP (5 μg) with human fibrinogen at different temperature for 30 min. (**D**) The pH-dependent effect of metalloproteinase SP on fibrinogen. Effect of pH on fibrinogenolytic activity was evaluated on 12% SDS-PAGE after incubation of metalloproteinase SP (5 μg) with human fibrinogen at different pH for 30 min. All data were expressed as mean ± SD.

2.5. *Effects of Snake Venom Monomer on the Chemical Constituents of Acute Pulmonary Embolism*

In this study, the effect of metalloproteinase SP in an acute pulmonary embolism animal model was explored by analyzing metabolomic changes among metalloproteinase SP, model, and blank groups. As shown in the OPLS-DA score plot in both positive and negative ion modes, the three groups could be clearly distinguished (Figure 7A,C). This suggests that metalloproteinase SP exhibited a significant effect on acute pulmonary embolism by changing the anomalous metabolic states. Therefore, the metalloproteinase SP may be able to improve the pathological process of acute pulmonary embolism. The metabolites identified herein are listed in Table 4.

Table 4. The significantly changed metabolites in mice plasma after treatment with metalloproteinase SP.

Peak NO	Metabolite Name	Formula	Rt (min)	Molecular Weight	p-Value	FD	Control	Metalloproteinase SP	(+) MS	(+) MS/MS
1	5-Aminopentanoic acid	C5H11NO2	0.850	117.0790	*↓	0.7883	1.205 ± 0.241	0.952 ± 0.179	118.0861	118.086, 100.075, 72.080, 57.033, 56.049, 55.054, 44.049, 43.018
2	o-Tyrosine	C9H11NO3	1.100	181.0739	*↓	0.7218	0.857 ± 0.255	0.615 ± 0.131	182.0812	182.081, 165.05, 136.075, 119.040
3	Piperidine	C5H11N	1.467	85.0891	**↓	0.7029	1.431 ± 0.445	1.006 ± 0.295	86.0964	86.096, 69.070, 57.070, 56.050, 55.054, 52.033, 44.050, 43.054,
4	L-Isoleucine	C6H13NO2	1.483	131.0946	*↓	0.7330	1.261 ± 0.316	0.924 ± 0.138	132.1016	132.101, 86.096, 69.069, 44.049, 41.038
5	L-Phenylalanine	C9H11NO2	1.917	165.0790	*↓	0.8219	1.314 ± 0.249	1.083 ± 0.237	166.0856	166.086, 149.059, 131.049, 120.080, 107.049, 103.054
6	L-Norleucine	C6H13NO2	2.083	131.0946	**↓	0.7168	0.860 ± 0.153	0.616 ± 0.095	132.1009	132.10, 97.653, 86.695, 69.069, 44.048, 30.338
7	Palmitoyl Ethanolamide	C18H37NO2	3.767	299.2824	**↑	1.3987	0.917 ± 0.208	1.283 ± 0.183	300.2892	300.289, 282.279, 85.100, 83.085, 71.085, 67.054, 62.060, 57.069, 43.054,
8	beta-Alanine	C3H7NO2	0.617	89.0477	*↓	0.7330	1.071 ± 0.195	0.751 ± 0.110	88.0406	88.040, 71.013, 59.013, 53.003, 44.997, 43.018, 42.034, 41.002
9	L-Tyrosine	C9H11NO3	0.900	181.0739	*↓	0.7772	0.795 ± 0.116	0.618 ± 0.127	180.0664	180.066, 163.040, 119.050 72.009, 93.034, 74.024
10	Benzyl glycinate	C9H11NO2	1.916	165.0790	*↓	0.8400	1.208 ± 0.269	1.015 ± 0.174	164.0716	164.071, 91.055, 77.039
11	p-Cresol glucuronide	C13H16O7	2.267	284.0896	**↓	0.6891	0.540 ± 0.116	0.372 ± 0.158	283.0818	283.081, 265.073, 175.026, 117.019, 107.049, 87.008, 71.014, 43.018
12	p-Cresol sulfate	C7H8O4S	2.633	188.0143	**↓	0.5838	0.714 ± 0.159	0.417 ± 0.128	187.0073	187.007, 107.50, 105.334, 77.039
13	10-HDoHE	C22H32O3	4.800	344.2351	**↓	0.6892	0.559 ± 0.120	0.385 ± 0.164	343.2274	343.227, 325.210, 281.227, 189.130, 161.130, 153.092, 137.095, 109.102, 59.013
14	Octadec-9-enoic Acid	C18H34O2	10.150	282.4680	**↑	1.6569	1.033 ± 0.174	1.713 ± 0.826	281.2478	281.248, 263.238,59.0139

Compared to control group, "↑ and ↓" means the relative content of ions which is significantly increased or decreased. * $p < 0.05$, ** $p < 0.01$.

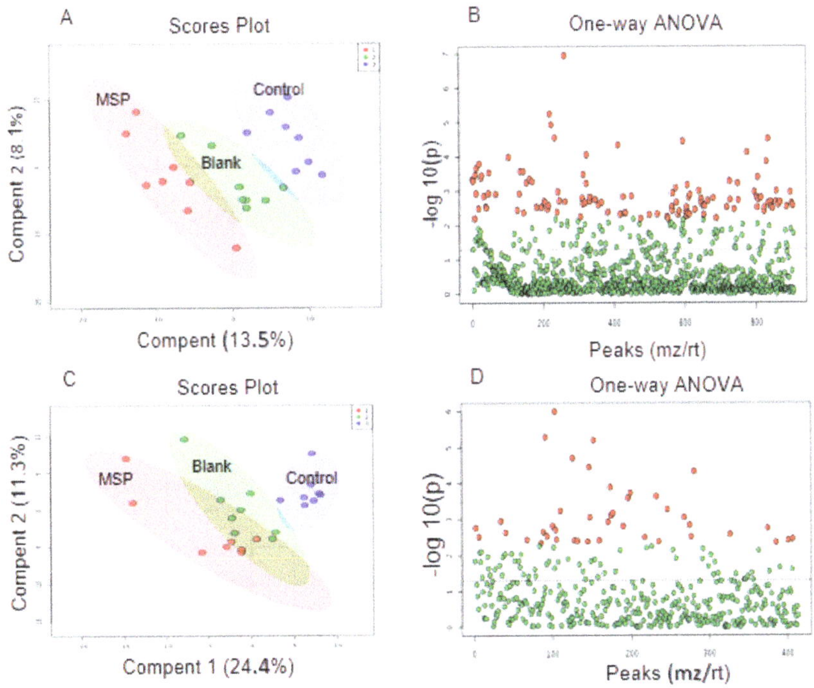

Figure 7. Score plot of plasma metabolomic profiling of three groups mice in positive ion mode (**A**) and negative ion mode (**C**). A total of 119 differential ions were screened by "adjusted *p*-value < 0.05, Fold change > 2" in positive ion mode (**B**). A total of 42 differential ions were screened by "adjusted *p*-value < 0.05, FC > 2" in negative ion mode (**D**).

The potential metabolic pathways and networks were analyzed by Metabo Analyst 4.0. Based on the Met PA analysis of Metabo Analyst 4.0, the related biomarkers primarily participated in the following pathways: tyrosine and tryptophan biosynthesis, phenylalanine metabolism, catecholamine biosynthesis, beta-alanine metabolism, aspartate metabolism, histidine metabolism, propanoate metabolism, pyrimidine metabolism, valine, leucine and isoleucine degradation, and tyrosine metabolism (Figure 8, Table 5).

Table 5. Metabolic pathways and differential metabolites.

NO	Pathway	Hits	Metabolites
1	Phenylalanine, tyrosine and tryptophan biosynthesis	2	L-Phenylalanine, L-Tyrosine
2	Phenylalanine metabolism	2	L-Phenylalanine, L-Tyrosine
3	Beta-Alanine Metabolism	1	Beta-Alanine
4	Aspartate Metabolism	1	Beta-Alanine
5	Histidine Metabolism	1	Beta-Alanine

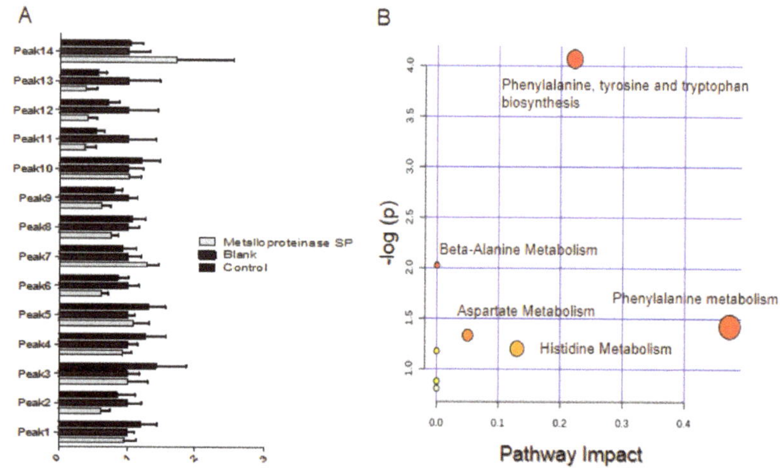

Figure 8. (**A**) Relative signal intensities for metabolic biomarkers in the blood. (**B**) Summary of pathway analysis with Metabo Analyst 4.0. Each ordinate in the plot represents a metabolic pathway. The ordinate is the original *p*-value obtained from the pathway analysis, and the abscissa is the influence value of pathways obtained from the topological analysis. The ordinate color and size of each circle was based on *p*-value and pathway impact value, respectively.

3. Discussion

The snake venom contains a large number of active proteins, peptides and small molecular weight compounds, which show anticoagulant, procoagulant, antithrombotic, antiplatelet aggregation, Fibrinogenolytic activity, analgesic, and other activities [15–18]. In the past few decades, various snake venom proteins have been extensively studied, and many snake venom proteins have been purified and characterized. Some snake venom preparations have been widely used in clinical settings as hemostatics, anticoagulants, antithrombotics, and thrombolytics. Examples include Ancrod, Batroxobin, and Acutase [18,19].

In the present study, metalloproteinase SP was purified form *Agkistrodon acutus* venom through ultracentrifugation, molecular exclusion chromatography on Sephadex TM G-75, anion exchange chromatography on DEAE-SphadexA-50, Sephadex TM G-75, and ultrafiltration. The molecular weight of metalloproteinase SP was determined by LC-MS/MS to be 22.9 kDa [20]. Metalloproteinase SP was proven to show anticoagulant, antithrombotic, antiplatelet aggregation, and fibrinogenolytic activities, and improve blood rheology. In vivo and in vitro experiments showed that metalloproteinase SP could cleave α, β, and γ chains of fibrinogen and further reduce fibrinogen content.

Fibrinogen plays a vital role blood coagulation, blood rheology, platelet aggregation, and thrombosis [21,22]. The coagulation process is traditionally classified into extrinsic, intrinsic, and common coagulation. In the common pathway of coagulation, fibrinogen forms the insoluble fibrin network under the stimulation of coagulation factors such as thrombin and FXa, which ultimately constitutes a fibrin clot that stops bleeding [23,24]. Fibrinogen mainly affects platelet aggregation, blood rheology, fibrinogen clot structure, atherosclerosis, plasmin activation, and fibrinogen decomposition (FgDP) to adjust the formation of thrombus [24–28]. Hence, the centrality of fibrinogen in platelet aggregation cannot be overemphasized. Fibrinogen has multiple GPIIb/IIa receptor binding sequences, such as γ400~411, α95~97, and α572~574 [29]. Fibrinogen functions as a bridge of platelet aggregation [29–31]. Moreover, fibrinogen is the most important factor affecting plasma viscosity. Increased plasma fibrinogen levels inevitably result in elevated plasma viscosity and shear stress, as well as endothelial cells and platelets activation. Activation of endothelial cells further promotes the expression or activation of various adhesion molecules and integrins, which leads to red blood cells adhesion and platelets aggregation and thrombosis [32–34]. Fibrin also possesses multiple plasminogen binding sites. Fibrinogen was cleaved by metalloproteinase SP to form fibrin, which may be through decomposing the α, β, and γ chain of fibrinogen, to reduce the content of the fibrinogen in the blood. Furthermore, it affected the blood coagulation pathway, prolonged the clotting time, directly inhibited the aggregation of platelets, indirectly regulated blood rheology, and indirectly regulated the formation of thrombus [35–38].

Based on the in vivo and in vitro experiments, the underlying mechanism of metalloproteinase SP on ADP-induced acute pulmonary embolism in mice was further investigated by untargeted mass spectrometry-based metabolomics profiling [39]. Metalloproteinase SP inhibited the formation of acute pulmonary embolism induced by ADP, which may be related to the amino acid metabolism and synthesis, including phenylalanine, tyrosine and tryptophan biosynthesis, phenylalanine metabolism, catecholamine biosynthesis, and eight other metabolic pathways [40,41].

Phenylalanine is one of the essential amino acids in humans that forms tyrosine under the action of phenylalanine hydroxylase in the liver [42,43]. Tyrosine synthesizes certain hormones and neurotransmitters such as dopamine (DA), norepinephrine (NE), epinephrine (E), and melanin in the nervous system and adrenal medulla. Thus, tyrosine plays an important role in regulating energy metabolism and scavenging free radicals [40]. Metalloproteinase SP regulated the biosynthesis of tyrosine and tryptophan, thereby alleviating lipid metabolism disorder and exert indirectly inhibited the formation of thrombus.

The content of catecholamines is closely related to blood coagulation and thrombosis [44]. Excessive levels of catecholamines could result in hypertension and acute myocardial infarction. Catecholamines act on platelet alpha (2)-adrenergic receptors to regulate platelet activation and aggregation, blood coagulation, and thrombosis.

The anticoagulant and antithrombotic activity of metalloproteinase SP may be ascribed to the regulation of histidine metabolism. Histidine is a semi-essential amino acid that is decarboxylated to form histamine under the action of histidine decarboxylase. Histamine exhibits diastolic vasoactive, antihypertensive and anti-inflammatory activities, and regulates a variety of allergic reactions. Histamine is clinically used for angina pectoris, cardiac insufficiency and other diseases [45]. Histamine has a strong vasodilator effect and is associated with a variety of allergic reactions and inflammation. Histidine may indirectly regulate inflammatory factors and inflammatory reaction by regulating the synthesis of histamine, thus inhibiting the formation of thrombus.

Metalloproteinase SP has many pharmacological activities, such as prolonging coagulation time, anti-platelet aggregation, inhibiting thrombosis, and decomposing fibrinogen. Metalloproteinase SP has high affinity with fibrinogen, and its anti-platelet aggregation and inhibition of thrombosis are closely related to its ability to decompose fibrinogen. Studies have shown that fibrinogen decomposition has an important relationship with the activation of fibrinolytic enzymes, and at the same time affects them. The experimental results showed that metalloproteinase SP can rapidly decompose fibrinogen and plays an anti-thrombotic role. However, fibrinogen content that is too low can easily lead to hemorrhage and coagulation disorders. At the same time, metalloproteinase has certain blood toxicity, which may lead to the reduction of platelet count and erythrocyte count. Thus, toxicity and safety of this venome component are needed to be further studied before its application in clinical.

4. Materials and Methods

4.1. Materials

SPF grade ICR mice (18–22 g) were purchased from the Comparative Medical Center of Yangzhou University (Jiangsu province, China, License No. SYXK (Su) 2018-0019). Sprague-Dawley rats (180–220 g) were purchased from Nanjing Qing long Shan Animal Center (Jiangsu province, China, License No. SYXK (su) 2018-0022). Animals were housed in a room of temperature of 22 ± 2 °C, relative humidity of 50 ± 5%, and exposed to 12 h dark/light cycle. The animals had free access to food and drinking water, and were allowed to acclimatize to their new environment for 1 week. The experiments were carried out in accordance with the guidelines and the regulations of the Ethical Committee of the China Pharmaceutical University. The protocols were approved by the Institutional Animal Care and Use Committee of the China Pharmaceutical University. The ethics approval number was AUC-37 (20180329) and the approval date was 29 March 2018.

4.2. *Separation and Purification of Metalloproteinase SP*

A total of 1000 mg venom powder was dissolved in 5 mM Tris-HCl solution (pH 7.4), and then centrifuged at 12,000 r/min for 15 min at 4 °C. After centrifugation, the supernatant was collected and loaded on a gel filtration column of Sephadex TM G-75. The transparent supernatant was further analyzed by molecular exclusion chromatography with Sephadex TM G-75 column (1 × 100 cm, GE Healthcare, Frankfurt, Germany), which was pre-conditioned with 5 mM Tris-HCl (pH 7.4). Aliquots separated samples (4 mL/tube) were collected at a flow rate of 0.5 mL/min at 15 °C. The target component was chosen according to its profile on sodium dodecyl sulfate poly-acrylamide gel electrophoresis (SDS-PAGE) and was pooled, lyophilized, and then stored at −20 °C for subsequent chromatographic analysis. The target component from Sephadex TM G-75 was solubilized in 4 mL of 5 mM Tris-HCl (pH 7.4), and centrifuged at 12,000 r/min for 15 min at 4 °C. The supernatant was loaded on the anion exchange chromatography column (DEAE-SphadexA-50, 2 × 150 cm GE Healthcare, Germany) previously equilibrated and eluted with 5 mM Tris-HCl (pH 7.4). The target proteins were sequentially washed out with 5 mM Tris-HCl (pH 7.4), 0–0.2 M NaCl in 5 mM Tris-HCl (pH 7.4), and 0.2 M NaCl in 5 mM Tris-HCl (pH 7.4). The 4 mL/tube samples were collected at a flow rate of 0.5 mL/min at 15 °C. The target component was chosen based on its SDS-PAGE profile and was re-chromatographed after lyophilization. The clear supernatant was subjected to further purification via molecular exclusion chromatography using Sephadex TM G-75 (1 × 50 cm, GE Healthcare, Germany), previously equilibrated and eluted with 5 mM Tris-HCl (pH 7.4). Samples of 3 mL/tube were collected at a flow rate of 0.5 mL/min at 15 °C. The target component was chosen according to its profile on sodium dodecyl sulfate poly-acrylamide gel electrophoresis (SDS-PAGE) and was pooled, lyophilized, and stored at −20 °C for further chromatographic analysis.

4.3. *Protein Quantification and Identification*

The protein concentration was determined by the Bradford Protein Concentration Assay Kit (Thermo Fisher Scientific Incorporation, MA, USA), according to the manufacturer's instructions [46]. The purified venome protein was separated by SDS-PAGE and identified by LC-MS/MS and a Uniprot database search (https://www.uniprot.org/uniprot/?query=agkistrodon+acutus&sort=score).

4.4. *Pharmacological Evaluation of Metalloproteinase SP*

4.4.1. Anti-Coagulant Activity

Blood Clotting Assays

Half an hour after metalloproteinase SP intravenous administration via the tail vein, blood samples were collected from the orbital plexus of mice. Afterward, the time for the blood coagulate was recorded. The clotting time was observed when microfibril was visible. The average time of each observation was 20 s. The collected blood samples were also spotted on glass slides and clotting times determined. Clotting was achieved when the fibrin mesh appeared and could be picked by glass dissecting needle [47,48].

Blood Parameters Test

Half an hour after clopidogrel sulfate and metalloproteinase SP administration via the tail vein, blood samples were collected from the orbital plexus of mice or from abdominal aorta of mice. Both venous and arterial blood were anticoagulated by heparin sodium. The anticoagulant whole blood was placed in the automatic blood cell analyzer (RT-7600Vet, Rayto, Shenzhen, China) for detection and analysis within 2 h.

Coagulation Function Test

One hour after the intravenous administration of test drugs or vehicles, blood samples were harvested from the orbital plexus. An aliquot of 600 μL blood was gently mixed with 66 μL citrate sodium anticoagulant. The changes of blood corpuscles in whole blood (anticoagulated by citrate sodium) were analyzed using blood cell analyzer RT-7600Vet (Rayto, Shenzhen China). After centrifugation at 3000 rpm/min for 15 min, the plasma and blood cells were collected and analyzed for four coagulation parameters using the semiautomatic blood coagulation analyzer LG-PABER-I (STEELLEX, Beijing, China) [49–51]. These parameters were: activated partial thromboplastin time (APTT), thrombin time (TT), prothrombin time (PT), and Fibrinogen time (FT). Standard analytical protocols were closely followed for all determinations.

4.4.2. Anti-Thrombotic Activity

The anti-thrombotic effect of metalloproteinase SP was evaluated using two thrombosis pathological models [52,53].

ADP-Induced Acute Pulmonary Thrombosis in Mice

After metalloproteinase SP and clopidogrel treatment for 1 h, the ADP solution (250 mg/kg) was intravenously administered to induce acute pulmonary embolism, which could probably cause the paralysis or unpredicted death of mice. The recovery time of the mice was then recorded. After establishment of the model, the whole blood cell changes were assayed and coagulation parameters determined as earlier mentioned.

Half an hour after metalloproteinase SP intravenous administration via the tail vein, blood samples were collected from the orbital plexus of mice of mice. Both venous and arterial blood were anticoagulated by heparin sodium. The anticoagulant whole blood was placed in the automatic blood cell analyzer (RT-7600Vet, Rayto, Shenzhen, China) for detection and analysis within 2 h.

Carrageenan-Induced Tail Thrombosis in Mice

One hour after intravenous administration of metalloproteinase SP and aspirin in each group, Carrageenan solution (20 mg/kg) was intraperitoneally injected to induce tail thrombosis. The mice were then placed in environmentally-friendly cages for 12 h and their tail lengths recorded.

4.4.3. Antiplatelet Aggregation Activity

In Vivo Antiplatelet Aggregation Assay

After 60 min of administration in each group, the rats were anaesthetized with 50 mg/kg chloral hydrate via abdominal injection. Blood samples were collected from abdominal aorta via cannulation. An aliquot of 8 mL blood was gently mixed with 880 μL citrate sodium anticoagulant (3.8%, *v/v*). Afterward, platelet-rich plasma (PRP) was obtained after centrifugation at 1000 rpm/min for 10 min. The residue was further centrifuged at 3000 rpm/min for 10 min to get platelet poor plasma (PPP). Prior to the antiplatelet aggregation assay, both PRP and PPP were incubated at 37 °C. An aliquot of 250 μL PPP was added into the measuring cup to calibrate the baseline. Thereafter, 225 μL of PRP was added and platelet aggregation stimulated with 25 μL different stimulants; thus, ADP (150 μM), AA (1 mg/mL), Collagen (50 μg/mL). Platelet aggregation was measured by aggregometer (AggRAM, Helena, USA), and the maximal aggregation rate was recorded within 5 min [54].

In Vitro Antiplatelet Aggregation Activity

The blank PRP plasma was harvested from control rats. Metalloproteinase SP (1 μg, 5 μg) was pre-incubated with 225 μL PRP at 37 °C for 5 min, and then stimulated with 25 μL different aggregating agents at the following final concentrations (ADP 15 μM). Platelet aggregation was assayed by a platelet aggregation instrument (AggRAM, Helena, TX, USA), and the maximum aggregation rate recorded in 5 min. 225 uL aliquot of PRP was measured into each cuvette and stimulated with metalloproteinase SP 25 uL (1 μg, 5 μg). Platelet aggregation was assayed by a platelet aggregation instrument, and the maximum aggregation rate recorded in 5 min [53].

4.4.4. Fibrinogenolytic Activity

Human fibrinogen (8 mg/mL) was incubated with 5 μg metalloproteinase SP at 37 °C for 0.25 h, 0.5 h, 1 h, 2 h, 3 h, 4 h, 6 h, 12 h, 24 h, 48h, 72 h, and 96 h [53]. Afterward, the samples mixed with loading buffer solution were heated at 100 °C for 10 min. All samples were transferred to SDS-PAGE system for fibrinogen degradation analysis.

Effect of Temperature and pH on Metalloproteinase SP Activity

Human fibrinogen (8 mg/mL) was incubated with 5 μg metalloproteinase SP at 0 °C, 10 °C, 20 °C, 30 °C, 40 °C, 50 °C, and 60 °C for 30 min [55]. Additionally, 5 μg metalloproteinase SP was incubated with buffer solutions of different pH (pH 3, 4, 5, 6, 7, 8, 9, 10, 11) at 37 °C for 30 min, and subsequently incubated with human fibrinogen (8 mg/mL) at 37 °C for another 30 min. All samples were subjected to SDS-PAGE to investigate the effect of temperature and pH on metalloproteinase SP activity.

Effects of Serine Protease and Metalloproteinase Inhibitors on Metalloproteinase SP Activity

Different inhibitors, benzamidine (5 mM), ethylenediaminetetraacetic acid (EDTA, 5 mM), butyl boronic acid (20 mM), Trasylol (400 IU), trypsin inhibitor (20 mM), and PMSF (5 mM) were incubated with 5 μg metalloproteinase SP at 37 °C for 30 min [54]. The samples were further incubated with human fibrinogen at 37 °C for 30 min. All samples were analyzed by SDS-PAGE to evaluate the effect of inhibitors on metalloproteinase SP activity.

4.5. Effects of Snake Venom Monomer on the Chemical Constituents of Acute Pulmonary Embolism Based on UHPLC-Q/TOF-MS Non-Targeted Metabolomics

An aliquot of 150 μL of methanol/acetonitrile (3:1) solution (containing 0.4 μg/mL L-2-chloropheylaanine and 10 μg/mL ketoprofen as the internal standard for the ESI+ and ESI− modes respectively) was added to 50 μL plasma followed by vigorous vortex-mixing for 30 s. The mixture was centrifuged at 13,000 r/min at 4 °C for 10 min to precipitate protein. The supernatant (150 μL) was divided into two portions and dried under a gentle stream of nitrogen gas at room temperature. One of the samples was reconstituted with 75 μL of 50% aqueous acetonitrile for ultra-performance liquid chromatography with quadrupole time-of-flight mass spectrometry (UPLC-Q-TOF-MS) analysis in positive mode while the other was for negative mode. The quality control (QC) samples were prepared to improve the data quality for metabolic profiling.

4.5.1. UHPLC/Q-TOF MS Analysis

Chromatographic evaluations were performed with an Agilent 1290 series (Agilent Corp, Santa Clara, CA, USA) HPLC system equipped with a binary pump, degasser, an autosampler, and a temperature-controlled column compartment. Chromatographic separations were achieved on an UPLC BEH C18 column (1.8 μm, 2.1 mm × 100 mm, Waters, Ireland) [25]. Mobile phase A was 0.1% formic acid aqueous solution and B was 0.1% formic acid-acetonitrile solution for the positive ion mode. Mobile phase A was 10 mM ammonium acetate solution and B was 10 mM ammonium acetate acetonitrile solution for the negative ion mode. This mobile phase system was run in a gradient elution program as follows: 90–99% A, at 0–1 min; 30–90% A, at 1–3 min; 15–30% A, at 3–8 min; 0–15% A, at 8–9 min; 0% A, at 9–10 min. The post-run time was 5 min. The oven temperature was set at 30 °C, the injection volume was 10 μL and the flow rate was 0.4 mL/min.

Separated components were detected by Agilent 6545A Q/TOF mass spectrometer (Agilent Corp, Santa Clara, CA, USA) equipped with an ESI interface. The operating parameters were as follows: drying N_2 gas flow rate, 8 L/min; temperature, 320 °C; nebulizer, 35 psig; capillary, 3000 V; skimmer, 65 V; OCT RFV, 750 V; fragmentor 100 V. The reference masses 254.2806 (ketoprofen) and 299.75 (L-Phenylalanine) were used for internal mass calibration during the runs in the positive and negative ion modes. At the scan rate of 1.50 spectra/s using fixed collision energies (0.00, 10.00, 20.00, 40.00 V) MS/MS data were acquired with isolation width MS/MS medium (~4 amu).

4.5.2. Data Processing and Identification of Differential Metabolites

LC-MS raw data files were converted to m/z Data format using DA reprocessor (Agilent Corp., Santa Clara, CA, USA). Peak finding, filtering, and alignment were subsequently carried out using open-source R-Package XCMS [25]. Metabo Analyst 4.0 (https://www.metaboanalyst.ca/) was then employed for the analysis of the data. The use of Metabo Analyst 4.0 can be broadly categorized into three major activities: (1) missing value processing, (2) data filtering, and (3) data normalization (i.e., sample normalization, data transformation, and data scaling) [55].

Using Metabo Analyst 4.0 unsupervised principal component analysis (PCA), a general picture of the relationship that exists among the data matrix was obtained. Thereafter, the supervised orthogonal partial least-squares discrimination analysis (OPLS-DA) was carried out to examine the metabolite differences between the treatment, blank, and model groups. The inclusion criterion of the metabolites was that, the fold change between the groups compared should be greater than 1.2 and p-value < 0.05 (one-way ANOVA).

Metabolites that met this criterion were considered as the differential metabolites. In the identification of the biomarkers, the following databases were used: METLIN (http://metlin.scripps.edu/), HMDB database (http://www.hmdb.ca/), Metabo Analyst4.0 (https://www.metaboanalyst.ca/), SMPDB (http://www.smpdb.ca/) and KEGG (http://www.genome.jp/kegg/) [55,56]

4.6. *Statistical Analysis*

Functional and enzymatic activities were assessed by two individual experiments in triplicate. All data were expressed as mean ± SD. The significance of differences was analyzed by one-way ANOVA followed by the Bonferroni correction. A value of $p < 0.05$ was considered statistically significant.

5. Conclusions

In this study, a novel metalloproteinase SP was isolated from Agkistrodon venom. In vivo, the metalloproteinase SP protein showed anticoagulant, antithrombotic, and antiplatelet aggregation activities and changed blood rheology. In vitro, metalloproteinase SP strongly inhibited platelet aggregation and cleaved the α, β, and γ chains of fibrinogen, but its activity was obviously affected by temperature or pH and could be reversed by EDTA. Finally, this study successfully built a mice model with acute pulmonary embolism to explore the antithrombotic mechanism of metalloproteinase SP. Assisted by LC-MS-Q-TOF-MS analysis, 14 representative biomarkers were identified, involving in phenylalanine, tyrosine and tryptophan biosynthesis, phenylalanine metabolism, catecholamine biosynthesis, and eight other metabolic pathways. All these findings indicated that metalloproteinase SP may affect blood coagulation and thrombus formation by decomposing fibrinogen and anti-platelet aggregation

Supplementary Materials: Supplementary Materials can be found at http://www.mdpi.com/1422-0067/20/17/4088/s1.

Author Contributions: J.H., F.H., and X.Y. made the major contributions to this study in the conception, design, drafting part of manuscript, and final revision. H.F. performed statistical analysis of the data and purified protein. All authors read and approved the final manuscript.

Funding: The work was supported by the National Natural Science Foundation of China (No. 6200020055A).

Acknowledgments: We thank Mr Yingchun Tan providing us snake venom. Thanks to the help and support provided by the National Platform for Natural Medicine Active Components and the National Key Laboratory of Pharmacodynamics and the Platform of Chinese Medicine College.

Conflicts of Interest: The authors declare no conflict of interest.

Abbreviations

ANOVA	Analysis of variance
ADP	Adenosine diphosphate
AA	Arachidonic acid
RBC	Red blood cell
PLT	Platelet
PCT	Plateletcrit
HCT	Hematocrit
MPV	Mean platelet volume
PDW	Platelet volume distribution width
P-LCR	Platelet -larger cell ratio

References

1. Raskob, G.E.; Angchaisuksiri, P.; Blanco, A.N.; Buller, H.; Gallus, A.; Hunt, B.J.; Hylek, E.M.; Kakkar, A.; Konstantinides, S.V.; McCumber, M. Thrombosis: A major contributor to global disease burden. *Thromb. Res.* **2014**, *134*, 931–938. [CrossRef] [PubMed]
2. Nieswandt, B.; Kleinschnitz, C.; Stoll, G. Ischaemic stroke: A thrombo-inflammatory disease. *J. Physiol.* **2011**, *589*, 4115–4123. [CrossRef] [PubMed]
3. Ortiz-Muñoz, G.; Mallavia, B.; Bins, A.; Headley, M.; Krummel, M.F.; Looney, M.R. Aspirin-triggered 15-epi-lipoxin A 4 regulates neutrophil-platelet aggregation and attenuates acute lung injury in mice. *Blood* **2014**, *124*, 2625–2635. [CrossRef]
4. Fry, B.G.; Roelants, K.; Champagne, D.E.; Scheib, H.; Tyndall, J.D.; King, G.F.; Nevalainen, T.J.; Norman, J.A.; Lewis, R.J.; Norton, R.S.; et al. The toxicogenomic multiverse: Convergent recruitment of proteins into animal venoms. *Annu. Rev. Genom. Hum. G.* **2009**, *10*, 483–511. [CrossRef]
5. Huang, Q.; Yang, Q.D.; Tan, X.L.; Feng, J.; Tang, T.; Xia, J.; Zhang, L.; Huang, L.; Bai, Y.P.; Liu, Y.H. Absence of association between atherosclerotic cerebral infarction and TNFSF4/TNFRSF4 single nucleotide polymorphisms rs1234313, rs1234314 and rs17568 in a Chinese population. *J. Int. Med. Res.* **2014**, *42*, 436–443. [CrossRef] [PubMed]
6. Mackessy, S.P. Thrombin-like enzymes in snake venoms. In *Toxins and Hemostasis*, 2nd ed.; Kini, R., Clemetson, K., Eds.; Springer: Dordrecht, Netherlands, 2010; Volume 39, pp. 2–6.
7. Dempfle, C.E.; Argiriou, S.; Kucher, K.; Müller-Peltzer, H.; Rübsamen, K.; Heene, D.L. Analysis of fibrin formation and proteolysis during intravenous administration of ancrod. *Blood* **2000**, *96*, 2793–2802.
8. Bocian, A.; Urbanik, M.; Hus, K.; Łyskowski, A.; Petrilla, V.; Andrejčáková, Z.; Petrillová, M.; Legáth, J. Proteomic analyses of Agkistrodon contortrix venom using 2D electrophoresis and MS techniques. *Toxins* **2016**, *8*, 372. [CrossRef] [PubMed]
9. Lomonte, B.; Tsai, W.C.; Ureña-Diaz, J.M.; Sanz, L.; Mora-Obando, D.; Sánchez, E.E.; Fry, B.G.; Gutiérrez, J.M.; Gibbs, H.L.; Sovic, M.G.; et al. Venomics of new world pit vipers: Genus-wide comparisons of venom proteomes across Agkistrodon. *J. Proteomics* **2014**, *96*, 103–116. [CrossRef] [PubMed]
10. Tang, S.S.; Wang, X.H.; Zhang, J.H.; Tang, B.S.; Qian, L.; Li, P.Y.; Luo, L.W. Biochemical properties and comparative pharmacology of a coagulant from Deinagkistrodon acutus snake venom. *Eur. J. Pharm. Sci.* **2013**, *49*, 90–98. [CrossRef] [PubMed]

11. Ding, B.; Xu, Z.; Qian, C.; Jiang, F.; Ding, X.; Ruan, Y.; Ding, Z.; Fan, Y. Antiplatelet aggregation and antithrombosis efficiency of peptides in the snake venom of *Deinagkistrodon acutusr*, isolation, identification, and evaluation. *Evid-Based Compl. Alt.* **2015**, *5*, 1–6.
12. Ming, X. A novel direct factor Xa inhibitory peptide with anti-platelet aggregation activity from Agkistrodon acutus venom hydrolysates. *Sci. Rep.* **2015**, *5*, 10846.
13. Zeng, F.; Shen, B.; Zhu, Z.; Zhang, P.; Ji, Y.; Niu, L.; Li, X.; Teng, M. Crystal structure and activating effect on RyRs of AhV_TL-I, a glycosylated thrombin-like enzyme from Agkistrodon halys snake venom. *Arch. Toxicol.* **2013**, *87*, 535–545. [CrossRef] [PubMed]
14. Castro, H.C.; Zingali, R.B.; Albuquerque, M.G.; Pujol-Luz, M.; Rodrigues, C.R. Snake venom thrombin-like enzymes: From reptilase to now. *Cell. Mol. Life Sci.* **2004**, *61*, 843–856. [CrossRef] [PubMed]
15. Ge, Y.H.; Chen, Y.Y.; Zhou, G.S.; Liu, X.; Tang, Y.P.; Liu, R.; Liu, P.; Li, N.; Yang, J.; Wang, J.; et al. Novel antithrombotic protease from marine worm sipunculus nudus. *Int. J. Mol. Sci.* **2018**, *19*, 3023. [CrossRef] [PubMed]
16. Kini, M.R. Anticoagulant proteins from snake venoms: Structure, function and mechanism. *Biochem. J.* **2006**, *397*, 377–387. [CrossRef]
17. Suntravat, M.; Nuchprayoon, I.; Pérez, J.C. Comparative study of anticoagulant and procoagulant properties of 28 snake venoms from families Elapidae, Viperidae, and purified Russell's viper venom-factor X activator (RVV-X). *Toxicon* **2010**, *56*, 544–553. [CrossRef]
18. Depta, J.P.; Bhatt, D.L. New approaches to inhibiting platelets and coagulation. *Annu. Rev. Pharmacol.* **2015**, *55*, 373–397. [CrossRef] [PubMed]
19. Koh, D.C.I.; Armugam, A.; Jeyaseelan, K. Snake venom components and their applications in biomedicine. *Cell Mol. Life Sci.* **2006**, *63*, 3030–3041. [CrossRef] [PubMed]
20. Niall, H.D. Automated edman degradation: The protein sequenator. *Method Enzymol.* **1973**, *27*, 942–1010.
21. Breen, P. Basics of coagulation pathways. *Int. Anesthesiol. Clin.* **2004**, *42*, 1–9. [CrossRef]
22. Corte, A.L.C.L.; Philippou, H.; Ariëns, R.A.S. Role of fibrin structure in thrombosis and vascular disease. *Adv. Protein Chem. Str.* **2011**, *75*–127.
23. Chernysh, I.N.; Nagaswami, C.; Weisel, J.W. Visualization and identification of the structures formed during early stages of fibrin polymerization. *Blood* **2011**, *117*, 4609–4614. [CrossRef] [PubMed]
24. Lowe, G.D.; Lee, A.J.; Rumley, A.; Price, J.F.; Fowkes, F.G. Blood viscosity and risk of cardiovascular events: The Edinburgh Artery Study. *Brit. J. Haematol.* **2015**, *96*, 168–173. [CrossRef]
25. Uitte de Willige, S.; de Visser, M.C.; Houwing-Duistermaat, J.J.; Rosendaal, F.R.; Vos, H.L.; Bertina, R.M. Genetic variation in the fibrinogen gamma gene increases the risk for deep venous thrombosis by reducing plasma fibrinogen gamma' levels. *Blood* **2005**, *106*, 4176–4183. [CrossRef] [PubMed]
26. Hacioglu, G.; Yalcin, O.; Bor-Kucukatay, M.; Ozkaya, G.; Baskurt, O.K. Red blood cell rheological properties in various rat hypertension models. *Clin. Hemorheol. Micro.* **2002**, *26*, 27–32.
27. Bennett, J.S. Platelet-fibrinogen interactions. *Ann. N. Y. Acad. Sci.* **2010**, *936*, 340–354. [CrossRef]
28. Weisel, J.W. The mechanical properties of fibrin for basic scientists and clinicians. *Biophys. Chem.* **2004**, *112*, 267–276. [CrossRef] [PubMed]
29. Collet, J.P.; Allali, Y.; Lesty, C.; Tanguy, M.L.; Silvain, J.; Ankri, A.; Blanchet, B.; Dumaine, R.; Gianetti, J.; Payot, L.; et al. Altered fibrin architecture is associated with hypofibrinolysis and premature coronary atherothrombosis. *Arterioscl. Throm. Vas.* **2006**, *26*, 2567–2573. [CrossRef]
30. Collet, J.P.; Allali, Y.; Lesty, C.; Tanguy, M.L.; Silvain, J.; Ankri, A.; Blanchet, B.; Dumaine, R.; Gianetti, J.; Payot, L.; et al. Fibrinogen Hershey IV: A novel dysfibrinogen with a gammaV411I mutation in the integrin alpha(IIb)beta(3) binding site. *J. Thromb. Haemost.* **2008**, *99*, 1008–1012.
31. Goncalves, I.I.; Hughan, S.C.; Schoenwaelder, S.M.; Yap, C.L.; Yuan, Y.; Jackson, S.P. Integrin alpha IIb beta 3-dependent calcium signals regulate platelet-fibrinogen interactions under flow involvement of phospholipase C gamma 2. *J. Biol. Chem.* **2003**, *278*, 34812. [CrossRef] [PubMed]
32. Wolberg, A.S.; Gabriel, D.A.; Hoffman, M. Analyzing fibrin clot structure using a microplate reader. *Blood Coagul Fibrin.* **2002**, *13*, 533–539. [CrossRef]
33. Mannila, M.N.; Lovely, R.S.; Kazmierczak, S.C.; Eriksson, P.; Samnegård, A.; Farrell, D.H.; Hamsten, A.; Silveira, A. Elevated plasma fibrinogen gamma' concentration is associated with myocardial infarction: Effects of variation in fibrinogen genes and environmental factors. *J. Thromb. Haemost.* **2010**, *5*, 766–773. [CrossRef] [PubMed]

34. Lowe, G.D. Blood viscosity and cardiovascular disease. *J. Thromb. Haemost.* **1992**, *67*, 494–498. [CrossRef]
35. Koenig, W.; Ernst, E. The possible role of hemorheology in atherothrombogenesis. *Atherosclerosis* **1992**, *94*, 93–107. [CrossRef]
36. Fredenburgh, J.C.; Nesheim, M.E. Lys-plasminogen is a significant intermediate in the activation of Glu-plasminogen during fibrinolysis in vitro. *J. Biol. Chem.* **1992**, *267*, 26150–26156. [PubMed]
37. Hsiao, G.; Shen, M.Y.; Lin, K.H.; Chou, C.Y.; Tzu, N.H.; Lin, C.H.; Chou, D.S.; Chen, T.F.; Sheu, J.R. Inhibitory activity of kinetin on free radical formation of activated platelets in vitro and on thrombus formation in vivo. *Eur. J. Pharmacol.* **2003**, *465*, 281–287. [CrossRef]
38. Chung, T.; Connor, D.; Joseph, J.; Emmett, L.; Mansberg, R.; Peters, M.; Ma, D.; Kritharides, L. Platelet activation in acute pulmonary embolism. *J. Thromb. Haemost.* **2007**, *5*, 918–924. [CrossRef] [PubMed]
39. Chen, H.; Miao, H.; Feng, Y.L.; Zhao, Y.Y.; Lin, R.C. Metabolomics in dyslipidemia. *Adv. Clin. Chem.* **2014**, *66*, 101–119. [PubMed]
40. Bujak, R.; García-Álvarez, A.; Rupérez, F.J.; Nuño-Ayala, M.; García, A.; Ruiz-Cabello, J.; Fuster, V.; Ibáñez, B.; Barbas, C. Metabolomics reveals metabolite changes in acute pulmonary embolism. *J. Proteome Res.* **2014**, *13*, 805–816. [CrossRef] [PubMed]
41. Zeleznik, O.A.; Poole, E.M.; Lindstrom, S.; Kraft, P.; Van, H.V.A.; Lasky-Su, J.A.; Harrington, L.B.; Hagan, K.; Kim, J.; Parry, B.A.; et al. Metabolomic analysis of 92 pulmonary embolism patients from a nested case-control study identifies metabolites associated with adverse clinical outcomes. *J. Thromb. Haemost.* **2017**, *16*, 500–507. [CrossRef] [PubMed]
42. Alkaitis, M.S.; Ackerman, H.C. Tetrahydrobiopterin supplementation improves phenylalanine metabolism in a murine model of severe malaria. *ACS Infect. Dis.* **2016**, *2*, 827–838. [CrossRef] [PubMed]
43. Wu, X.; Zhao, C.; Zhang, A.; Zhang, J.; Wang, X.; Sun, X.; Sun, Z.; Wang, X. High-throughput metabolomics used to identify potential therapeutic targets of Guizhi Fuling Wan against endometriosis of cold coagulation and blood stasis. *RSC Adv.* **2018**, *8*, 19238–19250. [CrossRef]
44. Schwencke, C.; Schmeisser, A.; Weinbrenner, C.; Braun-Dullaeus, R.C.; Marquetant, R.; Strasser, R.H. Trans regulation of the α2-adrenergic signal transduction pathway by chroni-blockade. *J. Cardiovasc. Pharm.* **2005**, *45*, 253–259. [CrossRef]
45. Li, R.J. The role of histidine in metabolism. *Prog. Physiol. Sci.* **1985**, *2*, 80–82.
46. Duan, J.; Liang, Z.; Yang, C.; Zhang, J.; Zhang, L.; Zhang, W.; Zhang, Y. Rapid protein identification using monolithic enzymatic microreactor and LC-ESI-MS/MS. *Proteomics* **2006**, *6*, 412–419. [CrossRef] [PubMed]
47. Wang, Y.; Hao, J.; Gao, W.; Liu, Z.; Wu, S.; Jing, S. Study on hemostatic activities of the rhizome of *Paris bashanensis*. *Pharm. Biol.* **2013**, *51*, 1321–1325.
48. Lu, W.J.; Lee, J.J.; Chou, D.S.; Chou, D.S.; Jayakumar, T.; Fong, T.H.; Hsiao, G.; Sheu, J.R. A novel role of andrographolide, an NF-kappa B inhibitor on inhibition of platelet activation: The pivotal mechanisms of endothelial nitric oxide synthase/cyclic GMP. *J. Mol. Med.* **2011**, *89*, 1261–1273. [CrossRef] [PubMed]
49. Lemini, C.; Jaimez, R.; Franco, Y. Gender and inter-species influence on coagulation tests of rats and mice. *Thromb. Res.* **2007**, *120*, 410–419. [CrossRef] [PubMed]
50. Fan, P.C.; Ma, H.P.; Hao, Y.; He, X.R.; Sun, A.J.; Jiang, W.; Li, M.X.; Jing, L.L.; He, L.; Ma, J.; et al. A new anti-fibrinolytic hemostatic compound 8-O-acetyl shanzhiside methylester extracted from *Lamiophlomis rotata*. *J. Ethnopharmacol.* **2016**, *187*, 232–238. [CrossRef] [PubMed]
51. Vesna, K.; Ivana, F.; Zorica, V. Effects of rutin and hesperidin and their Al (III) and Cu (II) complexes on in vitro plasma coagulation assays. *Molecules* **2011**, *16*, 1378–1388.
52. Sheu, J.R.; Hung, W.C.; Wu, C.H.; Lee, Y.M.; Yen, M.H. Antithrombotic effect of rutaecarpine, an alkaloid isolated from *Evodia rutaecarpa* on platelet plug formation in in vivo experiments. *British J. Haematol.* **2000**, *110*, 110–115. [CrossRef] [PubMed]
53. Simkhada, J.R.; Cho, S.S.; Mander, P.; Choi, Y.H.; Yoo, J.C. Purification, biochemical properties and antithrombotic effect of a novel Streptomyces enzyme on carrageenan-induced mice tail thrombosis model. *J. Thromb. Res.* **2012**, *129*, 170–182. [CrossRef]
54. Devaraja, S.; Girish, K.S.; Devaraj, V.R.; Kemparaju, K. Factor Xa-like and fibrin (ogen)olytic activities of a serine protease from *Hippasa agelenoidesspider* venom gland extract. *J. Thromb. Thrombolys* **2010**, *29*, 119–126. [CrossRef] [PubMed]

55. Majumdar, S.; Dutta, S.; Das, T.; Chattopadhyay, P.; Mukherjee, A.K. Antiplatelet and antithrombotic activity of a fibrin (ogen)olytic protease from Bacillus cereus strain FF01. *Int. J. Biol. Macromol.* **2015**, *79*, 477–489. [CrossRef] [PubMed]
56. Mais, E.; Alolga, R.N.; Wang, S.L.; Linus, L.O.; Yin, X.; Qi, L.W. A comparative UPLC-Q/TOF-MS-based metabolomics approach for distinguishing *Zingiber officinale* Roscoe of two geographical origins. *Food Chem.* **2017**, *240*, 239–244. [CrossRef]

International Journal of *Molecular Sciences*

MDPI

Article

Cellular and Molecular Effects of High-Molecular-Weight Heparin on Matrix Metalloproteinase 9 Expression

René Huber [1,†], Rozan Attili/Abedalkhader [1,†,‡], Daniela Küper [1,2], Lara Hauke [1], Bernadette Lüns [1], Korbinian Brand [1], Karin Weissenborn [3] and Ralf Lichtinghagen [1,†,*]

1 Institute of Clinical Chemistry, Hannover Medical School, 30625 Hannover, Germany; huber.rene@mh-hannover.de (R.H.); rozana@hebron.edu (R.A.); kueper.daniela@mh-hannover.de (D.K.); lara.hauke@gmx.de (L.H.); luens.bernadette@mh-hannover.de (B.L.); brand.korbinian@mh-hannover.de (K.B.)

2 Central Pharmacy, Hannover Medical School, 30625 Hannover, Germany

3 Department of Neurology, Hannover Medical School, 30625 Hannover, Germany; weissenborn.karin@mh-hannover.de

* Correspondence: lichtinghagen.ralf@mh-hannover.de; Tel.: +49-511-532-3940

† These authors contributed equally to this work.

‡ Current address: Institute of Pharmacy and Medical Laboratory Science, Department of Medical Laboratory, P.O. Box 40, 711 Hebron, West Bank, Palestine.

Received: 7 March 2019; Accepted: 28 March 2019; Published: 30 March 2019

Abstract: Blood sampling with different anticoagulants alters matrix metalloproteinase (MMP-) 9 expression, thus influencing its concentration and diagnostic validity. Here, we aimed to evaluate the effects of different anticoagulants on MMP-9 regulation. MMP-9 expression was assessed in response to ethylenediaminetetraacetic acid, citrate, and high-/low-molecular-weight heparin (HMWH, LMWH) in co-culture experiments using THP-1, Jurkat, and HT cells (representing monocytes, T, and B cells). Triple and double cell line co-culture experiments revealed that HMWH treatment of THP-1 and Jurkat led to a significant MMP-9 induction, whereas other anticoagulants and cell type combinations had no effect. Supernatant of HMWH-treated Jurkat cells also induced MMP-9 in THP-1 suggesting monocytes as MMP-9 producers. HMWH-induced cytokine/chemokine secretion was assessed in co-culture supernatant, and the influence of cytokines/chemokines on MMP-9 production was analyzed. These experiments revealed that Jurkat-derived IL-16 and soluble intercellular adhesion molecule (sICAM-) 1 are able to induce MMP-9 and IL-8 production by THP-1. As a consequence, the increased MMP-9 expression found in HMWH blood samples may be influenced by HMWH-dependent secretion of IL-16 and sICAM-1 by T cells resulting in an increased production of MMP-9 and IL-8 by monocytes. IL-8, in turn, may support MMP-9 and its own expression in a positive autocrine feedback loop.

Keywords: blood sampling; anticoagulants; MMP-9; high-molecular-weight heparin; IL-16; sICAM-1; IL-8; monocytes; T cells

1. Introduction

Matrix metalloproteinases (MMPs) are a major group of endopeptidases degrading extracellular matrix (ECM) and basement membrane components [1] involved in the regulation of cellular events such as proliferation, differentiation, and migration [2]. In humans, 23 zinc-dependent MMPs have been identified [3]. Emerging evidence indicate an important role of these enzymes in normal and pathological processes including tissue remodeling, wound healing, inflammation, arthritis, cardiovascular diseases, cancer, and neurological diseases [3–5]. Of particular importance is MMP-9, also known as gelatinase B or type IV collagenase, since it has recently become a subject of growing interest in diverse human pathologies such as stroke. Stroke is regarded as the one of the leading causes of death in the world [6] and MMP-9 levels in the blood have been suggested to represent a suitable biomarker supporting its prognosis [7]. It was previously shown by our group and others that the way of blood sampling may affect subsequent analytical tests and that the kind of anticoagulant used, especially heparin, may alter the amount of MMP-9 (and other proteases or protease inhibitors) measured [8–11]. The molecular mechanism underlying this phenomenon, however, is still not well elucidated. Nevertheless, currently, MMP-9 assessment in blood is performed using heparin, ethylenediaminetetraacetic acid (EDTA), citrate, or serum samples [12]. The clinical chemist who has to evaluate a patient's blood MMP-9 concentration usually does not know about the patient's medical treatment, and an impact of concurrent treatment upon measured MMP-9 levels has not been considered in practice, so far.

High-molecular-weight heparin (HMWH) is a highly sulfated glycosaminoglycan. Due to its high negative charge density, HMWH prevents clotting and is widely used as an anticoagulant in heparin blood samples [13]. Low-molecular-weight heparin (LMWH), another class of anticoagulant derived from unfractionated (i.e., high-molecular-weight) heparin, has a number of advantages over HMWH, which has led to its increasing use for a number of thromboembolic indications [13,14]. The interaction of HMWH, LMWH, and other commonly used anticoagulants such as EDTA [15] and citrate [16] with human blood cells and their potential modulation of cellular processes are not well defined, neither in general nor in respect of MMP-9 expression [17].

The aim of this study was to characterize the influence of direct and indirect effects of different anticoagulants on the regulation of MMP-9 and to identify the molecular mechanism mediating these effects so as to assess the impact of the respective substances on the suitability of MMP-9 as a biomarker.

2. Results

2.1. Direct Stimulation of Different Cell Types with Anticoagulants Has No Influence on MMP-9 Expression

Stimulation of individual cell line cultures (THP-1, Jurkat, HT cells) with 3.2 mg/well EDTA (dissolved in medium), 220 μL/well citrate, or 50 international units (IU)/well HMWH revealed that MMP-9 expression and secretion are not influenced directly by any of these anticoagulants in monocytic cells (Figure 1a), T cells (Figure 1b), or B cells (Figure 1c) since no significant induction of MMP-9 mRNA could be shown by qPCR. This indicates that the reported MMP-9 induction in heparin blood sampling [8] is mediated via an indirect mechanism.

2.2. Significant Induction of MMP-9 Expression by HMWH in A Co-Culture Including THP-1, Jurkat, and HT Cells

The analysis of MMP-9 mRNA expression in a co-culture of THP-1, Jurkat, and HT cells demonstrated that MMP-9 expression increased significantly over time after addition of HMWH (approximately 7-fold after 24 h; Figure 2a). Equivalent results were obtained when the amounts of secreted MMP-9 protein were measured in the culture supernatant (approximately 3.5-fold induction after 24 h; Figure 2b). In contrast, stimulation with other anticoagulants such as EDTA or citrate (Figure 2a) had no MMP-9-inducing effect in this co-culture model. These results suggest that both MMP-9 mRNA and protein expression in one of the cell types used depends on an interaction with another cell type present in the mixture in response to HMWH, potentially by cell-to-cell contacts or indirectly via the stimulation with a secreted mediator.

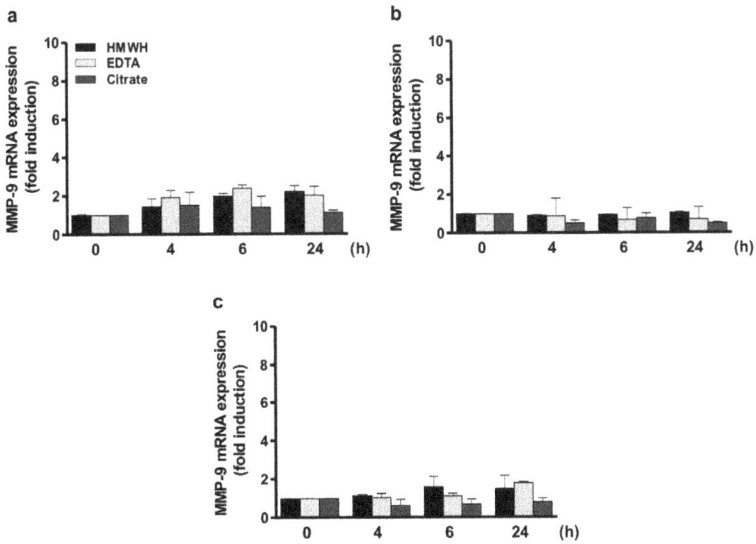

Figure 1. No significant induction of MMP-9 expression by HMWH, EDTA, or citrate in individual cell line cultures of THP-1, Jurkat, and HT cells. 2×10^5 THP-1 (**a**), Jurkat (**b**), or HT cells (**c**) per well were starved overnight and then stimulated with 50 IU/well HMWH, 3.2 mg/well EDTA, or 220 µL/well citrate up to 24 h. MMP-9 mRNA expression was determined using qPCR (QuantiTect Custom Assay; housekeeping gene: GAPDH); mean ± SD, $n = 3$.

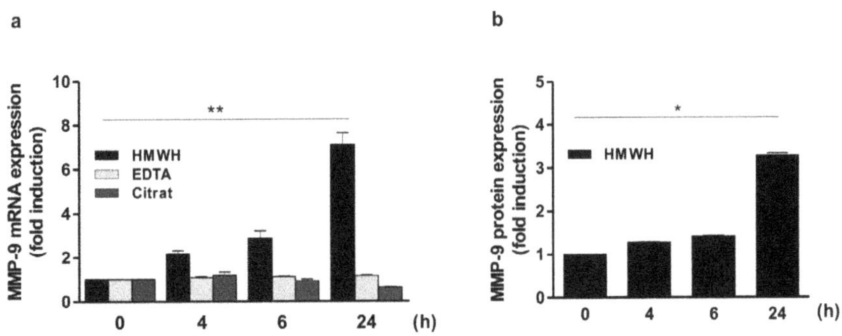

Figure 2. Induction of MMP-9 expression by HMWH in a THP-1, Jurkat, and HT cells containing co-culture. 7×10^5 THP-1, Jurkat, and HT cells per well each (i.e., a total of 2.1×10^6 cells/well) were starved overnight and then stimulated with 50 IU/well HMWH, 3.2 mg/well EDTA, or 220 µL/well citrate up to 24 h. MMP-9 mRNA (**a**) and protein (**b**) expression were determined using qPCR (QuantiTect Custom Assay; housekeeping gene: GAPDH) and ELISA (MMP-9 Quantikine Kit); mean ± SD, $n = 3$ (measured in duplicates). Kruskal–Wallis test, * $p \leq 0.05$; ** $p \leq 0.01$.

2.3. Significant Induction of MMP-9 Expression by HMWH in the THP-1 and Jurkat Co-Culture

To determine whether the impact of HMWH on MMP-9 expression depends on an interplay of the three cell types used or a cooperation of two cell lines, MMP-9 expression was further assessed in co-culture approaches consisting of two cell types each. Thus, cell line mixtures including HMWH-stimulated monocytes and T cells, monocytes and B cells, as well as T and B cells were performed. In control approaches, the mixtures were alternatively treated with EDTA or citrate. As expected, no increase in MMP-9 mRNA expression was observed in the cultures of THP-1 and Jurkat, THP-1 and HT, as well as Jurkat and HT cells following control stimulation with EDTA or citrate (data not shown). Moreover, there was also no significant induction of MMP-9 levels following HMWH stimulation in a mixture of THP-1 and HT cells or HT and Jurkat cells (Figure 3a). In contrast, HMWH stimulation of a combination of THP-1 and Jurkat cells led to a significantly increased MMP-9 mRNA expression over time (approximately 8-fold after 24 h; Figure 3a). These results could also be confirmed on the protein level (approximately 3-fold induction after 24 h; Figure 3b).

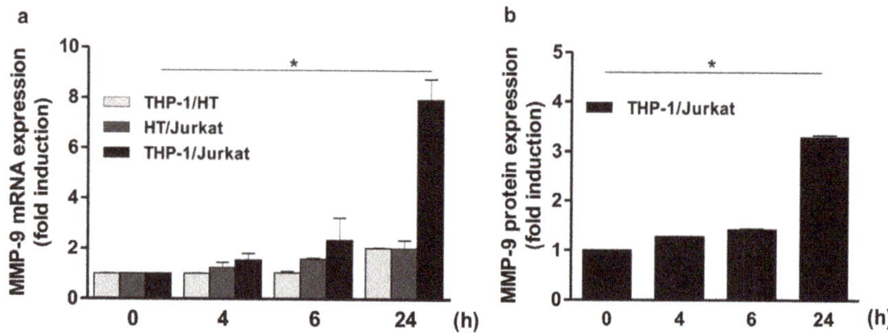

Figure 3. Induction of MMP-9 expression by HMWH in a THP-1 and Jurkat cells containing co-culture. 1×10^6 THP-1 and HT, HT and Jurkat, or THP-1 and Jurkat cells per well (i.e., a total of 2×10^6 cells/well) were starved overnight and then stimulated with 50 IU/well HMWH up to 24 h. MMP-9 mRNA (**a**) and protein (**b**) expression were determined using qPCR (QuantiTect Custom Assay; housekeeping gene: GAPDH) and ELISA (MMP-9 Quantikine Kit); mean ± SD, n = 3 (measured in duplicates). Kruskal–Wallis test, * $p \leq 0.05$.

2.4. *Significant Induction of MMP-9 Expression in THP-1 Cells in Response to Culture Supernatant Derived from HMWH-Treated Jurkat Cells*

In the next step, it was analyzed whether the increased MMP-9 production depends on cell-to-cell-contacts or is rather based on an indirect effect, e.g., via soluble mediator(s). Since monocytes/macrophages are able to produce large amounts of MMP-9 [18], we hypothesized that in this context, the T cells are responsible for the secretion of soluble factor(s) activating the MMP-9 production in monocytes. Therefore, the MMP-9 expression in THP-1 cells following incubation with culture supernatant derived from HMWH-stimulated Jurkat cells was analyzed. In the respective experiments, a significant induction of MMP-9 mRNA up to 24 h could be shown (approximately 8-fold after 24 h) whereas the supernatant of HMWH-treated HT cells had only a slight (though significant) effect on MMP-9 (up to 3-fold; Figure 4a). Stimulation of THP-1 cells with the supernatant of EDTA- or citrate treated Jurkat and HT cells did not increase MMP-9 mRNA expression considerably (data not shown). THP-1 cells were also incubated with the supernatant from Jurkat cells which have been incubated with human plasma derived from heparin-containing monovettes and MMP-9 was also significantly increased in this experimental setting, albeit to a lesser extent and characterized by a divergent temporal development (peaking at 4 h with approximately 5.5-fold induction; Figure 4b). These data support the suggestion that the monocytes are the main producers of MMP-9 and that factor(s) present in the supernatant of HMWH-stimulated T cells are able to significantly induce MMP-9 expression in monocytes.

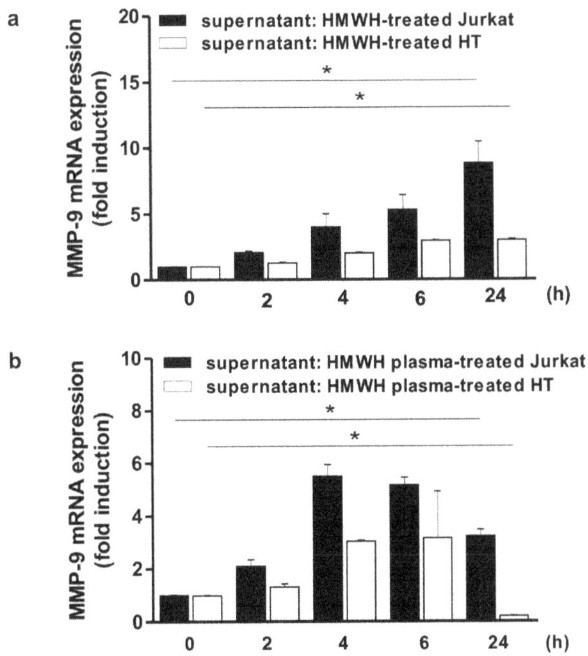

Figure 4. Induction of MMP-9 expression in THP-1 cells by incubation with supernatant from HMWH- and HMWH plasma-stimulated Jurkat cells. 2×10^6 THP-1 cells/well were starved overnight and subsequently stimulated up to 24 h with the supernatant of Jurkat or HT cells which have been treated for 24 h with 50 IU/well HMWH (**a**) or HMWH plasma (derived from a heparin monovette) (**b**). MMP-9 mRNA expression was determined using qPCR (QuantiTect Custom Assay; housekeeping gene: GAPDH); mean ± SD, $n = 3$ (measured in duplicates). Kruskal–Wallis test, * $p \leq 0.05$.

2.5. High-Molecular-Weight Heparin Versus Low-Molecular-Weight Heparin

To further investigate the stimulatory effect of other types of heparin on MMP-9 expression, the effect of two different types of LMWH (i.e., Clexane and Fragmin) was analyzed. As demonstrated in Figure 5, the stimulation of THP-1 and Jurkat with LMWH had no increasing effect on MMP-9 mRNA amounts, neither using Clexane nor Fragmin, in contrast to the significant approximately 10-fold effect shown by HMWH after 24 h. These results indicate that induction of MMP-9 expression also depends on the type of heparin used.

2.6. Identification of Soluble Mediators Secreted in Response to HMWH

The basic idea of the following experiment was to identify the (combination of) T cell-derived mediator(s) which is able to increase MMP-9 synthesis in monocytes. Therefore, all three individual cell lines as well as the relevant double (THP-1 and Jurkat) and triple (THP-1, Jurkat, and HT) cell line mixtures were cultivated and stimulated with HMWH. Cell culture supernatants were then profiled for the expression of multiple cytokines and chemokines. As shown in Table 1, HMWH-stimulated THP-1 released 3 types of mediators, macrophage migration inhibitory factor (MIF), interleukin 1 receptor antagonist (IL-1RA), and CC-chemokine ligand 5 (CCL5). Jurkat secreted IL-13, IL-16, MIF, sICAM-1, and Serpin E1, whereas HT cells produced IL-13, MIF, tumor necrosis factor (TNF), SICAM-1, and Serpin E1. In the co-cultures of HMWH-stimulated THP-1 and Jurkat or THP-1, Jurkat and HT cells, virtually all identified factors could be detected, additionally supplemented with IL-8 as a factor specifically occurring in cell line mixtures. This suggested that IL-16—as the only T cell-specific soluble mediator present under all MMP-9-inducing conditions—might be involved (either individually or in a combination with other mediators) in the induction of MMP-9 secretion from monocytes.

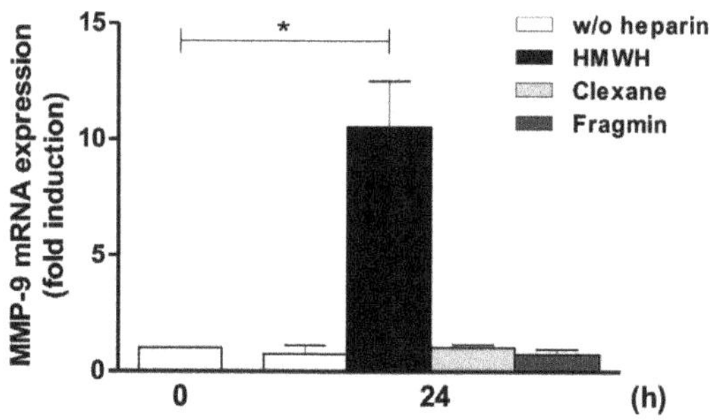

Figure 5. No Induction of MMP-9 expression by LMWH in a THP-1 and Jurkat cells containing co-culture. 1×10^6 THP-1 and Jurkat cells (i.e., a total of 2×10^6 cells/well) were starved overnight. Afterwards, the co-culture was stimulated with HMWH (as a positive control), Clexane, or Fragmin (50 IU/well each) for 24 h. Then, MMP-9 mRNA expression was determined using qPCR (QuantiTect Custom Assay; housekeeping gene: GAPDH); mean ± SD, n = 3 (measured in duplicates). T test, * $p \leq 0.05$.

Table 1. Cytokines/chemokines expressed by HMWH-treated THP-1, Jurkat, and HT cells as determined by Proteome Profiler Human XL Cytokine Array. T cell-specific IL-16 and co-culture-specific IL-8 are highlighted in bold.

Cell Line(s)	Identified Cytokines/Chemokines
THP-1	IL-1RA, CCL5, MIF
Jurkat	MIF, IL-13, **IL-16**, sICAM-1, Serpin E1
HT	MIF, IL-13, sICAM-1, Serpin E1, TNF
THP-1, Jurkat, and HT	IL-1RA, CCL5, MIF, IL-13, **IL-16**, sICAM-1, Serpin E1, **IL-8**
THP-1 and Jurkat	IL-1RA, CCL5, MIF, IL-13, **IL-16**, sICAM-1, Serpin E1, **IL-8**

Interestingly, following stimulation with LMWH, no secreted cytokines/chemokines could be detected in co-cultures of THP-1 and Jurkat cells (data not shown) supporting the consideration that only HMWH (but not LMWH) is able to activate the production of MMP-9-inducing cytokine(s) from T cells.

2.7. Regulation of MMP-9 Expression in THP-1 by Jurkat-Derived IL-16 and sICAM-1

In the next step, THP-1 cells were stimulated with IL-16 to assess its influence on MMP-9 mRNA synthesis. Unexpectedly, this approach did not result in a significant upregulation of MMP-9, and equivalent negative results were obtained when THP-1 were stimulated with IL-13, MIF, or Serpin E1, i.e., the other factors found in the supernatant of HMWH-treated T cells. Only in the case of sICAM-1 a significant (but only approximately 3-fold) MMP-9 induction was observed (data not shown). This indicates that IL-16 alone is not the driver of MMP-9 production in monocytes and suggests that sICAM-1—potentially in combination with other secreted proteins—might contribute to the stimulatory effect on MMP-9 expression.

To reveal whether an interaction between two or more factors is responsible for this effect, (T cell-specific) IL-16, sICAM-1 (due to its mild significant effect on MMP-9) and (cell mixture-specific) IL-8 were applied in different combinations. A slight but significant effect on MMP-9 was observed following stimulation of THP-1 cells with IL-16 and sICAM-1 or IL-16 and IL-8 with about 3– to 4-fold induction levels after 24 h (Figure 6). Our results suggest that the T cell-derived mediators IL-16 and sICAM-1 are acting in combination with the presumably monocyte-derived chemokine IL-8 thus mediating an induction of MMP-9. To further confirm these results, monocytes were incubated with the triple combination of these mediators. In fact, a synergistic influence of IL-16, sICAM-1, and IL-8 was confirmed (7-fold, i.e., as strong as the positive control TNF; Figure 6).

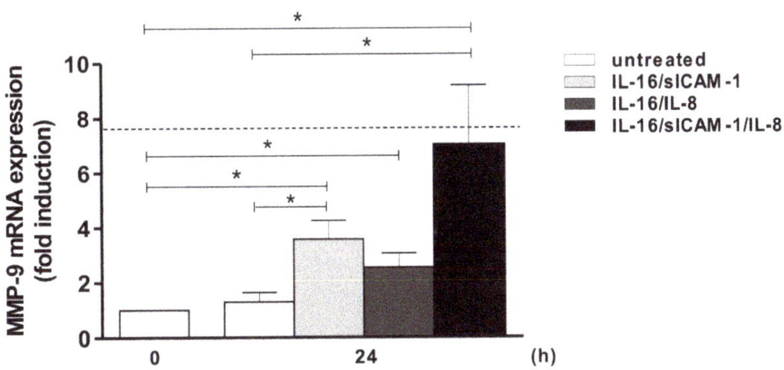

Figure 6. Induction of MMP-9 expression by IL-16, sICAM-1, and IL-8 in THP-1 cells. 2×10^6 THP-1 cells/well were starved overnight. Afterwards, cells were stimulated with the mediators identified via Proteome Profiler Array: IL-16 and sICAM-1, IL-16 and IL-8, or IL-16, sICAM-1, and IL-8 for 24 h (5 ng/mL each). The dashed line indicates MMP-9 mRNA expression following TNF stimulation for 2 h as a positive control. MMP-9 mRNA expression was determined using qPCR (QuantiTect Custom Assay; housekeeping gene: GAPDH); mean ± SD, $n = 3$ (measured in duplicates). T test, * $p \leq 0.05$.

In contrast, the analysis of further cytokine combinations including MIF and IL-13 did not yield a significant increase in MMP-9 mRNA (data not shown) indicating that these proteins do not play an influential role in the regulation of MMP-9.

2.8. Regulation of IL-8 Expression in THP-1 by IL-16 and Autocrine Activation

To further assess whether the supporting chemokine IL-8 is induced by the same mediator cocktail as MMP-9, or whether other factors are involved, monocytic cells were stimulated individually or with combinations of IL-16, sICAM-1, and IL-8 and the resulting IL-8 mRNA expression was measured. Initially, stimulation of THP-1 with individual factors as demonstrated for IL-16 and IL-8 itself resulted in a significant increase in the IL-8 measurement (3- to 4-fold induction; Figure 7). Moreover, further trials using IL-16 and sICAM-1 or IL-16, sICAM-1, and IL-8 also enhanced IL-8 expression with about 4- to 6-fold induction levels (thus roughly reaching a dimension also observed using the positive control TNF; Figure 7). With respect to the effect of IL-16, sICAM-1, and IL-8 on MMP-9 expression shown above, the results presented here substantiate the assumption that the T cell-derived factors IL-16 and sICAM-1 provoke the secretion of monocyte-derived IL-8. This combination, in turn, is able to induce both a continuing IL-8 production by monocytes and an increased MMP-9 expression.

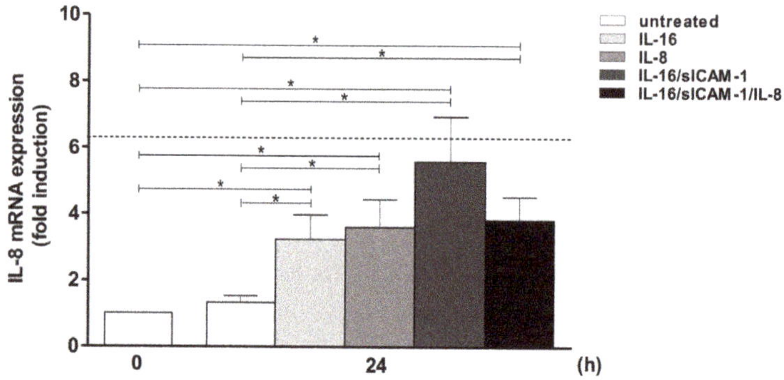

Figure 7. Induction of IL-8 expression by IL-16, sICAM-1, and IL-8 in THP-1 cells. 2×10^6 THP-1 cells/well were starved overnight. Afterwards, cells were stimulated with IL-16, IL-8, IL-16 and sICAM-1, or IL-16, sICAM-1, and IL-8 for 24 h (5 ng/mL each). The dashed line indicates IL-8 mRNA expression following TNF stimulation for 2 h as a positive control. IL-8 mRNA expression was determined using qPCR (housekeeping gene: GAPDH); mean ± SD, $n = 3$ (measured in duplicates). T test, * $p \leq 0.05$.

3. Discussion

Like other MMPs, MMP-9 is a member of the broad ECM remodeling network [19,20] and plays a crucial role in physiological processes such as proliferation, differentiation, and wound healing, but also in tumor progression, metastasis, and angiogenesis [2,21] due to its ability to efficiently degrade (amongst others) gelatin, collagens (IV, V, VII, X, XIV), elastin, aggrecan, and fibronectin [22]. Its structure comprises an N-terminal pro-domain, the catalytically active metalloproteinase domain (including three fibronectin type II repeats and a zinc binding motif), and a hemopexin domain [20]. MMP-9 is initially synthesized as a latent pro-enzyme (the zymogen) which is activated via the so-called cysteine switch, i.e., the proteolytic cleavage of the pro-peptide [23]. To avoid deleterious effects in normal tissue, the enzymatic activity of MMP-9 is negatively regulated by tissue inhibitor of metalloproteinases (TIMP-) 3 and (to a lesser extent) 1, either by association with pro-MMP-9 or the inhibition of mature MMP-9 [24]. It has been reported that MMP-9 production is induced by TNF, interferon (IFN-) α, IL-1β, and several growth factors (e.g., epidermal growth factor (EGF), platelet-derived growth factor (PDGF)) [21]. Its mRNA expression is positively regulated via IκB kinase (IKK-), protein kinase C (PKC-), Ras-, or mitogen-activated protein kinases (MAPK; especially p38-, extracellular signal-regulated kinase (ERK-), and Jun N-terminal kinase (JNK-)) dependent signaling pathways, resulting in a nuclear factor κB (NF-κB-), specificity protein (Sp-) 1-, or activator protein (AP-) 1-driven gene expression. The repression of MMP-9 transcription is mainly Janus kinase (Jak)/signal transducer and activator of transcription (STAT-) dependent in response to IFN-β, IL-4, and IL-10 [21]. Consequently, MMP-9 expression is pharmacologically controlled using agents negatively affecting the MMP-9-regulating pathways or transcriptions factors, or both, thus especially reducing the activity of NF-κB or AP-1. For instance, MMP-9 expression can be significantly reduced via the suppression of NF-κB as demonstrated in breast cancer cells following the application of the isothiocyanate Sulforaphane [25] and in monocytic THP-1 cells treated with the labdane diterpenoid Andrographolide [26]. Consistently, MMP-9 can be induced in THP-1 cells using NF-κB-activating agents such as the synthetic triacylated lipopeptide Pam3CSK4 [27]. In a variety of other studies, the application of small molecule inhibitors for several kinases provoked a significant reduction of MMP-9 expression in different cell types. This has been demonstrated for inhibitors of PKC (e.g., RO318220, Calphostin C, Safingol), p38-MAPK (SB203580), MEK (PD098059), ERK (Rosiglitazone), and other signaling molecules (reviewed in [28]). Alternatively, signaling pathway activation can be prevented via neutralizing antibodies either against activating mediators (such as TNF), the respective receptors such as EGFR, or involved signaling molecules (e.g., MEK, Ras, PKC) [28].

In addition to the approaches to address MMP-9 as a target in neoplastic, vascular, and other diseases [5,29], it is also regarded as a promising diagnostic or prognostic factor, e.g., in inflammatory bowel disease [30] or stroke [7], rendering MMP-9 an interesting potential biomarker [30]. In the framework of a recently published study [8], we found a considerable relation between induction of MMP-9 expression and different anticoagulants used. Moreover, this relation affected both MMP-9 protein baseline values and kinetics, which was clearly shown in heparin samples with or without proteinase inhibitors. Consequently, the way of blood sampling influences the measurement of MMP-9. This is also reflected by further studies also showing that blood sampling influences the concentration and the diagnostic validity of MMP-9 [31] and other circulating MMPs (1, 2, 3, 7, and 8) [32]. Therefore, we aimed to characterize the effects of common anticoagulants on MMP-9 expression and to identify the cell types contributing to this effect to assess the impact of these anticoagulants in clinical applications and the suitability of MMP-9 as a biomarker. Our initial analyses showed that direct stimulation of THP-1, Jurkat, and HT cells (representing the major leukocyte cell types in the blood, i.e., monocytes, T cells, or B cells) with the common anticoagulants EDTA, citrate, or HMWH had no effect on the amounts of MMP-9 mRNA or protein. This indicates that its expression in blood cells is not induced directly by anticoagulants, but rather influenced via an indirect mechanism, i.e., an interaction of two or more cell types in the blood.

Based on this consideration, we performed an experiment with a mixture of the three cell lines and incubated them with EDTA, citrate, or HMWH. A strong (approximately 7-fold and 3.5-fold, respectively) and significant induction of MMP-9 mRNA and secreted protein in the mixture was only observed in response to HMWH, whereas EDTA and citrate had no effect on MMP-9 levels. These results are in good agreement to the literature since it has been reported that stimulation with heparin increases MMP-9 and TIMP levels in leukocytes and platelets [10], whereas citrate and EDTA appeared to be less effective [8]. We concluded that the MMP-9 expression in at least one of the included cell types depends on an activating interaction with another involved cell type in response to HMWH, either directly by cell-to-cell contacts or indirectly via the stimulation with a secreted mediator. To address this question, MMP-9 levels were analyzed in co-culture approaches including two different cell types each, i.e., THP-1/Jurkat, THP-1/HT, and Jurkat/HT. In contrast to all other combinations, stimulation of THP-1 and Jurkat cells with HMWH led to a significant increase in MMP-9 mRNA expression indicating that the interplay between monocytes and T cells is involved in the upregulation of MMP-9 in heparinized blood samples. The results may also reflect the tendency of leukocytes to generate the members of the MMP family not ubiquitously, but in distinct gene expression patterns [33] which has also been predicted bioinformatically for other cell types, e.g., fibroblasts [19].

In the next step, we analyzed whether the increased MMP-9 production depends on direct or indirect effects. With respect to the known ability of monocytes and macrophages to produce large amounts of MMP-9, e.g., in the course of tissue invasion [18], while T cells are important producers of cytokines [34], it was reasonable to speculate that during the interaction of monocytes and T cells, the latter are responsible for the secretion of one or more soluble mediator(s) to which monocytes react with increased MMP-9 production. Therefore, we measured the MMP-9 expression in THP-1 cells in response to stimulation with culture supernatant derived from HMWH-stimulated Jurkat cells. Indeed, a significant MMP-9 induction was observed under these conditions. In order to elucidate whether it is further possible to increase MMP-9 production under clinical-like circumstances, THP-1 were also incubated with the supernatant from Jurkat cells which have been activated by human plasma derived from heparin-containing monovettes. Again, even under these conditions, the MMP-9 mRNA amounts were enhanced in THP-1 cells stimulated with the transferred supernatant. Together, these results support the suggestion that in response to HMWH, T cells release monocyte-activating factor(s) while monocytes stimulated in this manner are the relevant producers of MMP-9. This assumption also fits to earlier reports demonstrating that monocytes are the main producers of MMP-9 among leukocytes [33].

To identify the T cell-derived mediator(s) involved in this process, cytokines and chemokines secreted by T cells in response to HMWH were monitored. These analyses revealed that HMWH-challenged Jurkat cells are characterized by the production of IL-13, IL-16, MIF, sICAM-1, and Serpin E1. Since most of these factors were also produced by THP-1 (MIF) or HT cells (IL-13, MIF, sICAM-1, and Serpin E1) in response to HMWH, IL-16 proved to be the only T cell-specific factor in this setting. Thus, IL-16, which is constitutively expressed but not secreted in T cells on the mRNA and pro-protein level [35], appeared to be a promising candidate cytokine, especially in consideration of other studies that have already shown that IL-16 is able to stimulate human monocytes, e.g., to increase the expression of pro-inflammatory cytokines [36]. A subsequent stimulation of THP-1 cells with IL-16, however, did not result in an enhanced MMP-9 expression, although its MMP-9 inducing capability has been demonstrated in vascular smooth muscle cells [37]. This suggested that further factors are required to achieve that effect. Among the mixtures of cytokines/chemokines then tested, only the combination consisting of IL-16, sICAM-1 (which showed a slight but significant effect on MMP-9 when individually applied), and IL-8 (which was the only mediator specifically expressed in co-culture experiments) exhibited a monocytic MMP-9 expression equivalent to the positive control (2 h TNF) and, more important, to that observed in the co-culture of THP-1 and Jurkat cells following HMWH stimulation. These circumstances indicated that T cell-derived IL-16 and sICAM-1, which had together a significant but relatively low effect on MMP-9, require the supporting presence of IL-8 to exploit their full activating potential on MMP-9. In this context, monocytes were the presumed source of IL-8, since they are known to produce large amounts of IL-8 in response to certain stimuli such as TNF [38]. Other factors, however, such as CCL5 or MIF appear not to be involved, presumably due to the absence of further stimulating agents, e.g., TNF or IL-1β, which have been reported to be necessary for their MMP-9-inducing activity [39,40]. To monitor whether IL-8 can be induced in monocytic cells by the same cocktail as MMP-9 (or at least one mediator included), THP-1 cells were stimulated individually or with combinations of IL-16, sICAM-1, and IL-8. Indeed, THP-1 cells expressed significantly increased amounts of IL-8 chemokine following a stimulation with IL-16 or IL-8 individually—thus confirming the previously reported autocrine regulation of IL-8 [41]—as well as combinations of IL-16 and sICAM-1 or IL-16, sICAM-1, and IL-8. In the latter cases, the IL-8 levels observed were roughly comparable to those found in the presence of the positive control TNF. Considered together with the result that IL-16, sICAM-1, and IL-8 are potent drivers of MMP-9 expression, our data suggest a model in which the presence of HMWH may lead to the secretion of T cell-derived IL-16 and sICAM-1 which induces monocytic IL-8 production. The combination of these factors appears to yield both a continuing IL-8 secretion as well as an enhanced MMP-9 production by monocytes (Figure 8). However, with respect to our experimental conditions, the participation of other cell types and further mediators could not be excluded and has to be addressed by future studies. It should also be mentioned that the mechanism presented here may also have an impact on the expression of further proteases thus potentially affecting their validity as biomarkers (such as MMP-2, -11, or -13 [42]).

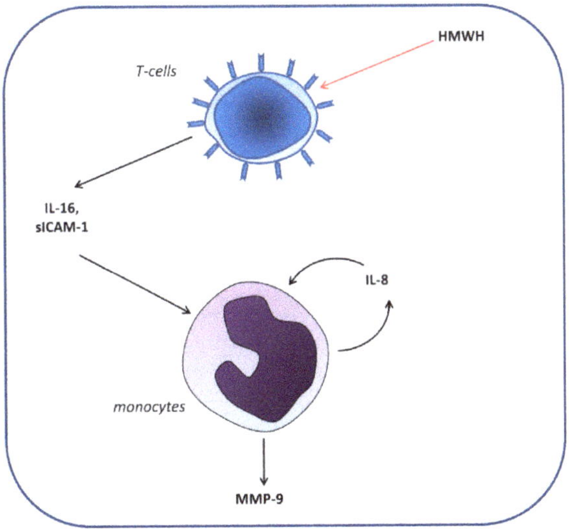

Figure 8. Proposed model of MMP-9 induction in monocytes by HMWH-treated T cells. Our data suggest that in response to HMWH, T cells secrete the mediators IL-16 and sICAM-1, which induce monocytic IL-8 production. Together, these factors yield a continuing IL-8 secretion as well as an enhanced MMP-9 production by monocytes.

In some of our experiments, TNF was used as a positive control in the cytokine stimulation experiments since it remarkably increases MMP-9 [43] and IL-8 expression [38] by monocytes. Thus, it appears surprising that the presence of HT cells had no influence on MMP-9 levels in THP-1- and HT-containing co-cultures although they secrete TNF in response to HMWH (Table 1). This might be ascribed to the absence of TNF in the case of the triple cell line culture. Whether this absence is due to an altered expression time course, ineffectively low TNF levels (i.e., under the detection limit of the proteome profiler membrane), or a suppression of HT-derived TNF expression in this cell mixture, remains to be established. In the pure THP-1/HT culture, in which TNF expression could be detected, the relatively low TNF amounts produced by HMWH-challenged HT cells (as represented by only slightly stained THF-representing dots on the detection membrane; data not shown) might be insufficient to evoke MMP-9 expression in THP-1 cells over time.

In contrast to HMWH, we did not detect a significant cytokine/chemokine secretion or MMP-9 expression in co-cultured THP-1 and Jurkat cells in response to LMWH. Although the effects of HMWH and LMWH on cytokine/chemokine or growth factor expression appear to differ among different conditions and cell lines [44], thus providing (at least in part) contradictory results, it has been observed that HMWH enhances the expression of CCL5 in endometrial carcinoma cells [44]. In heparinized plasma samples, elevated levels of methyl-accepting chemotaxis protein (MCP), stem cell factor (SCF), PDGFβ, vascular endothelial growth factor (VEGF), CXC chemokine ligand (CXCL) 11, and Serpin E1 [45] or IL-6, -8, -17, granulocyte/macrophage-colony stimulating factor (GM-CSF), and macrophage inflammatory protein (MIP-) 1 in comparison to citrate or EDTA plasma, or both, were detected [46]. In contrast, LMWH has been reported to significantly reduce the expression of pro-inflammatory cytokines and chemokines in several cell types. For instance, LMWH application inhibited the secretion of IL-4 and TNF by mast cells [47] and the expression of IL-6 and -8 in epithelial cells [48] as well as IL-4, -5, -13, and TNF in T cells, effects that fit well to the anti-inflammatory features of LMWH [49]. The remarkable difference in cytokine induction between HMWH and LMWH described in the literature and observed in our study may be attributed to the molecular structure of the respective heparins. Natural heparin is exclusively produced by mast cells and predominantly isolated on an industrial scale from porcine intestinal mucosa [50]. Both the unfractionated heparin (i.e., HMWH) and its fractionated derivative LMWH (produced by chemical or enzymatic processing of HMWH [50]) are glycosaminoglycans exhibiting a common molecular structure, i.e., repeating disaccharide units consisting of variably sulfated glucosamine, glucuronic acid, or iduronic acid [51]. Due to the general structural similarity, the anticoagulant mechanism is comparable: both heparins interact with antithrombin and suppress the activity of factor Xa and—to a lesser extent—thrombin [50]. However, in contrast to the long-chained HMWH with an average molecular weight of 16 kDa, LMWH is characterized by considerably shorter disaccharide chains resulting in an average molecular weight of 3.5–6 kDa [51], a reduced charge per chain, and less unspecific interactions with proteins [52]. In consequence, HMWH's length, charge, and tendency towards unspecific interactions may play a crucial role in the activation of T cells. For instance, an interaction with specific surface receptors is to be assumed, since it has been shown that unfractionated heparin is able to interact with the high mobility group box 1 protein receptor on macrophages [53] as well as G-protein-coupled receptor GPR56 on a variety of human cell lines [54]. Moreover, in the case of EDTA and citrate, the—in comparison to HMWH—negligible length and charge may be crucial for the absence of an MMP-9-inducing effect of these anticoagulants under the conditions tested in our study. Taking these points into account, one can speculate that receptor crosslinking may play a role during HMWH-induced activation of T cells, an effect which could not be provided by LMWH, EDTA, or citrate.

Our results also possess some clinical implications. For instance, HMWH-containing monovettes should be avoided when MMP-9 levels have to be assessed as a biomarker to circumvent its overvaluation. More important, heparin is also administered systemically during therapeutic approaches, e.g., for the treatment of venous thromboembolism (VTE) [13] or the prevention of myocardial infraction during the acute phase of unstable angina [55]. Known tissue-damage-associated adverse effects of heparin such as increased tendency towards bleeding [56] may be influenced or intensified by its MMP-9 promoting capacity. LMWH which according to our data does not appear to possess an influence on the expression of MMP-9 nor any cytokine/chemokine monitored may represent a preferable "gentle" alternative to HMWH. In another study, even a reduction of MMP-9 concentration and activity has been reported in the plasma of LMWH-treated patients with abdominal aortic aneurysm [57], an effect which might be connected to the LMWH-dependent downregulation of IL-8 as shown in enoxaparin-treated epithelial cells [48]. Accordingly, LMWH is increasingly used as an anticoagulant, e.g., for VTE prevention [58,59].

In summary, our study suggests that the increased MMP-9 expression previously found in HMWH blood samples may be ascribed to an HMWH-induced secretion of IL-16 and sICAM-1 by T cells resulting in an increased production of MMP-9 by monocytes, a process potentially supported by IL-8 in an autocrine feedback loop (Figure 8). This implies that HMWH shall be used cautiously in clinical applications.

4. Materials and Methods

4.1. Collection and Storage of Human Heparin Plasma

A total of three venous blood samples from healthy individuals were collected at the Department of Clinical Chemistry at Hannover Medical School. Informed consent was obtained from all donors before blood sampling. The experiments were carried out in accordance with the relevant guidelines and regulations and approved by the Hannover Medical School ethics committee in accordance with the Declaration of Helsinki (No. 388-2008, 02.12.2008). Blood sampling was performed using commercially available heparin- (i.e., HMWH) containing monovette tubes (Sarstedt, Nümbrecht, Germany). Blood samples were immediately centrifuged at 1900× g for 10 min and after removal of the supernatant, plasma samples were aliquoted and stored at −80 °C until assay.

4.2. Cell Lines and Cell Culture Conditions

For the analyses, THP-1 (human acute monocytic leukemia cell line; ACC-16), Jurkat (human T cell leukemia cell line, ACC-282), and HT cells (human B cell lymphoma cell line, ACC-567) were used (DSMZ, Braunschweig, Germany), representing monocytes, T cells, and B cells, respectively. All cell lines were cultured in RPMI 1640 medium containing 300 mg/L L-glutamine (PAA, Linz, Austria), supplemented with 10% fetal calf serum (FCS), 100 U/mL penicillin, and 100 mg/mL streptomycin (Biochrom, Berlin, Germany) at 37 °C in a humidified atmosphere containing 5% CO_2 [60].

Cell culture experiments were performed in 6-well plates (Sarstedt) with 2 mL medium/well. For stimulation experiments, cells were incubated without antibiotics. For starvation, the FCS concentration in the culture medium was reduced to 1% and cells were incubated overnight. Starved cells without further treatment/stimulation served as negative controls to avoid FCS-induced disturbances of the respective measurements.

4.3. Cell Culture Experiments

4.3.1. Individual and Co-Culture Experiments

In the initial experiments, 2×10^6 THP-1, Jurkat, or HT cells/well were starved in individual cultures overnight. Afterwards, cells were stimulated with 3.2 mg EDTA (Sigma Aldrich, Darmstadt, Germany), 10 µL HMWH (=50 IU; Ratiopharm, Ulm, Germany) or LMWH (Clexane; Sanofi-Aventis, Frankfurt, Germany or Fragmin; Pfizer, Berlin, Germany), or 220 µL citrate (Sarstedt) per well and harvested after 0, 4, 6, and 24 h for the analysis of MMP-9 mRNA expression.

In double co-culture experiments, THP-1 and Jurkat, THP-1 and HT, or Jurkat and HT cells were cultivated together (in total 2×10^6 cells/well, i.e., 1×10^6 cells per cell line). Following starvation overnight, cells were stimulated with the respective anticoagulants and harvested at the indicated time points. In addition, supernatants of the respective double co-cultures were collected for the analysis of secreted proteins.

Furthermore, in a triple co-culture experiment a mixture of the three cell lines used was seeded with 2.1×10^6 cells/well (i.e., 0.7×10^6 cells per cell line) and starved overnight. Then, the respective anticoagulant was added and the cells were harvested and the supernatant collected at the indicated time points.

4.3.2. Cell Supernatant Transfer Experiments

In a further series of experiments, starved monocytic THP-1 cells were stimulated with the supernatant of anticoagulant-treated Jurkat or HT cells and the MMP-9 expression was analysed as indicated. Alternatively, starved THP-1 cells were incubated with supernatant from starved Jurkat or HT cells which have been stimulated for 24 h with human plasma derived from blood samples collected in HMWH-containing monovettes.

4.3.3. Cytokine/Chemokine Stimulation Experiments

Following a starvation phase overnight, 2×10^6 THP-1 cells/well were stimulated with 5 ng/mL MIF, 5 ng/mL IL-13, 5 ng/mL IL-16, 5 ng/mL sICAM-1, 25 ng/mL Serpin E1, or 5 ng/mL IL-8 (Peprotech, Hamburg, Germany), either individually or in combination, for 24 h. The respective concentrations were derived from the respective product sheet by selecting a low effective concentration and then validated by dose response experiments. As a positive control, starved THP-1 cells were stimulated for 2 h with 25 ng/mL human TNF (Sigma Aldrich). Following stimulation, cells were harvested and the MMP-9 or IL-8 mRNA expression was measured.

4.4. RNA Extraction, cDNA Synthesis, and qPCR

Cells were lysed and RNA was isolated using the RNeasy Mini Kit (Qiagen, Hilden, Germany) according to the manufacturer's instructions [61]. RNA was eluted in 50 µL RNase-free water and concentrations were determined using the NanoDrop ND-1000 photometer at 260/280 nm (Peqlab, Erlangen, Germany). For first-strand complementary DNA (cDNA) synthesis, 1 µg total RNA per sample were reverse transcribed using the SuperScript™ II Reverse Transcriptase kit (Invitrogen, Carlsbad, CA, USA) and 250 ng random primers in a total volume of 100 µL as previously described [62]. Following reverse transcription, cDNA samples were immediately frozen and stored at −80 °C until further use. Quantitative polymerase chain reaction (qPCR) was performed using the LightCycler 480 II (Roche Diagnostics, Mannheim, Germany) as described in [8]. In short, MMP-9 analyses were based on the QuantiTect® Custom Assay (Qiagen; forward primer: 5′-TCCAGTACCGAGAGAAAG-3′, reverse primer: 5′-CAGGATGTCATAGGTCACGTAG-3′, hybridization probe: 5′-non-fluorescent quencher-GGAGTGAGTTGAACCAG-6-carboxyfluoroscein-3′). Suitability of glyceraldehyde-3-phosphate dehydrogenase (GAPDH), actin, and β2-microglobulin as housekeeping genes was assessed in 5 randomly selected cDNA samples. In all cases, expression levels were roughly comparable among tested samples. GAPDH was finally selected as housekeeping gene since it exhibited the highest expression levels. GAPDH mRNA determination was performed using sequence-specific fluorescent resonance energy transfer probes (forward primer: 5′-TGCTGAGTATGTCGTGGAGTC-3′, reverse primer: 5′-GGATGCAGGGATGATGTTCT-3′; donor hybridization probe: 5′-GACAACTTTGGTATCGTGGAAGGACTCATGACCACA-fluorescein thiocyanate-3′, acceptor hybridization probe: 5′-Cy5.5-CTGAGCGTGGCTATTCCTTCGTGACTACTG-phosphate-3′). 5 µL cDNA were mixed with 4 pmol hybridization probes, 10 pmol forward and reverse primers, and 10 µL LightCycler 480 Probes Master (Roche). H_2O was added to a final volume of 20 µL. PCR conditions were selected as follows: initial hot start incubation (95 °C, 5 min), denaturation (95 °C, 10 s), annealing (56 °C, 30 s for MMP-9; 59 °C, 30 s for GAPDH), and extension (72 °C, 30 s).

IL-8 mRNA expression was determined using the SYBR Select Master Mix (Invitrogen) and the LightCycler 480 (forward primer: 5′ -TCCTGTTTCTGCAGCTCTGG-3′, reverse primer: 5′ -GGCCACTCTCAATCACTCTC-3′). An amount of 5 µL cDNA was mixed with 10 pmol forward and reverse primers, and 10 µL SYBR Select Master Mix. H_2O was added to a final volume of 20 µL. PCR conditions were selected as follows: enzymatic degradation of contaminating uracil-containing DNA (50 °C, 2 min), initial hot start incubation (95 °C, 2 min), denaturation (95 °C, 10 s), annealing (59 °C, 15 s) extension (72 °C, 20 s) [38].

For the calculation of cDNA concentration, plasmid standards were generated and used in serial dilutions ranging from 1 mg/µL to 100 fg/µL as previously described [63].

4.5. Enzyme-Linked Immunosorbent Assay

Measurement of total secreted MMP-9 protein in the cell culture supernatant was performed using the Quantikine Kit (R&D Systems, Wiesbaden, Germany). Amounts of secreted total MMP-9 protein were assessed in thawed cell culture supernatant (stored at −150 °C until use) according to the manufacturer's instructions as described before [8]. The mean intra- and inter-assay coefficients of variation were given as 2.3% and 7.5%, respectively. The minimum detectable concentration was given as 0.156 µg/L.

4.6. Proteome Profiler Array

To identify secreted mediators, the Proteome Profiler Human XL Cytokine Array (based on the sandwich ELISA principle; R&D Systems) was carried out according to the manufacturer's instructions [64]. In short, an array membrane (consisting of capture antibodies spotted in duplicate on nitrocellulose) was incubated with 1.5 mL diluted supernatant (derived from HMWH-stimulated individual cell lines or co-cultures) and a cocktail of biotin-coupled capture antibodies (15 μL) overnight at 4 °C. Afterwards, the membrane was washed and incubated with 1.5 mL Streptavidin-HRP dilution for 30 min. Subsequently, membrane-bound cytokines/chemokines were detected using the chemiluminescent detection reagent and Amersham Hyperfilm ECL films (GE Healthcare, Freiburg, Germany).

4.7. Statistical Analysis

Data were described as means ± standard deviation (SD). The Kruskal–Wallis test was applied to analyze differences in mRNA or protein levels among several time points. The unpaired T test was used to analyze differences among mRNA levels following LMWH or cytokine stimulation. Analyses were performed using IBM SPSS Statistics, version 24 (IBM Deutschland GmbH, Ehningen, Germany).

Author Contributions: R.A., D.K., L.H., and B.L. performed the experiments. R.H., R.A., and R.L. wrote the paper. R.H. and R.L. conceived the study and designed the experiments. K.B. and K.W. contributed to the design of the study. R.H., R.A., D.K., L.H., B.L., K.B., K.W., and R.L. analyzed the results and approved the final version of the manuscript.

Funding: This research was funded by the *Deutsche Vereinte Gesellschaft für Klinische Chemie und Laboratoriumsmedizin* (DGKL; Stiftung für Pathobiochemie und Molekulare Diagnostik; grant number 2015/16 to RA and RL) and the *Doktor Robert Pfleger Stiftung Bamberg* (grant number RDR/chu to RH and RL).

Conflicts of Interest: The authors declare no conflict of interest.

Abbreviations

AP	Activator protein
cDNA	Complementary DNA
CCL	CC-chemokine ligand
CXCL	CxC-chemokine ligand
DSMZ	Deutsche Sammlung von Mikroorganismen und Zellkulturen
ECL	Enhanced chemiluminescence
ECM	Extracellular matrix
EDTA	Ethylenediaminetetraacetic acid
EGF	Epidermal growth factor
ELISA	Enzyme-linked immunosorbent assay
ERK	Extracellular signal-regulated kinase
FCS	Fetal calf serum
GAPDH	Glyceraldehyde-3-phosphate dehydrogenase
GM-CSF	granulocyte/macrophage-colony stimulating factor
GPR	G-protein-coupled receptor
HMGB	High mobility group box
HMWH	High-molecular-weight heparin
HRP	Horseradish peroxidase
IFN	Interferon
IKK	IκB kinase
IL	Interleukin
IL-1RA	IL-1 receptor antagonist
IU	International unit
Jak	Janus kinase
JNK	Jun N-terminal kinase
LMWH	Low-molecular-weight heparin
MAPK	Mitogen-activated protein kinasesmethyl-accepting chemotaxis protein
MCP	Methyl-accepting chemotaxis protein
MIF	Macrophage migration inhibitory factor
MIP	Macrophage inflammatory protein
MMP	Matrix metalloproteinase

NF-κB	Nuclear factor κB
PDGF	Platelet-derived growth factor
PKC	Protein kinase C
qPCR	Quantitative polymerase chain reaction
SCF	Stem cell factor
SD	Standard deviation
sICAM	Soluble intercellular adhesion molecule
Sp	Specificity protein
STAT	Signal transducer and activator of transcription
TIMP	Tissue inhibitor of metalloproteinases
TNF	Tumor necrosis factor
VEGF	Vascular endothelial growth factor
VTE	Venous thromboembolism

References

1. Nelson, A.R.; Fingleton, B.; Rothenberg, M.L.; Matrisian, L.M. Matrix metalloproteinases: Biologic activity and clinical implications. *J. Clin. Oncol.* **2000**, *18*, 1135–1149. [CrossRef]
2. Egeblad, M.; Werb, Z. New functions for the matrix metalloproteinases in cancer progression. *Nat. Rev. Cancer* **2002**, *2*, 161–174. [CrossRef]
3. Lemaitre, V.; D'Armiento, J. Matrix metalloproteinases in development and disease. *Birth Defects Res. C Embryo Today* **2006**, *78*, 1–10. [CrossRef]
4. Yong, V.W. Metalloproteinases: Mediators of pathology and regeneration in the CNS. *Nat. Rev. Neurosci.* **2005**, *6*, 931–944. [CrossRef] [PubMed]
5. Hu, J.; Van den Steen, P.E.; Sang, Q.X.; Opdenakker, G. Matrix metalloproteinase inhibitors as therapy for inflammatory and vascular diseases. *Nat. Rev. Drug Discov.* **2007**, *6*, 480–498. [CrossRef]
6. Gronberg, N.V.; Johansen, F.F.; Kristiansen, U.; Hasseldam, H. Leukocyte infiltration in experimental stroke. *J. Neuroinflamm.* **2013**, *10*, 115. [CrossRef]
7. Sapojnikova, N.; Kartvelishvili, T.; Asatiani, N.; Zinkevich, V.; Kalandadze, I.; Gugutsidze, D.; Shakarishvili, R.; Tsiskaridze, A. Correlation between MMP-9 and extracellular cytokine HMGB1 in prediction of human ischemic stroke outcome. *Biochim. Biophys. Acta* **2014**, *1842*, 1379–1384. [CrossRef] [PubMed]
8. Rababah, M.; Worthmann, H.; Deb, M.; Tryc, A.B.; Ma, Y.T.; El Bendary, O.M.; Hecker, H.; Goldbecker, A.; Heeren, M.; Brand, K.; et al. Anticoagulants affect matrix metalloproteinase 9 levels in blood samples of stroke patients and healthy controls. *Clin. Biochem.* **2012**, *45*, 483–489. [CrossRef] [PubMed]
9. Mannello, F.; Luchetti, F.; Canonico, B.; Papa, S. Effect of anticoagulants and cell separation media as preanalytical determinants on zymographic analysis of plasma matrix metalloproteinases. *Clin. Chem.* **2003**, *49*, 1956–1957. [CrossRef]
10. Mannello, F.; Jung, K.; Tonti, G.A.; Canestrari, F. Heparin affects matrix metalloproteinases and tissue inhibitors of metalloproteinases circulating in peripheral blood. *Clin. Biochem.* **2008**, *41*, 1466–1473. [CrossRef] [PubMed]
11. Meisser, A.; Cohen, M.; Bischof, P. Concentrations of circulating gelatinases (matrix metalloproteinase-2 and -9) are dependent on the conditions of blood collection. *Clin. Chem.* **2005**, *51*, 274–276. [CrossRef] [PubMed]
12. Gerlach, R.F.; Demacq, C.; Jung, K.; Tanus-Santos, J.E. Rapid separation of serum does not avoid artificially higher matrix metalloproteinase (MMP)-9 levels in serum versus plasma. *Clin. Biochem.* **2007**, *40*, 119–123. [CrossRef]
13. Gray, E.; Mulloy, B.; Barrowcliffe, T.W. Heparin and low-molecular-weight heparin. *Thromb. Haemost.* **2008**, *99*, 807–818. [PubMed]
14. Pineo, G.F.; Hull, R.D. Economic and practical aspects of thromboprophylaxis with unfractionated and low-molecular-weight heparins in hospitalized medical patients. *Clin. Appl. Thromb. Hemost.* **2009**, *15*, 489–500. [CrossRef] [PubMed]
15. Banfi, G.; Salvagno, G.L.; Lippi, G. The role of ethylenediamine tetraacetic acid (EDTA) as in vitro anticoagulant for diagnostic purposes. *Clin. Chem. Lab. Med.* **2007**, *45*, 565–576. [CrossRef] [PubMed]

16. Van den Besselaar, A.M.; Chantarangkul, V.; Tripodi, A. A comparison of two sodium citrate concentrations in two evacuated blood collection systems for prothrombin time and ISI determination. *Thromb. Haemost.* **2000**, *84*, 664–667. [PubMed]
17. Yoffe, J.R.; Taylor, D.J.; Woolley, D.E. Mast-cell products and heparin stimulate the production of mononuclear-cell factor by cultured human monocyte/macrophages. *Biochem. J.* **1985**, *230*, 83–88. [CrossRef] [PubMed]
18. Webster, N.L.; Crowe, S.M. Matrix metalloproteinases, their production by monocytes and macrophages and their potential role in HIV-related diseases. *J. Leukoc. Biol.* **2006**, *80*, 1052–1066. [CrossRef] [PubMed]
19. Wollbold, J.; Huber, R.; Pohlers, D.; Koczan, D.; Guthke, R.; Kinne, R.W.; Gausmann, U. Adapted Boolean network models for extracellular matrix formation. *BMC Syst. Biol.* **2009**, *3*, 77. [CrossRef]
20. Page-McCaw, A.; Ewald, A.J.; Werb, Z. Matrix metalloproteinases and the regulation of tissue remodelling. *Nat. Rev. Mol. Cell Biol.* **2007**, *8*, 221–233. [CrossRef]
21. Van den Steen, P.E.; Dubois, B.; Nelissen, I.; Rudd, P.M.; Dwek, R.A.; Opdenakker, G. Biochemistry and molecular biology of gelatinase B or matrix metalloproteinase-9 (MMP-9). *Crit. Rev. Biochem. Mol. Biol.* **2002**, *37*, 375–536. [CrossRef]
22. Murphy, G.; Knäuper, V.; Atkinson, S.; Butler, G.; English, W.; Hutton, M.; Stracke, J.; Clark, I. Matrix metalloproteinases in arthritic disease. *Arthritis Res.* **2002**, *4*, S39–S49. [CrossRef] [PubMed]
23. Kessenbrock, K.; Plaks, V.; Werb, Z. Matrix metalloproteinases: Regulators of the tumor microenvironment. *Cell* **2010**, *141*, 52–67. [CrossRef]
24. Lambert, E.; Dasse, E.; Haye, B.; Petitfrere, E. TIMPs as multifacial proteins. *Crit. Rev. Oncol. Hematol.* **2004**, *49*, 187–198. [CrossRef]
25. Lee, Y.R.; Noh, E.M.; Han, J.H.; Kim, J.M.; Hwang, B.M.; Kim, B.S.; Lee, S.H.; Jung, S.H.; Youn, H.J.; Chung, E.Y.; et al. Sulforaphane controls TPA-induced MMP-9 expression through the NF-kappaB signaling pathway, but not AP-1, in MCF-7 breast cancer cells. *BMB Rep.* **2013**, *46*, 201–206. [CrossRef] [PubMed]
26. Lee, W.R.; Chung, C.L.; Hsiao, C.J.; Chou, Y.C.; Hsueh, P.J.; Yang, P.C.; Jan, J.S.; Cheng, Y.W.; Hsiao, G. Suppression of matrix metalloproteinase-9 expression by andrographolide in human monocytic THP-1 cells via inhibition of NF-kappaB activation. *Phytomedicine* **2012**, *19*, 270–277. [CrossRef]
27. Al-Rashed, F.; Kochumon, S.; Usmani, S.; Sindhu, S.; Ahmad, R. Pam3CSK4 Induces MMP-9 Expression in Human Monocytic THP-1 Cells. *Cell Physiol. Biochem.* **2017**, *41*, 1993–2003. [CrossRef]
28. St-Pierre, Y.; Couillard, J.; Van Themsche, C. Regulation of MMP-9 gene expression for the development of novel molecular targets against cancer and inflammatory diseases. *Expert Opin. Ther. Targets* **2004**, *8*, 473–489. [CrossRef] [PubMed]
29. Overall, C.M.; Lopez-Otin, C. Strategies for MMP inhibition in cancer: Innovations for the post-trial era. *Nat. Rev. Cancer* **2002**, *2*, 657–672. [CrossRef]
30. Manfredi, M.A.; Zurakowski, D.; Rufo, P.A.; Walker, T.R.; Fox, V.L.; Moses, M.A. Increased incidence of urinary matrix metalloproteinases as predictors of disease in pediatric patients with inflammatory bowel disease. *Inflamm. Bowel Dis.* **2008**, *14*, 1091–1096. [CrossRef]
31. Jung, K.; Lein, M.; Laube, C.; Lichtinghagen, R. Blood specimen collection methods influence the concentration and the diagnostic validity of matrix metalloproteinase 9 in blood. *Clin. Chim. Acta* **2001**, *314*, 241–244. [CrossRef]
32. Jung, K.; Klotzek, S.; Stephan, C.; Mannello, F.; Lein, M. Impact of blood sampling on the circulating matrix metalloproteinases 1, 2, 3, 7, 8, and 9. *Clin. Chem.* **2008**, *54*, 772–773. [CrossRef] [PubMed]
33. Bar-Or, A.; Nuttall, R.K.; Duddy, M.; Alter, A.; Kim, H.J.; Ifergan, I.; Pennington, C.J.; Bourgoin, P.; Edwards, D.R.; Yong, V.W. Analyses of all matrix metalloproteinase members in leukocytes emphasize monocytes as major inflammatory mediators in multiple sclerosis. *Brain* **2003**, *126*, 2738–2749. [CrossRef] [PubMed]
34. Raphael, I.; Nalawade, S.; Eagar, T.N.; Forsthuber, T.G. T cell subsets and their signature cytokines in autoimmune and inflammatory diseases. *Cytokine* **2015**, *74*, 5–17. [CrossRef] [PubMed]
35. Cruikshank, W.W.; Kornfeld, H.; Center, D.M. Interleukin-16. *J. Leukoc. Biol.* **2000**, *67*, 757–766. [CrossRef] [PubMed]
36. Mathy, N.L.; Scheuer, W.; Lanzendorfer, M.; Honold, K.; Ambrosius, D.; Norley, S.; Kurth, R. Interleukin-16 stimulates the expression and production of pro-inflammatory cytokines by human monocytes. *Immunology* **2000**, *100*, 63–69. [CrossRef]

37. Park, S.L.; Hwang, B.; Lee, S.Y.; Kim, W.T.; Choi, Y.H.; Chang, Y.C.; Kim, W.J.; Moon, S.K. p21WAF1 Is Required for Interleukin-16-Induced Migration and Invasion of Vascular Smooth Muscle Cells via the p38MAPK/Sp-1/MMP-9 Pathway. *PLoS ONE* **2015**, *10*, e0142153. [CrossRef] [PubMed]
38. Gunther, J.; Vogt, N.; Hampel, K.; Bikker, R.; Page, S.; Muller, B.; Kandemir, J.; Kracht, M.; Dittrich-Breiholz, O.; Huber, R.; et al. Identification of two forms of TNF tolerance in human monocytes: Differential inhibition of NF-kappaB/AP-1- and PP1-associated signaling. *J. Immunol.* **2014**, *192*, 3143–3155. [CrossRef] [PubMed]
39. Robinson, S.C.; Scott, K.A.; Balkwill, F.R. Chemokine stimulation of monocyte matrix metalloproteinase-9 requires endogenous TNF-alpha. *Eur. J. Immunol.* **2002**, *32*, 404–412. [CrossRef]
40. White, D.A.; Fang, L.; Chan, W.; Morand, E.F.; Kiriazis, H.; Duffy, S.J.; Taylor, A.J.; Dart, A.M.; Du, X.J.; Gao, X.M. Pro-inflammatory action of MIF in acute myocardial infarction via activation of peripheral blood mononuclear cells. *PLoS ONE* **2013**, *8*, e76206. [CrossRef]
41. Browning, D.D.; Diehl, W.C.; Hsu, M.H.; Schraufstatter, I.U.; Ye, R.D. Autocrine regulation of interleukin-8 production in human monocytes. *Am. J. Physiol. Lung Cell Mol. Physiol.* **2000**, *279*, L1129–L1136. [CrossRef] [PubMed]
42. Ren, F.; Tang, R.; Zhang, X.; Madushi, W.M.; Luo, D.; Dang, Y.; Li, Z.; Wei, K.; Chen, G. Overexpression of MMP Family Members Functions as Prognostic Biomarker for Breast Cancer Patients: A Systematic Review and Meta-Analysis. *PLoS ONE* **2015**, *10*, e0135544. [CrossRef] [PubMed]
43. Leber, T.M.; Balkwill, F.R. Regulation of monocyte MMP-9 production by TNF-alpha and a tumour-derived soluble factor (MMPSF). *Br. J. Cancer* **1998**, *78*, 724–732. [CrossRef] [PubMed]
44. Doster, A.; Schwarzig, U.; Zygmunt, M.; Rom, J.; Schutz, F.; Fluhr, H. Unfractionated Heparin Selectively Modulates the Expression of CXCL8, CCL2 and CCL5 in Endometrial Carcinoma Cells. *Anticancer Res.* **2016**, *36*, 1535–1544. [PubMed]
45. Biancotto, A.; Feng, X.; Langweiler, M.; Young, N.S.; McCoy, J.P. Effect of anticoagulants on multiplexed measurement of cytokine/chemokines in healthy subjects. *Cytokine* **2012**, *60*, 438–446. [CrossRef] [PubMed]
46. Patil, R.; Shukre, S.; Paranjape, R.; Thakar, M. Heparin and EDTA anticoagulants differentially affect the plasma cytokine levels in humans. *Scand. J. Clin. Lab. Investig.* **2013**, *73*, 452–455. [CrossRef] [PubMed]
47. Baram, D.; Rashkovsky, M.; Hershkoviz, R.; Drucker, I.; Reshef, T.; Ben-Shitrit, S.; Mekori, Y.A. Inhibitory effects of low molecular weight heparin on mediator release by mast cells: Preferential inhibition of cytokine production and mast cell-dependent cutaneous inflammation. *Clin. Exp. Immunol.* **1997**, *110*, 485–491. [CrossRef]
48. Shastri, M.D.; Stewart, N.; Horne, J.; Peterson, G.M.; Gueven, N.; Sohal, S.S.; Patel, R.P. In-vitro suppression of IL-6 and IL-8 release from human pulmonary epithelial cells by non-anticoagulant fraction of enoxaparin. *PLoS ONE* **2015**, *10*, e0126763. [CrossRef]
49. Shastri, M.D.; Stewart, N.; Eapen, M.; Peterson, G.M.; Zaidi, S.T.; Gueven, N.; Sohal, S.S.; Patel, R.P. Opposing effects of low molecular weight heparins on the release of inflammatory cytokines from peripheral blood mononuclear cells of asthmatics. *PLoS ONE* **2015**, *10*, e0118798. [CrossRef] [PubMed]
50. Onishi, A.; St Ange, K.; Dordick, J.S.; Linhardt, R.J. Heparin and anticoagulation. *Front. Biosci. (Landmark Ed.)* **2016**, *21*, 1372–1392.
51. Liu, J.; Linhardt, R.J. Chemoenzymatic synthesis of heparan sulfate and heparin. *Nat. Prod. Rep.* **2014**, *31*, 1676–1685. [CrossRef] [PubMed]
52. Crowther, M.; Lim, W. Low molecular weight heparin and bleeding in patients with chronic renal failure. *Curr. Opin. Pulm. Med.* **2007**, *13*, 409–413. [CrossRef] [PubMed]
53. Li, L.; Ling, Y.; Huang, M.; Yin, T.; Gou, S.M.; Zhan, N.Y.; Xiong, J.X.; Wu, H.S.; Yang, Z.Y.; Wang, C.Y. Heparin inhibits the inflammatory response induced by LPS and HMGB1 by blocking the binding of HMGB1 to the surface of macrophages. *Cytokine* **2015**, *72*, 36–42. [CrossRef]
54. Chiang, N.Y.; Chang, G.W.; Huang, Y.S.; Peng, Y.M.; Hsiao, C.C.; Kuo, M.L.; Lin, H.H. Heparin interacts with the adhesion GPCR GPR56, reduces receptor shedding, and promotes cell adhesion and motility. *J. Cell Sci.* **2016**, *129*, 2156–2169. [CrossRef]
55. Theroux, P.; Waters, D.; Qiu, S.; McCans, J.; de Guise, P.; Juneau, M. Aspirin versus heparin to prevent myocardial infarction during the acute phase of unstable angina. *Circulation* **1993**, *88*, 2045–2048. [CrossRef]
56. Cavender, M.A.; Sabatine, M.S. Bivalirudin versus heparin in patients planned for percutaneous coronary intervention: A meta-analysis of randomised controlled trials. *Lancet* **2014**, *384*, 599–606. [CrossRef]

57. Grzela, T.; Brawura-Biskupski-Samaha, R.; Jelenska, M.M.; Szmidt, J. Low molecular weight heparin treatment decreases MMP-9 plasma activity in patients with abdominal aortic aneurysm. *Eur. J. Vasc. Endovasc. Surg.* **2008**, *35*, 159–161. [CrossRef] [PubMed]
58. Kalyani, B.S.; Roberts, C.S. Low molecular weight heparin: Current evidence for its application in orthopaedic surgery. *Curr. Vasc. Pharmacol.* **2011**, *9*, 19–23. [CrossRef]
59. Lyman, G.H. Thromboprophylaxis with low-molecular-weight heparin in medical patients with cancer. *Cancer* **2009**, *115*, 5637–5650. [CrossRef]
60. Huber, R.; Panterodt, T.; Welz, B.; Christmann, M.; Friesenhagen, J.; Westphal, A.; Pietsch, D.; Brand, K. C/EBPbeta-LAP*/LAP Expression Is Mediated by RSK/eIF4B-Dependent Signalling and Boosted by Increased Protein Stability in Models of Monocytic Differentiation. *PLoS ONE* **2015**, *10*, e0144338. [CrossRef]
61. Huber, R.; Kunisch, E.; Gluck, B.; Egerer, R.; Sickinger, S.; Kinne, R.W. Comparison of conventional and real-time RT-PCR for the quantitation of jun protooncogene mRNA and analysis of junB mRNA expression in synovial membranes and isolated synovial fibroblasts from rheumatoid arthritis patients. *Z. Rheumatol.* **2003**, *62*, 378–389. [CrossRef] [PubMed]
62. Haas, S.C.; Huber, R.; Gutsch, R.; Kandemir, J.D.; Cappello, C.; Krauter, J.; Duyster, J.; Ganser, A.; Brand, K. ITD- and FL-induced FLT3 signal transduction leads to increased C/EBPbeta-LIP expression and LIP/LAP ratio by different signalling modules. *Br. J. Haematol.* **2010**, *148*, 777–790. [CrossRef] [PubMed]
63. Schwettmann, L.; Wehmeier, M.; Jokovic, D.; Aleksandrova, K.; Brand, K.; Manns, M.P.; Lichtinghagen, R.; Bahr, M.J. Hepatic expression of A disintegrin and metalloproteinase (ADAM) and ADAMs with thrombospondin motives (ADAM-TS) enzymes in patients with chronic liver diseases. *J. Hepatol.* **2008**, *49*, 243–250. [CrossRef] [PubMed]
64. Gutsch, R.; Kandemir, J.D.; Pietsch, D.; Cappello, C.; Meyer, J.; Simanowski, K.; Huber, R.; Brand, K. CCAAT/enhancer-binding protein beta inhibits proliferation in monocytic cells by affecting the retinoblastoma protein/E2F/cyclin E pathway but is not directly required for macrophage morphology. *J. Biol. Chem.* **2011**, *286*, 22716–22729. [CrossRef] [PubMed]

International Journal of *Molecular Sciences*

MDPI

Article

Biomarker Analysis of Orally Dosed, Dual Active, Matrix Metalloproteinase (MMP)-2 and MMP-9 Inhibitor, AQU-118, in the Spinal Nerve Ligation (SNL) Rat Model of Neuropathic Pain

Mei Yee Kwan [1], Anthony Choo [1], Taleen Hanania [1], Afshin Ghavami [1], Jose Beltran [1], John Shea [1], Amidi Barboza [1], Andrew Hu [1], Marcie Fowler [2], Venugopal Rao Neelagiri [3] and Irving Sucholeiki [3,*]

1 PsychoGenics Inc., 215 College Road, Paramus, NJ 07652, USA; Mei.Kwan@psychogenics.com (M.Y.K.); amchoo@gmail.com (A.C.); Taleen.Hanania@psychogenics.com (T.H.); Afshin.Ghavami@psychogenics.com (A.G.); Jose.Beltran@psychogenics.com (J.B.); John.Shea@psychogenics.com (J.S.); Amidi.Barboza@psychogenics.com (A.B.); andreweiy2000@yahoo.com (A.H.)

2 United States Army of Surgical Research, Joint Base San Antonio (JBSA), Fort Sam Houston, TX 78234, USA; Marcie.fowler.phd@gmail.com

3 Aquilus Pharmaceuticals Inc., 3H Gill Street, Suite 300, Woburn, MA 01801, USA; neelagiriv@yahoo.com

* Correspondence: sucholeiki@aquiluspharma.com

Received: 23 December 2018; Accepted: 11 February 2019; Published: 14 February 2019

Abstract: There is an unmet medical need for the development of non-addicting pain therapeutics with enhanced efficacy and tolerability. The current study examined the effects of AQU-118, an orally active inhibitor of metalloproteinase-2 (MMP-2) and MMP-9, in the spinal nerve ligation (SNL) rat model of neuropathic pain. Mechanical allodynia and the levels of various biomarkers were examined within the dorsal root ganglion (DRG) before and after oral dosing with AQU-118. The rats that received the SNL surgery exhibited significant mechanical allodynia as compared to sham controls. Animals received either vehicle, positive control (gabapentin), or AQU-118. After SNL surgery, the dorsal root ganglion (DRG) of those rats dosed with vehicle had elevated messenger RNA (mRNA) expression levels for MMP-2, IL1-β & IL-6 and elevated protein levels for caspase-3 while exhibiting decreased protein levels for myelin basic protein (MBP) & active IL-β as compared to sham controls. Rats orally dosed with AQU-118 exhibited significantly reduced mechanical allodynia and decreased levels of caspase-3 in the DRG as compared to vehicle controls. Results demonstrate that oral dosing with the dual active, MMP-2/-9 inhibitor, AQU-118, attenuated mechanical allodynia while at the same time significantly reduced the levels of caspase-3 in the DRG.

Keywords: matrix metalloproteinase; MMP-2; MMP-9; inhibitor; allodynia; caspase-3; neuropathic; pain; dorsal root ganglion; spinal nerve ligation

---

1. Introduction

The population suffering from neuropathic pain is large; neuropathic pain is a significant problem for patients and the healthcare system. This type of pain is difficult to treat effectively with currently available pharmacological agents, in part due to the fact that the aetiologies that drive the neuropathic pain states are varied, complex, and not well-understood. Indeed, many of the currently approved neuropathic pain medications, and in particular anticonvulsants such as gabapentin and pregabalin, do not have a direct effect on the pathophysiology of the disorder but rather modulate neurotransmission [1]. Since only one in four patients with neuropathic pain experiences greater than a 50% reduction in pain following treatment, novel therapeutic approaches are an urgent priority [2]. One tactic for treating neuropathic pain is through the inhibition of specific protein targets involved in the initiation or maintenance of pain sensitization within the spinal cord.

Surgically-induced rodent models of neuropathic pain cause the levels of various inflammatory proteins such as cytokines and, in particular, certain matrix metalloproteinases (MMPs), to become elevated within the spinal nerves [3–9]. MMPs are zinc-dependent endopeptidases belonging to a larger family of proteases known as the metzincin superfamily. MMPs are involved in degrading the extracellular matrix as well as processing of growth factors, cytokines, chemokines, adhesion molecules, and a variety of other enzymes. Increased expression of MMPs or an imbalance of different MMPs has been observed in a variety of pathological conditions, including various neurodegenerative diseases [10,11].

It has been postulated that in injured dorsal root ganglion (DRG) primary sensory neurons, MMP-9 can induce early neuropathic pain via interleukin-1β cleavage and microglia activation, and MMP-2 induces delayed neuropathic pain via IL-1β cleavage and astrocyte activation [12]. MMP-9 also promotes Schwann cell-mediated myelin basic protein (MBP) degradation and macrophage infiltration in the spinal nerve and activation of astrocytes in the spinal cord [13]. Based upon this evidence, we sought to develop an orally active MMP-2 and MMP-9 inhibitor to treat neuropathic pain. Our first compound in this arena is AQU-118, a potent small molecule inhibitor of both MMP-2 and MMP-9 with an in vitro IC50 of 3 nM and 9 nM, respectively [14]. AQU-118 attenuates neuropathic pain behaviors in the SNL and chronic constriction injury of the infraorbital nerve (CCI-IoN) rodent models as well as in the naloxone precipitated morphine withdrawal rodent model [14,15]. While these models demonstrated the analgesic efficacy of AQU-118 in neuropathic pain, they did not address the mechanistic or biochemical correlates associated with this activity. The present study analyzes the effects of AQU-118 on mechanical allodynia in the SNL-rat model and then identifies underlying molecular effects of oral dosing with AQU-118 on various inflammatory biomarkers associated with neuropathic pain.

Although many biomarkers are associated with neuropathic pain, we focused on those whose expression and/or protein levels have been reported to change within the dorsal root ganglion (DRG) or dorsal horn following spinal nerve injuries [12,13,16–18]. We examined transcript expression levels of MMP-2, MMP-9, IL-1β, IL-6 and protein levels of MBP (20 kDa and 15 kDa molecular weight bands), the pro and active forms of IL-1β, IL-6 and caspase-3 from rat DRG samples. We first determined if any protein or mRNA expression changes were related to the spinal nerve ligation (SNL) surgery by comparing DRG samples from rats who received the SNL with control animals that underwent a sham surgery. Any biomarker(s) that were observed to significantly change in the DRG after SNL-surgery were then examined further to determine if they would return to pre-SNL levels (i.e., sham) upon oral dosing with AQU-118 and its corresponding reduction in mechanical allodynia. The ultimate goals were to better understand the mechanism of action for the antiallodynic effects of AQU-118 and to find biomarkers that could be used in a clinical setting to help monitor the effectiveness of AQU-118. This study presents for the first time biomarker data for an orally administered, dual active MMP-2/MMP-9 inhibitor in the SNL rat model for neuropathic pain. These results further validate the oral effectiveness of AQU-118 and support its continued preclinical development and future clinical testing as a novel treatment for general neuropathic pain.

2. Results

2.1. Attenuation of von Frey Mechanical Allodynia by Oral Administration of AQU-118 in the Spinal Nerve Ligation (SNL)-Rat Model

Due to the need for sufficient spinal tissue (i.e., DRG) to perform both the gene expression and protein biomarker work, the SNL-rat model studies were designed to incorporate higher numbers of rodents among certain groups than would typically be needed for measuring only mechanical allodynia. Except for the positive control ($n = 10$, gabapentin, 100 mg/kg, 0.58 mmoles/kg), a larger number of rodents were used for the sham ($n = 20$), vehicle (0.5% methyl cellulose, $n = 40$) and drug groups ($n = 20$, AQU-118, 160 mg/kg, 0.34 mmoles/kg) (Table 1). Two weeks after surgery, rats with L-5 spinal nerve ligation displayed significant mechanical allodynia as compared to pre-operative testing (Figure 1). Oral dosing of AQU-118 beginning on day 1 caused an increase in the paw withdrawal threshold (PWT) as compared to the vehicle control group (Figure 1). No statistically significant effect on contralateral PWT was observed with oral dosing of AQU-118 which was comparable to both the vehicle and positive control (gabapentin) arms (Figure 2).

Table 1. Protocol for SNL study using male Sprague-Dawley rats.

Group	#Rats	Route	Dose [1,5] (mg/kg)	Compound	Dosing Days	von-Frey Testing Days [5]
1	20	P.O.	NA [2]	Sham	NA [2]	1,3,5
2	40	P.O.	NA [3]	Vehicle [3]	1–5	1,3,5
3	10	P.O.	100	Gabapentin	1,3,5 [4]	1,3,5
4	20	P.O.	160	AQU-118	1–15	1,3,5

[1] Once per day dosing via gastric gavage (P.O.), 60 min before von-Frey testing. [2] NA not applicable. [3] 0.5% methyl cellulose at a dose volume equivalent to the test compound (5 mL/kg). [4] Gabapentin was dissolved in saline and administered 60 min prior to von-Frey testing. [5] A base-line measure was taken via von-Frey before surgery and then two weeks after surgery (Day 0). The animals were then balanced and assigned to treatment groups based on their post-operative PWT values. Animals with a VF score over 4.5 g were excluded from the study.

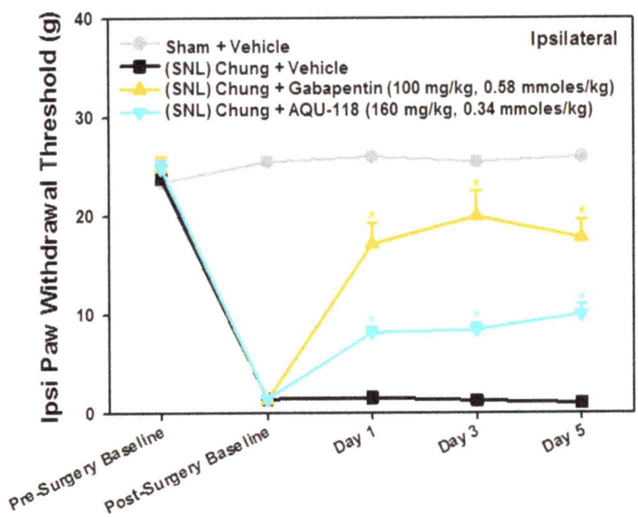

Figure 1. Mechanical Response Threshold: Spinal Nerve Ligation (SNL) Chung Model (Chung), Ipsilateral (Ipsi) Paw Withdrawal. Paw withdrawal thresholds following SNL-surgery for ipsilateral hind paws. Data are presented as mean ± SEM. * $p < 0.05$ vs. (SNL) Chung + Vehicle group on the same day. Post-Surgery Baseline was performed 15 days after SNL-surgery.

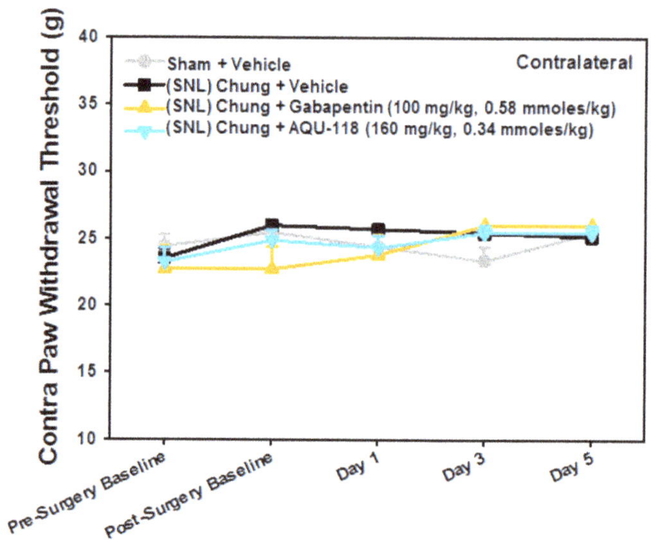

Figure 2. Mechanical Response Threshold: Spinal Nerve Ligation (SNL) Chung Model (Chung), Contralateral (Contra) Paw Withdrawal. Contralateral paw withdrawal thresholds following surgery and administration of AQU-118. Data are presented as mean ± SEM.

2.2. Transcript Expression Changes in the DRG 20 Days after SNL-Surgery (D5) between Vehicle and Sham Groups

Comparisons of MMP-2, MMP-9, IL-1β and IL-6 mRNA expression levels in the left (ipsilateral) DRG between vehicle and sham groups were measured and all were found to be significantly elevated except for MMP-9 (Figures 3 and 4). An apparent correlation was observed between the mRNA expression levels of MMP-2 and IL-1β (Figure 5). However, no such correlation was observed between MMP-2 and IL-6 mRNA levels (Figure 5).

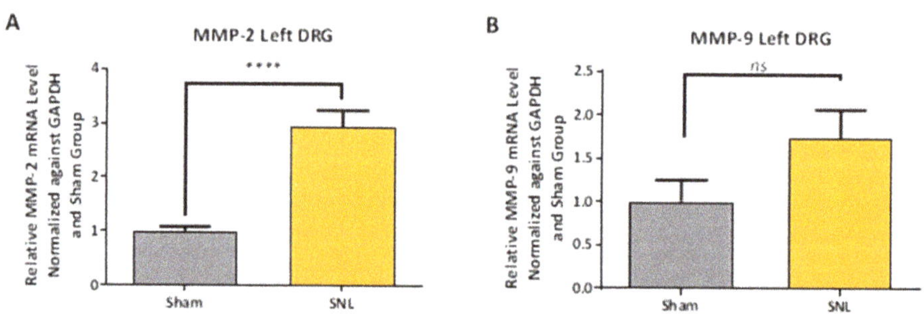

Figure 3. MMP2 (**A**) and MMP9 (**B**) mRNA expression levels in ipsilateral DRG for sham and SNL vehicle groups. MMP-2 mRNA expression levels were found to be significantly elevated after SNL-surgery while MMP-9 mRNA was found to be not significant. Data presented as the average from n = 10 per group ± SEM and analyzed with a two-tailed unpaired t-test (ns, not significant; ****, $p < 0.0001$).

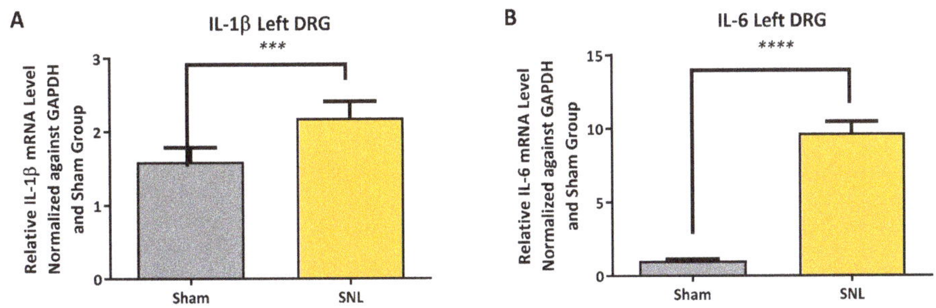

Figure 4. IL-1β (**A**) and IL-6 (**B**) mRNA expression levels in ipsilateral DRG for sham and SNL vehicle groups. Both IL-1β and IL-6 mRNA expression levels were found to be significantly elevated after SNL-surgery in vehicle group compared to sham. Data presented as the average from $n = 10$ per group ± SEM and analyzed a with two-tailed unpaired t-test. (***, $p < 0.001$; ****, $p < 0.0001$).

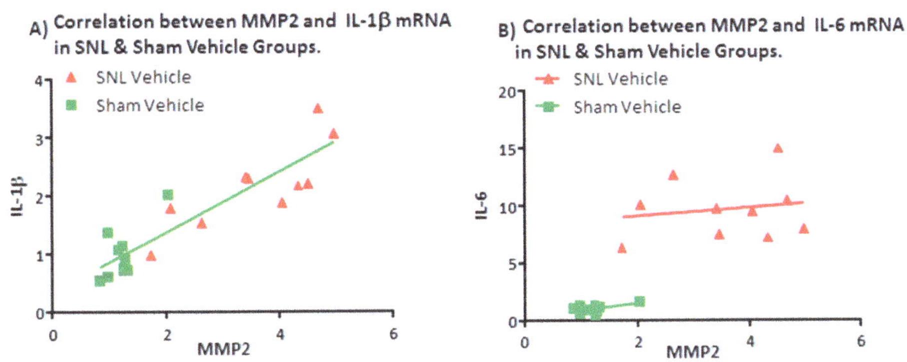

Figure 5. Correlation between MMP2 and (**A**) IL-1β mRNA and (**B**) IL-6 mRNA expression levels in left DRG of SNL & sham Vehicle groups. Pearson correlation coefficients (with 95% confidence) were calculated for sham vehicle ($r = 0.7372$; two-tailed p value = 0.015) and SNL vehicle ($r = 0.8135$; two-tailed p value = 0.0042) for MMP2 vs IL-1β mRNA. Similar analysis of MMP2 vs. IL-6 mRNA showed no clear correlation with IL-6 transcript level: sham vehicle ($r = 0.4912$; two tailed p value = 0.1494) and SNL vehicle ($r = 0.1622$; two tailed p value = 0.6543).

2.3. Protein Level Changes in the DRG 20 Days after SNL-Surgery (D5) between Vehicle and Sham Groups

IL-1β (Pro and cleaved form), MBP, IL-6 and caspase-3 protein levels were measured via western blotting of tissue obtained from the left (ipsilateral) DRG and compared between sham and vehicle groups. All protein targets were normalized to glyceraldehyde-3-phosphate dehydrogenase (GAPDH) protein levels. The results show that cleaved IL-1β was found to significantly decrease in the ipsilateral DRG of the vehicle group as compared to sham while no significant changes were observed in contralateral DRG (Figure 6). Caspase-3 was found to significantly increase in the ipsilateral DRG while both the higher and lower molecular weight bands of MBP were found to significantly decrease in the vehicle group as compared to the sham (Figure 7). The pro forms of IL-1β and IL-6 were not detected in the DRG by western blotting.

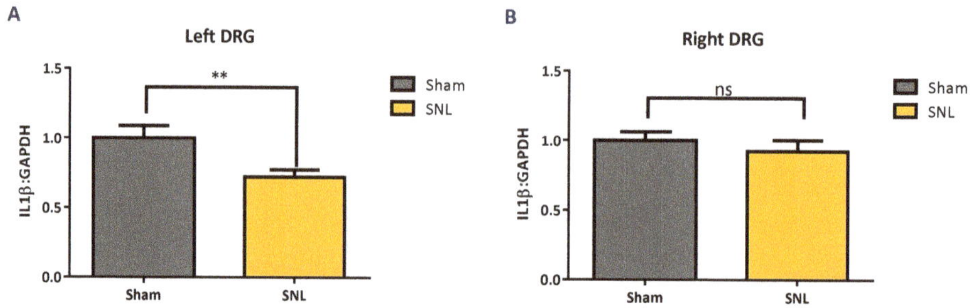

Figure 6. Comparison of cleaved IL-1β protein level in ipsilateral and contralateral DRG for sham and vehicle groups. Cleaved, active IL-1β protein levels in left (ipsilateral) DRG (**A**) significantly decreased in the vehicle group (n = 10) compared to the sham (n = 10). (**B**) No difference between the sham (n = 10) and vehicle group (n = 10) in right (contralateral) DRG. Data presented as the average ± SEM. An unpaired t-test was performed: ns, not significant; **, $p < 0.01$.

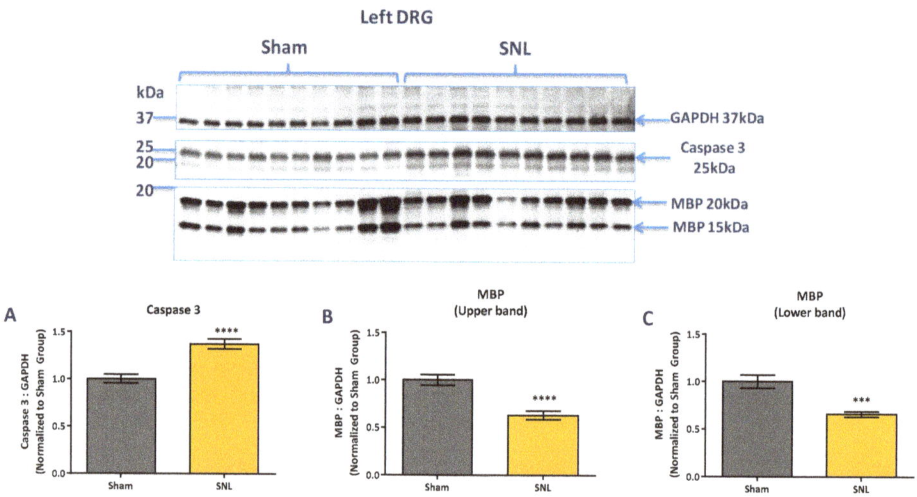

Figure 7. Comparison of Caspase-3 and MBP protein levels in in the ipsilateral DRG for sham and vehicle groups. (**A**) Caspase-3 protein levels (via western blot) in the left (ipsilateral) DRG significantly increased in the vehicle group compared to the sham. (**B**) MBP levels (via western blot) significantly decreased for both the upper band (20 kDa) and (**C**) lower band (15 kDa) in the vehicle group of the left (ipsilateral) DRG as compared to the sham. Protein level is presented normalized to GAPDH and relative to the sham group. Data presented as the average ± SEM. An unpaired t-test was performed: ***, $p < 0.001$; ****, $p < 0.0001$.

2.4. Protein Level Changes in the DRG 20 Days after SNL-Surgery (D5) between AQU-118 and Vehicle Groups

MBP (both higher and lower molecular weight bands) and caspase-3 protein levels were measured via western blotting of tissue obtained from the left (ipsilateral) DRG of the AQU-118 group ($n = 20$) and compared with the vehicle ($n = 20$). All protein targets were normalized to glyceraldehyde-3-phosphate dehydrogenase (GAPDH) protein levels. In the samples from AQU-118 treated animals, caspase-3 levels were significantly decreased in the ipsilateral DRG but there were no significant changes observed in the levels of MBP as compared to samples obtained from vehicle-treated animals (Figure 8). The DRG samples were tested for MMP-2 mRNA expression levels and retested for caspase-3 protein levels in the AQU-118 ($n = 10$) and vehicle ($n = 10$) groups. (Figure 9). Comparison of these samples showed that the AQU-118 ($n = 10$) group exhibited a statistically significant decrease in the caspase-3 protein level as compared to the vehicle group ($n = 10$), while still exhibiting elevated MMP-2 mRNA expression levels (Figure 9).

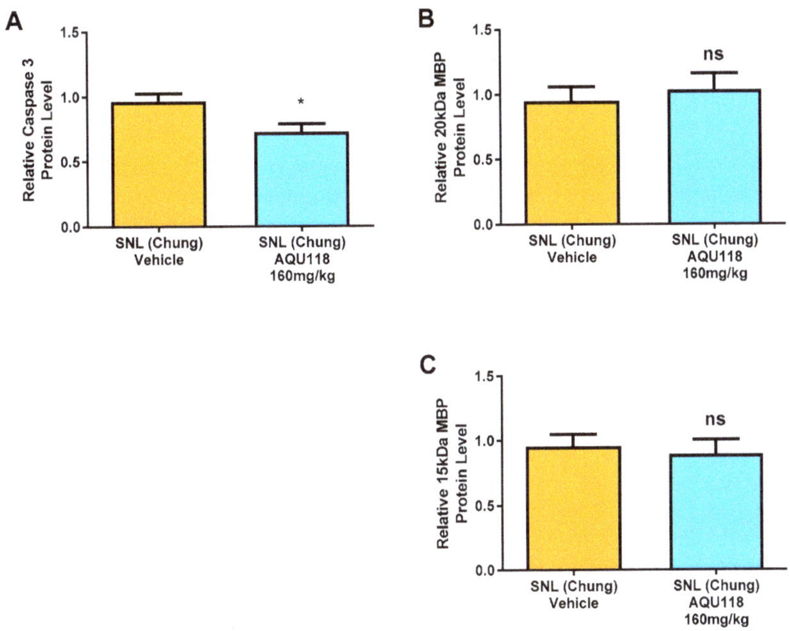

Figure 8. Comparison of caspase-3 and MBP in the ipsilateral DRG for vehicle and AQU-118 groups. (**A**) Caspase-3 protein levels in the left (ipsilateral) DRG significantly decreased in the AQU-118 group ($n = 20$) compared to the vehicle ($n = 20$). (**B**) There were no significant changes in MBP levels in both the upper band (20 kDa) and (**C**) lower band (15 kDa) in AQU-118 group ($n = 20$) of the left (ipsilateral) DRG as compared to the vehicle group ($n = 20$). Protein level is presented normalized to GAPDH and relative to the vehicle group. Data presented as the average ± SEM. An unpaired t-test was performed: ns, not significant; *, $p < 0.05$.

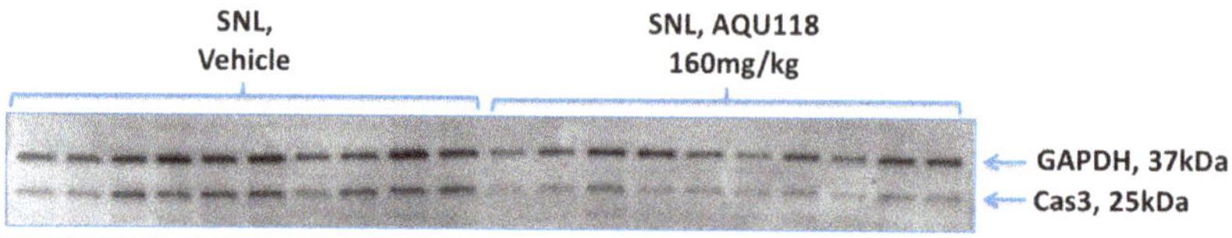

MMP2 mRNA Expression Level	0.80	2.90	3.00	1.90	2.00	2.70	3.30	3.50	0.50	2.80	1.50	1.50	4.30	3.30	3.30	3.10	1.80	3.80	3.60	3.90
Caspase 3 Protein Level	0.67	0.82	1.58	0.74	1.04	0.91	0.70	1.02	0.63	1.19	0.53	0.74	0.72	0.29	0.40	0.52	0.66	0.13	0.69	0.63

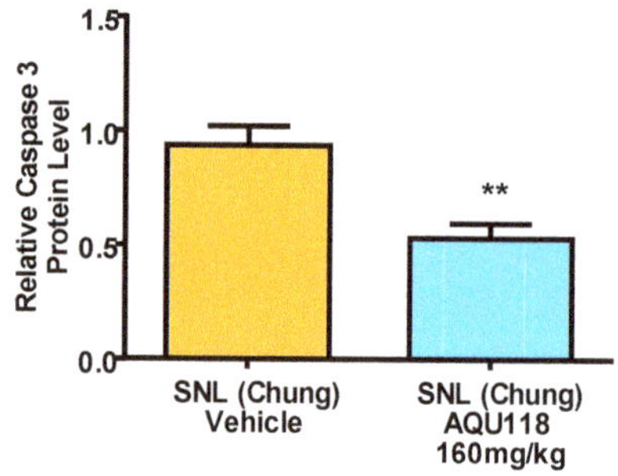

Figure 9. Comparison of MMP2 mRNA expression levels versus caspase-3 in a subset of ipsilateral DRG tissue for both vehicle and AQU-118 groups. Caspase 3 protein and MMP-2 mRNA expression levels were performed in the ipsilateral DRG tissue (via western blot) for both the AQU-118 (n = 10) and vehicle (n = 10) groups and compared to each other. The protein level is presented normalized to GAPDH and relative to the vehicle group (no samples were excluded in analysis). Data presented as the average ± SEM. An unpaired t-test was performed: **, $p < 0.01$.

3. Discussion

We have shown, for the first time, that oral dosing with AQU-118 attenuates mechanical allodynia and reduces caspase-3 protein levels in the DRG, likely via inhibition of MMP-2/-9. These findings are significant because this is the first analysis of the molecular and biochemical factors that mediate the behavioral effects seen with oral dosing with AQU-118, including the significantly reduced mechanical allodynia in both the rat SNL and CCI-IoN models of neuropathic pain as well as in the naloxone precipitated morphine withdrawal rodent model [14,15]. In this study we specifically looked at changes in biomarker levels after SNL surgery and then after dosing with AQU-118. We confirmed that after SNL surgery there is an increase in mechanical allodynia (Figure 1), which corresponds to an elevation in both the mRNA MMP-2 expression levels (Figure 3A) and caspase-3 protein levels (Figure 7A) in the DRG as compared to the sham. Once per day (SID) oral dosing with AQU-118 produces a statistically significant decrease in mechanical allodynia (Figure 1), which corresponds to a decrease in the caspase-3 protein levels (Figures 8 and 9) as compared to the vehicle group.

The significant increase in MMP-2 but not MMP-9 mRNA expression levels, 25 days after SNL surgery, is consistent with the work of Ji and coworkers. They found that MMP-9 was upregulated in the DRG soon after SNL but declined several days later; this decline in MMP-9 was soon followed by a sharp increase in MMP-2 that remained elevated up to 21 days after SNL-surgery [12]. Our observation of an apparent linear correlation between the MMP-2 mRNA expression levels and that of the expression level of IL-1 β but not IL-6 (Figure 5A,B) is also consistent with the hypothesis that MMP-2 induces neuropathic pain symptoms through activation of IL-1 β and astrocyte activation [12]. However, one would also expect a concomitant elevation in the level of cleaved IL-1 β protein in the DRG which we did not observe. Instead, a significant decrease in the level of cleaved IL-1 β was observed in the ipsilateral DRG of the vehicle group as compared to the sham. (Figure 6A,B). Also, while there was an up-regulation in mRNA IL-6 in the DRG (Figure 4B), no IL-6 protein was detected [19]. Repeated attempts were made to measure IL-6 using multiple kits and reagents; while IL-6 protein was easily detected using commercially available protein standards, no IL-6 protein was measured in the DRG samples [19]. Although protein levels cannot reliably be predicted from mRNA levels, the absence of detectable IL-6 protein in the rat DRG samples in this study was surprising [20]. Several explanations could account for this result. The levels of IL-6 in the tested samples may have been below the limits of detection of the assays due to low levels of translation. Alternatively, cytokine-producing cells (i.e., glial, astrocytes & Schwann cell) are not evenly distributed throughout the DRG, which would produce variable results in quantification or detection of proteins depending upon the specific sampling location [21].

Another protein that decreased in the DRG after SNL surgery was MBP. This is consistent with the work of Shubayev and coworkers who observed a decrease in both the higher and lower molecular weight bands of MBP after L5 spinal nerve crush (SNC) [13]. They also observed that the decrease in MBP coincided with an initial increase in both MMP-9 mRNA expression and MMP-9 activity in the DRG [13]. However, while Shubayev and coworkers found that daily intraperitoneal (ip) injection of a potent broad spectrum MMP inhibitor (GM6001) was able to reverse the decrease in MBP, we saw no such reversal with oral dosing with AQU-118. Potential explanations for this disparity include the possibility that the amount of MMP inhibitor required for reversal of MBP breakdown may be higher than what was used in our current study, or that there is a significant time delay before a reversal in MBP levels occur.

While we did not observe a reversal in MBP decline after oral dosing with AQU-118, we did observe a decrease in caspase-3 protein levels in the DRG that coincided with a decrease in mechanical allodynia. This is similar to what was observed by Shubayev and coworkers, who measured a significant decrease in the level of caspase-3 in the DRG at the same time as a decrease in mechanical allodynia was noted following daily injections of GM6001 [13]. Given the fact that we observe similar decreases in both mechanical allodynia and caspase-3 protein levels, but no reversal in the levels of MBP, it is possible that either there is a significant delay between the reduction of caspase-3/apoptosis and remyelination, or that MBP breakdown is not directly connected to a caspase-3 dependent apoptotic pathway.

Previous studies have observed a strong correlation between mechanical allodynia and neuronal apoptosis [18,22–25]. For example, Sekiguki and co-workers reported that the level of mechanical allodynia observed within various surgically induced rodent models (nerve root compression versus spinal nerve compression) of neuropathic pain correlated with the level of DRG neuronal apoptosis [18]. It is well known that caspase-3 functions as the key executioner of apoptosis in neuronal cells [26–28]. Inhibition of caspase 3 has been found to reduce pain behaviors in rodent models of neuropathic pain [23–25]. Additionally, hypoxia-reoxygenation, which induces apoptosis in brain endothelial cells, also results in an increase in the expression of MMP-2, MMP-9, and caspase-3, and when these cells are exposed to a non-selective MMP inhibitor the levels of both caspase-3 and apoptosis significantly decrease [29]. This phenomenon has also been observed more recently in an in-vitro and in-vivo rat model of traumatic brain injury (TBI), supporting the idea that inhibition of MMP-2 reduces neuronal apoptosis, at least in part, through decreasing caspase-3 levels or activity [30]. These studies reinforce our findings that inhibition of MMP -2/-9 via AQU-118 reduces caspase 3 protein levels and nerve damage-induced mechanical allodynia. Clearly more studies are required to determine what intermediaries are involved in AQU-118's ability to attenuate mechanical allodynia via caspase-3. The objective of this research is to identify biomarkers that can be used for monitoring the effects of AQU-118 in the clinic as well as to potentially screen patients with neuropathic pain who would be most likely to benefit from treatment with AQU-118.

4. Materials and Methods

4.1. Testing of Orally Administered AQU-118 in the SNL-Rat Model of Neuropathic Pain

4.1.1. Animals

The use of animals was approved by the Institutional Animal Care and Use Committee (IACUC) at PsychoGenics, Inc. (IACUC 243_0214, Pain in rats, approved 9 May 2014 and IACUC 234_0314 Surgical procedures addendum for rats, approved 24 March 2014). A total of 90 male Sprague Dawley rats (200–225 g) were obtained from Envigo (Indianapolis, IN, USA). Upon receipt, rats were assigned unique identification numbers and were group housed with 3 rats per cage in polycarbonate cages with micro-isolator filter tops. All rats were examined, handled, and weighed prior to initiation of the study to assure adequate health and suitability. During the course of the study, 12/12 light/dark cycles were maintained. Rodent chow and water were provided *ad libitum* for the duration of the study.

Spinal Nerve Ligation (SNL) Chung Model [31,32]: Under general anesthesia with continuous inhalation of isoflurane, surgery was performed with aseptic procedures in a surgery unit. Sterile ophthalmic ointment was used to lubricate the eyes. Animals were observed continuously for the level of anesthesia, testing for the animal's reflex response to a tail or paw pinch and closely monitoring the animal's breathing. A heating pad was used to maintain body temperature at 37 °C while the animals recovered from anesthesia. The skin at the area of the lower lumber and sacral level of the rat was shaved and disinfected with betadine and alcohol. A left longitudinal incision at the level next to the vertebral column was made, and the left paraspinal muscles were separated. The transverse process of L6 was removed and nearby connective tissue cleaned to expose L5 and L6 spinal nerves [31,32]. After the nerves were isolated and clearly visualized, 4-0 silk threads were used to ligate the left L5. The muscles were sutured with 4-0 silk threads and the wound closed by staples. All rats received an analgesic (Carprofen, 5 mg/kg, subcutaneous, s.c.) immediately before and the next day after surgery. Each rat was monitored until awake and moving freely around the recovery chamber. Animals were then single-housed for the duration of the study.

4.1.2. Von Frey Test for Mechanical Allodynia

Withdrawal from a mechanical stimulus was measured by applying von Frey (VF) filaments (Stoelting, Wood Dale, IL, USA) of ascending bending force to the plantar surface of the hind paws, ipsilateral and contralateral to the surgical manipulation. Filaments ranged from 0.69 to 26 g (0.692, 1.202, 1.479, 2.041, 3.63, 6, 8, 10, 15 and 26). Each filament was applied 3 times to determine withdrawal. A positive response was defined as withdrawal from the von Frey filament. Confirmation of the paw withdrawal threshold (PWT) was tested by assessing the response to the filament above and below the withdrawal response. Rats were brought to the experimental room and allowed to habituate in the room for one hour prior to testing, and acclimated to the observation chambers for 15 minutes prior to taking PWT measurements.

4.2. Pre-Operative Baseline Testing

Prior to surgery, all rats were tested using the VF test. Rats that had an ipsilateral PWT of less than 12 g were excluded from the study.

4.3. Post-Operative Testing

Two weeks following surgery, baseline VF responses were taken and animals were balanced and assigned to treatment groups based on their post-operative PWT values. Animals with a VF score over 4.5 g were excluded from the study. The animals were divided into four groups. One group (n = 20) which only had sham surgery was assigned as the sham group (Table 1). The next three groups consisted of those rodents that had undergone SNL surgery and were assigned as the vehicle group (n = 40), AQU-118 group (n = 20) and positive control (i.e., gabapentin) group (n = 10) (Table 1).

Rats were dosed with either AQU-118 of vehicle (via oral gavage once per day for 5 days. Gabapentin was administered (via oral gavage) only on von Frey test days (Day 1, 3, and 5). Von Frey PWT values were measured at 60 min following compound administration (Table 1).

4.3.1. Compound Formulation/Administration

AQU-118 (160 mg/kg, 0.34 mmoles/kg, once per day dosing, SID) was dissolved in 0.5% methylcellulose (400 centipoise, cps) to give a fine suspension and administered orally (P.O.). for 5 days, 60 min prior to VF testing, at a dose volume of 5 mL/kg. AQU-118 was prepared fresh daily. The vehicle (0.5% methylcellulose,400 cps) was administered at a dose volume equivalent to the test compound (5 mL/kg). Gabapentin (100 mg/kg, 0.58 mmoles/kg, SID; Toronto Research Chemicals, North York, ON, Canada) was dissolved in saline and administered P.O. on Days 1, 3, and 5 of the treatment periods, 60 min prior to VF testing, at a dose volume of 1 mL/kg. Gabapentin was prepared daily.

4.3.2. Tissue Collection & Preparation

After VF test on Day 5, rats were anesthetized by CO_2. An incision was made in the right atrium and an automated pump was used to perfuse the animal with ice cold PBS. The spinal cord (L4–L6) and DRGs (L4, L5, and L6) from both sides (into separate left and right tubes) were collected, flash frozen, and stored in a −80 °C freezer until analysis. Frozen DRG tissue samples were homogenized in RIPA buffer (75 µL per sample) using a handheld pestle homogenizer, 30 s on ice. Samples were cleared by centrifugation at 14,000 rpm for 15 min at 4 °C. Supernatants were collected and protein amount was determined using Bio-Rad DC protein kit (Biorad Laboratories, Hercules, CA, USA). For DRG samples that were analyzed by both qPCR and Western, frozen DRG samples were crushed in mini liquid nitrogen cooled mortar. Ten percent of the tissue powder was used in RNA/transcript analysis. The remaining tissue was lysed in RIPA buffer and processed as indicated above.

4.4. RNA and cDNA Preparation for qPCR

Sample preparation was performed via a modification of a published literature procedure [33]. DRG tissue samples were homogenized 2 × 1 min at 25 Hz in 750 µL of QIAzol Lysis Reagent with Tissue Lyser (Qiagen, Valencia, CA, USA) and 5 mm stainless steel beads (Qiagen). Disrupted samples were incubated at room temperate for 5 min. For RNA extraction, manufacturer protocol for RNeasy 96 Universal Tissue Kit (Qiagen) for RNA isolation was followed. Briefly, 150 µL of Chloroform was added and samples were shaken vigorously for 15 s followed by 3 min incubation at room temperature. The aqueous phase was separated from the organic phase by centrifugation at 6000× *g* at 4 °C for 15 min. The aqueous phase was then transferred to a new 96-well block and total RNA was precipitated with equal volume of 70% ethanol and then transferred to a RNeasy 96-well plate followed by centrifugation at 6000× *g* at room temperate for 4 min. Total RNA bound to column membranes was treated with RNase-Free DNase set (Qiagen) for 30 min, followed by 3 washing steps with RW1 and RPE buffers (provided with RNeasy 96 Universal Tissue Kit). RNA was eluted with 20 µL RNase-Free water.

RNA was quantified using Nanodrop 8000. Total RNA (0.5 µg of RNA) was reverse transcribed into cDNA with 3.2 µg random hexamers (Roche Applied Science, Indianapolis, IN, USA), 1 mM each dNTP (Roche Applied Science), 20U Protector RNase Inhibitor (Roche Applied Science), 1xTranscriptor Reverse Transcription reaction buffer and 10U Transcriptor Reverse Transcriptase (Roche Applied Science) in 20 µL total volume. The reactions were allowed to proceed at room temperature for 10 min, 55 °C for 30 min and were then inactivated at 85 °C for 5 min in a GeneAmp PCR Systems 9700 thermal cycler (Applied Biosystems, Foster City, CA, USA). cDNA samples were diluted 10-fold with RNase-Free water for qPCR analysis.

4.5. qPCR Analysis

All qPCR reagents and TaqMan Expression Assays were purchased from ThermoFisher Scientific (Waltham, MA, USA). Briefly, 5 μL of diluted cDNA was amplified with qPCR primer and probe sets in 1x TaqMan Fast Advanced Master Mix in 20 μL final reaction volume. Reactions were run on Applied Biosystems 7900HT Fast Block System with the following parameters: 95 °C for 20 s; 40 cycles of (95 °C/3 s; 60 °C/30 s). Each cDNA sample was run in triplicate. The Ct value for serially diluted pooled (SNL) cDNA was plotted against the log value of dilution factor and the slope of the linear regression was determined. Only assays with r2 values greater than 0.95 were used in this study. PCR efficiency was calculated as follows:

$$\text{PCR Efficiency} = 10 - 1/\text{slope}$$

The relative mRNA expression for each target was calculated following the method of Pfaffl [34]:

$$\Delta Ct_{sample} \text{ for Target} = \text{Ct Target control} - \text{Ct Target sample}$$

$$\Delta Ct_{sample} \text{ for GAPDH} = \text{Ct GAPDHcontrol} - \text{Ct GAPDH sample}$$

$$\text{Normalizing Target to GAPDH for each sample} = \text{PCR Efficiency Target}^{(\Delta \text{Ctsample for Target})} / \text{PCR Efficiency GAPDH}^{(\Delta \text{Ctsample for GAPDH})}$$

Normalized target expression was expressed relative to the average of the sham group.

4.6. Western Blotting

The procedure was performed following a published literature procedure [35]. IL-1β antibodies were purchased from three different vendors and validated using known standards of which only the antibody purchased from EMD Millipore passed and was used in the study. IL-6 antibodies were purchased from three different vendors of which two passed and were used in the study (R & D Systems and ThemoFisher) [19]. Caspase-3 antibody was purchased from Santa Cruz Biotechnology and MBP was purchased from Santa Cruz Biotechnology, both of which were validated and were used in the study. GAPDH was purchased from Cell Signalling Technology. Protein samples were denatured in Laemmli buffer/2-mercaptoethanol for 5 min at 95 °C. Denatured protein samples were separated by SDS-PAGE. After electrophoresis, proteins were transferred from gel to LFP-PVDF membranes by electroblotting. Non-specific binding of antibodies was blocked with 5% *w*/*v* dried milk in 1× TBST for one hour. After a brief rinse in TBST, the blots were probed with primary antibody (each diluted at 1:1000 in 1% *w*/*v* milk with 1× TBST) at 4 °C overnight. Protein-PVDF blots were washed once for 15 min followed by 3 more 5-min washes with TBST. Protein-PVDF blots were then incubated with the appropriate secondary antibody (diluted in 1:1000 prepared with 1% milk in TBST) for 1 h at room temperature. Protein-PVDF blots were washed once for 15 minutes followed by one more wash for 5 min. Antibody binding was detected using the ECL Plus Western Chemifluorescence Detection Kit (ThermoFisher Scientific). The detection solution was made fresh according to manufacturer's directions and dispensed onto membranes. After 5 min incubation, the protein-PDVF membranes were scanned using Typhoon 9410 scanner (GE Healthcare Bioscience, Pittsburgh, PA, USA) using 457 nm blue laser for excitation and 520 nm emission filter at 400 V. The scanned images from the Typhoon were analyzed with ImageQuant TL software version 7.0 (GE Healthcare Bioscience, Piscataway, NJ, USA). Band intensities were determined using the Rolling Ball method. Each protein target was first normalized to the in-lane housekeeping protein GAPDH (from the same gel). The normalized protein target for each sample was presented as a ratio relative to the sham vehicle group.

4.7. Statistical Analyses

Data at all time points post-surgery were analyzed by two-way repeated measures analysis of variance (RM ANOVA) with time as the within-subjects factor and treatment as the between-subjects factor. This was followed by Fisher PLSD post-hoc comparisons where appropriate (Appendix A). Pre-surgery baseline paw withdrawal data were analyzed by one-way ANOVA. An effect was considered significant if $p < 0.05$. Data are presented as the mean ± standard error of the mean (SEM) {* $p < 0.05$, ** $p < 0.01$, *** $p < 0.0001$}.

5. Conclusions

The results of this study demonstrate for the first time that attenuation of mechanical allodynia via the oral dosing of the dual active MMP-2/MMP-9 inhibitor, AQU-118, in the SNL rat model of neuropathic pain is likely due to the regulation of apoptotic pathways, and specifically through reduction of caspase-3. We have confirmed previous reports that 25 days after SNL surgery, mRNA levels of MMP-2 but not of MMP-9 are elevated in the DRG. We have also been able to confirm previous reports that MBP declines and caspase-3 protein levels increase 25 days after SNL surgery. This study observed a linear relationship between levels of mRNA MMP-2 and that of IL-1β but not of IL-6. While increases in MBP or that of IL-1 β were not observed after daily oral dosing with AQU-118, their participation in attenuating mechanical allodynia cannot be ruled out at this point. While identifying a protein (caspase-3) that changes at the site of nerve damage that coincides with a reduction of mechanical allodynia is a good first step in helping elucidate a mechanism of action for AQU-118, more research will need to be performed to determine if this observation is applicable to non-rodent animal models of neuropathic pain. Additionally, while our initial focus has been on AQU-118's effect on attenuating neuropathic pain, it would be of interest to see if other neurological disorders that implicate a caspase-3 mediated apoptotic pathway would benefit from the intervention of an orally active MMP-2/-9 inhibitor.

Author Contributions: A.C., A.B. and A.H. performed the surgery and/or behavioral portions of the SNL rat model. M.Y.K. performed the biomarker portion of the SNL-rat model and edited the manuscript. J.B. and J.S. performed the biomarker portion of the SNL-rat model. A.G. supervised the biomarker portion of the SNL-rat model. T.H. supervised the surgical and behavioral portions of the SNL rat model. M.F. assisted in the design of the biomarker portion of the SNL-rat model and wrote and edited the manuscript. V.R.N. chemically synthesized and purified AQU-118. I.S. conceived and chemically synthesized AQU-118, designed the behavioral and biomarker portions of the SNL-rat model and wrote the manuscript. All authors read and approved the final manuscript.

Funding: This work was supported by a Small Business Innovation Research (SBIR) grant (#1R44DE022207-02) from the National Institutes of Health (NIH)/National Institute of Dental & Craniofacial Research (NIDCR).

Conflicts of Interest: I.S. owns stock in Aquilus Pharmaceuticals, Inc. All other authors declare they have no competing interests.

Appendix A

Means Table for BASELINE Post-op
Effect: Treatment

	Count	Mean	Std. Dev.	Std. Err.
Sham + Vehicle	20	25.450	2.460	0.550
Chung + Vehicle	40	1.470	0.462	0.073
Chung + Gabapentin (100 mg/kg)	10	1.370	0.579	0.183
Chung + AQU-118 (160 mg/kg)	20	1.515	0.576	0.129

Figure A1. Statistical analysis of post-operative von-Frey (VF) baseline paw withdrawal threshold (PWT) of the ipsilateral hind paws. All Chung surgical groups showed a significant decrease in the PWT compared to the sham surgical group. No significant difference in post-operative baseline PWT was detected between the individual Chung surgical groups.

Fisher's PLSD for BASELINE Post-op
Effect: Treatment
Significance Level: 5 %

	Mean Diff.	Crit. Diff.	P-Value	
Sham + Vehicle, Chung + Vehicle	23.980	0.676	<0.0001	S
Sham + Vehicle, Chung + Gabapentin (100 mg/kg)	24.080	0.956	<0.0001	S
Sham + Vehicle, Chung + AQU-118 (160 mg/kg)	23.935	0.781	<0.0001	S
Chung + Vehicle, Chung + Gabapentin (100 mg/kg)	0.100	0.873	0.8204	
Chung + Vehicle, Chung + AQU-118 (160 mg/kg)	-0.045	0.676	0.8950	
Chung + Gabapentin (100 mg/kg), Chung + AQU-118 (160 mg/kg)	- 0.145	0.956	0.7638	

Figure A2. Fisher's protected least significant difference (PLSD) test of post-operative VF baseline PWT of the ipsilateral hind paws.

Two Way Repeated Measures ANOVA (One Factor Repetition)
Dependent Variable: VF

Source of Variation	DF	SS	MS	F	P
Group	3	26090.609	8696.870	746.529	<0.001
Rat ID(Group)	86	1001.878	11.650		
Time	4	10431.824	2607.956	318.583	<0.001
Group x Time	12	9127.455	760.621	92.916	<0.001
Residual	344	2816.027	8.186		
Total	449	55212.380	122.967		

Figure A3. Ipsilateral Two-Way RM ANOVA statistical analysis for behavioral portion (presented in Figure 1).

Two Way Repeated Measures ANOVA (One Factor Repetition)
Dependent Variable: VF

Source of Variation	DF	SS	MS	F	P
Group	3	45.509	15.170	1.066	0.368
Rat ID(Group)	86	1223.915	14.232		
Time	4	175.450	43.862	4.374	0.002
Group x Time	12	173.299	14.442	1.440	0.146
Residual	344	3449.710	10.028		
Total	449	5093.024	11.343		

Figure A4. Contralateral Two-Way RM ANOVA statistical analysis for behavioral portion (presented in Figure 2).

References and Note

1. Harden, N.; Cohen, M. Unmet needs in the management of neuropathic pain. *J. Pain Symptom Manag.* **2003**, *25*, S12–S17. [CrossRef]
2. Nightingale, S. The neuropathic pain market. *Nat. Rev. Drug Discov.* **2012**, *11*, 101–102. [CrossRef] [PubMed]
3. La Fleur, M.; Underwood, J.L.; Rappolee, D.A.; Werb, Z. Basement membrane and repair of injury to peripheral nerve: Defining a potential role for macrophages, matrix metalloproteinases, and tissue inhibitor of metalloproteinases-1. *J. Exp. Med.* **1996**, *184*, 2311–2326. [CrossRef] [PubMed]
4. DeLeo, J.A.; Colburn, R.W.; Nichols, M.; Malhotra, A. Interleukin-6-mediated hyperalgesia/allodynia and increased spinal IL-6 expression in a rat mononeuropathy model. *J. Interferon Cytokine Res.* **1996**, *16*, 695–700. [CrossRef] [PubMed]
5. Sweitzer, S.M.; Colburn, R.W.; Rutkowski, M.; DeLeo, J.A. Acute peripheral inflammation induces moderate glial activation and spinal IL-1 beta expression that correlates with pain behaviour in the rat. *Brain Res.* **1999**, *829*, 209–221. [CrossRef]
6. George, A.; Schmidt, C.; Weishaupt, A.; Toyka, K.V.; Sommer, C. Serial determination of tumor necrosis factor-alpha content in rat sciatic nerve after chronic constriction injury. *Exp. Neurol.* **1999**, *160*, 124–132. [CrossRef] [PubMed]

7. Siebert, H.; Dippel, N.; Mader, M.; Frank, W.; Bruck, W. Matrix metalloproteinase expression and inhibition after sciatic nerve axotomy. *J. Neuropathol. Exp. Neurol.* **2001**, *60*, 85–93. [CrossRef] [PubMed]
8. Schafers, M.; Sorkin, L.S.; Geis, C.; Shubayev, V.I. Spinal nerve ligation induces transient upregulation of tumor necrosis factor 1 and 2 in injured and adjacent uninjured dorsal root ganglia in the rat. *Neurosci. Lett.* **2003**, *347*, 179–182. [CrossRef]
9. Mika, J.; Korostynski, M.; Kaminska, D.; Wawrzczak-Bargiela, A.; Osikowicz, M.; Makuch, W.; Przewlocki, R.; Przewlocka, B. Interleukin-1 alpha has antiallodynic and antihyperalgesic activities in a rat neuropathic pain model. *Pain* **2008**, *138*, 587–597. [CrossRef]
10. Brkic, M.; Balusu, S.; Libert, C.; Vandenbroucke, R.E. *Friends or Foes: Matrix Metalloproteinases and Their Multifaceted Roles in Neurodegenerative Diseases, Mediators of Inflammation*; Hindawi Publishing Corp.: Cairo, Egypt, 2015; 27p.
11. Tokito, A.; Jougasaki, M. Matrix metalloproteinases in non-neoplastic disorders. *Int. J. Mol. Sci.* **2016**, *17*, 1178. [CrossRef]
12. Kawasaki, Y.; Xu, Z.Z.; Wang, X.; Park, J.Y.; Zhuang, Z.Y.; Tan, P.H.; Gao, Y.J.; Roy, K.; Corfas, G.; Lo, E.H.; et al. Distinct roles of matrix metalloproteases in the early- and late-phase development of neuropathic pain. *Nat. Med.* **2008**, *14*, 331–336. [CrossRef] [PubMed]
13. Kobayashi, H.; Chattopadhyay, S.; Kato, K.; Dolkas, J.; Kikuchi, S.; Myers, R.R.; Shubayev, V.I. MMPs initiate Schwann cell-mediated MBP degradation and mechanical nociception after nerve damage. *Mol. Cell Neurosci.* **2008**, *39*, 619–627. [CrossRef] [PubMed]
14. Henry, M.A.; Fairchild, D.D.; Patil, M.J.; Hanania, T.; Hain, H.S.; Davis, S.F.; Malekiani, S.A.; Hu, A.; Sucholeiki, R.; Nix, D.; Sucholeiki, I. Effect of a novel, orally active matrix metalloproteinease-2 and-9 inhibitor in spinal and trigeminal rat models of neuropathic pain. *J. Oral Fac. Pain Headache* **2015**, *29*, 286–296.
15. Sucholeiki, I. Compounds and Methods for the Treatment of Pain and Other Disorders. U.S. Patent PCT US2011/026848, WO 2012/118498 A1, 9 July 2012.
16. Dubovy, V.; Brazda, V.; Klusakova, I.; Svzenska, I.H. Bilateral elevation of interleukin-6 protein and mRNA in both lumbar and cervical dorsal root ganglia following unilateral chronic compression injury of the sciatic nerve. *J. Neuroinflamm.* **2013**, *10*, 824. [CrossRef] [PubMed]
17. Kim, D.S.; Figueroa, K.W.; Li, K.W.; Boroujerdi, A.; Yolo, T.; Luo, Z.D. Profiling of dynamically changed gene expression in dorsal root ganglia post peripheral nerve injury and a critical role of injury-induced glial fibrillary acetic protein in maintenance of pain behaviors. *Pain* **2009**, *143*, 114–122. [CrossRef] [PubMed]
18. Sekiguchi, M.; Sekiguchi, Y.; Konno, S.; Kobayashi, H.; Homma, Y.; Kikuchi, S. Comparison of neuropathic pain and neuronal apoptosis following nerve root or spinal nerve compression. *Eur. Spine J.* **2009**, *18*, 1978–1985. [CrossRef] [PubMed]
19. IL-6 antibodies (21 kDa) for Western blot were commercially obtained from ThermoFisher, R & D systems and Santa Cruz Biotechnology. Antibodies from R & D Systems and ThermoFisher passed validation with known controls and were then used to measure the level in DRG.
20. Gygi, S.P.; Rochon, Y.; Franza, R.B.; Aebersold, R. Correlation between protein and mRNA abundance in yeast. *Mol. Cell. Biol.* **1999**, *19*, 1720–1730. [CrossRef]
21. Nascimento, R.S.; Santiago, M.F.; Marques, S.A.; Allodi, S.; Martinez, A.M.B. Diversity among satellite glial cells in dorsal root ganglia of the rat. *Braz. J. Med. Biol. Res.* **2008**, *41*, 1011–1017. [CrossRef]
22. Whitesides, G.T.; Munglani, R. Cell death in the superficial dorsal horn in a model of neuropathic pain. *J. Neurosci. Res.* **2001**, *64*, 168–173. [CrossRef]
23. Joseph, E.K.; Levine, J.D. Caspase signaling in neuropathic and inflammatory pain in the rat. *Eur. J. Neurosci.* **2004**, *20*, 2896–2902. [CrossRef]
24. Scholz, J.; Broom, D.C.; Youn, D.-H.; Mills, C.D.; Kohono, T.; Suter, M.R.; Moore, K.A.; Decosterd, I.; Coggeshall, R.E.; Woolf, C.J. Blocking caspase activity prevents trans-synaptic neuronal apoptosis and the loss of inhibition in Lamina II of the dorsal Horn after peripheral nerve injury. *J. Neurosci.* **2005**, *25*, 7317–7323. [CrossRef] [PubMed]
25. Gradl, G.; Herlyn, P.; Finke, B.; Bierer, P.; Wree, A.; Witt, M.; Mittlmeier, T.; Vollmar, B. A pan-caspase inhibitor reduces myocyte apoptosis and neuropathic pain in rats with chronic constriction injury of the sciatic nerve. *Anesth. Analg.* **2013**, *116*, 216–223. [CrossRef] [PubMed]
26. Elmore, S. Apoptosis: A review of programmed cell death. *Toxicol. Pathol.* **2007**, *35*, 495–516. [CrossRef] [PubMed]

27. Kavanagh, E.; Rodhe, J.; Burguillos, M.A.; Venero, J.L.; Joseph, B. Regulation of caspase-3 processing by cIAP2 controls the switch between pro-inflammatory activation and cell death in microglia. *Cell Death Dis.* **2014**, *5*, e1565. [CrossRef] [PubMed]
28. Amelio, M.D.; Cavallucci, V.; Cecconi, F. Neuronal caspase-3 signaling: Not only cell death. *Cell Death Differ.* **2010**, *17*, 1104–1114. [CrossRef] [PubMed]
29. Lee, S.R.; Lo, E.H. Induction of caspase-mediated cell death by matrix metalloproteinases in cerebral endothelial cells after hypoxia-reoxygenation. *J. Cereb. Blood Flow Metab.* **2004**, *24*, 720–727. [CrossRef] [PubMed]
30. Abdul-Muneer, P.M.; Conte, A.A.; Haldar, D.; Long, M.; Patel, R.K.; Santhakumar, V.; Overall, C.M.; Pfister, B.J. Traumatic brain injury induced matrix metalloproteinase 2 cleaves CXCL12α(stromal cell derived factor 1α) and causes neurodegeneration. *Brain Behav. Immun.* **2017**, *59*, 190–199. [CrossRef]
31. Kim, S.H.; Chung, J.M. An experimental model for peripheral neuropathy produced by segmental spinal nerve ligation in the rat. *Pain* **1992**, *50*, 355–363. [CrossRef]
32. Chung, J.M.; Kim, H.K.; Chung, K. Segmental spinal nerve ligation model of neuropathic pain. *Methods Mol. Med.* **2004**, *99*, 35–45.
33. Menalled, L.B.; Kudwa, A.E.; Miller, S.; Fitzpatrick, J.; Watson-Johnson, J.; Keating, N.; Ruiz, M.; Mushlin, R.; Alosio, W.; McConnell, K.; et al. Comprehensive behavioral and molecular characterization of a new knock-in mouse model of Huntington's disease: zQ175. *PLoS ONE* **2012**, *7*, e49838. [CrossRef]
34. Pfaffl, M.W. A new mathematical model for relative quantification in real-time RT–PCR. *Nucleic Acids Res.* **2001**, *29*, e45. [CrossRef] [PubMed]
35. Menalled, L.B.; Kudwa, A.E.; Oakeshott, S.; Farrar, A.; Paterson, N.; Filippov, I.; Miller, S.; Kwan, M.; Olsen, M.; Beltran, J.; et al. Genetic deletion of transglutaminase 2 does not rescue the phenotypic deficits observed in R6/2 and zQ175 mouse models of Huntington's disease. *PLoS ONE* **2014**, *9*, e995. [CrossRef] [PubMed]

International Journal of
Molecular Sciences

MDPI

Article

Matrix Metalloproteinase-3 is Key Effector of TNF-α-Induced Collagen Degradation in Skin

Ursula Mirastschijski [1], Blaž Lupše [1], Kathrin Maedler [1], Bhavishya Sarma [1], Arlo Radtke [2], Gazanfer Belge [2], Martina Dorsch [3], Dirk Wedekind [3], Lisa J. McCawley [4], Gabriele Boehm [5], Ulrich Zier [1], Kazuhiro Yamamoto [6], Sørge Kelm [1] and Magnus S. Ågren [7,*]

1 Center for Biomolecular Interactions Bremen, Department of Biology and Biochemistry, University of Bremen, 28359 Bremen, Germany; mirastsc@uni-bremen.de (U.M.); blazpikalupse@gmail.com (B.L.); kmaedler@uni-bremen.de (K.M.); bhavishyasarma@yahoo.co.in (B.S.); u.x.zier@gmail.com (U.Z.); skelm@uni-bremen.de (S.K.)

2 Faculty of Biology and Biochemistry, University of Bremen, 28359 Bremen, Germany; arlo.radtke@uni-bremen.de (A.R.); belge@uni-bremen.de (G.B.)

3 Institute for Laboratory Animal Science, Hannover Medical School, 30625 Hannover, Germany; Dorsch.Martina@mh-hannover.de (M.D.); Wedekind.Dirk@mh-hannover.de (D.W.)

4 Department of Biomedical Engineering, Vanderbilt University, Nashville, TN 37232-6840, USA; lisa.mccawley@Vanderbilt.edu

5 Department of General, Visceral and Oncologic Surgery, Klinikum Bremen-Mitte, 28177 Bremen, Germany; Gabriele.Boehm@gesundheitnord.de

6 Institute of Ageing and Chronic Disease, University of Liverpool, Liverpool L69 3BX, United Kingdom; kazuhiro.yamamoto@liverpool.ac.uk

7 Digestive Disease Center and Copenhagen Wound Healing Center, Bispebjerg Hospital, University of Copenhagen, 2400 Copenhagen, Denmark

* Correspondence: magnus.agren@mail.dk; Tel.: +45-3863-5954

Received: 10 October 2019; Accepted: 17 October 2019; Published: 22 October 2019

Abstract: Inflammatory processes in the skin augment collagen degradation due to the up-regulation of matrix metalloproteinases (MMPs). The aim of the present project was to study the specific impact of MMP-3 on collagen loss in skin and its interplay with the collagenase MMP-13 under inflammatory conditions mimicked by the addition of the pro-inflammatory cytokine tumor necrosis factor-α (TNF-α). Skin explants from MMP-3 knock-out (KO) mice or from transgenic (TG) mice overexpressing MMP-3 in the skin and their respective wild-type counterparts (WT and WTT) were incubated ex vivo for eight days. The rate of collagen degradation, measured by released hydroxyproline, was reduced ($p < 0.001$) in KO skin explants compared to WT control skin but did not differ ($p = 0.47$) between TG and WTT skin. Treatment with the MMP inhibitor GM6001 reduced hydroxyproline media levels from WT, WTT and TG but not from KO skin explants. TNF-α increased collagen degradation in the WT group ($p = 0.0001$) only. More of the active form of MMP-13 was observed in the three MMP-3 expressing groups (co-incubation with receptor-associated protein stabilized MMP-13 subforms and enhanced detection in the media). In summary, the innate level of MMP-3 seems responsible for the accelerated loss of cutaneous collagen under inflammatory conditions, possibly via MMP-13 in mice.

Keywords: extracellular matrix; inflammation; cytokines; proteinases; interstitial collagens

1. Introduction

More than half a century ago, Gross and Lapière discovered collagenase-1, the first matrix metalloproteinase (MMP) [1]. Since then, research on MMP has yielded vast information on these zinc-dependent endopeptidases [2,3]. Expanding knowledge on MMP function and its impact on physiological and pathological processes has increased the interest in MMP substrates and has fueled further research [4]. Stromelysin-1 (MMP-3) belongs to the MMP family and is induced during development [5], wound repair [6], inflammation [7,8] and cancer [9]. Apart from cleaving extracellular matrix proteins, e.g., E-cadherin, laminins and type IV collagen, MMP-3 activates cytokines, growth factors [3,4] and other MMP members, e.g., the collagenases MMP-1, MMP-8 and MMP-13 and gelatinase B (MMP-9) [3]. In contrast to the collagenases MMP-1, MMP-8 and MMP-13, which are the main effectors of type I collagen degradation during wound repair, MMP-3 is incapable of cleaving native type I collagen. MMP-3 has been implicated in many inflammatory diseases, e.g., ultraviolet-B irradiation and photoaging [10], arthritis [11], after lung injury [8], vascular disease [12] and intestinal inflammation [13]. MMP-3-deficient mice show diminished inflammatory responses to various stimuli [14,15] and reduced cutaneous wound contraction [16]. Pro-inflammatory tumor necrosis factor-α (TNF-α) is an important mediator of inflammatory processes. TNF-α induces a variety of MMPs, e.g., MMP-1, MMP-3, MMP-8, MMP-9 and MMP-13 [4]. Apart from TNF-α converting enzyme (TACE) [17], MMP-3 was shown to activate TNF-α [18,19], yielding a 17 kDa peptide [20]. Activated TNF-α induces the downstream expression and secretion of MMP-3 [21,22]. In our previous study, human skin explant cultures challenged with TNF-α reacted with a concomitant and increased MMP-1, MMP-3 and collagen degradation. In this human model, we hypothesized that the increased collagen degradation was due to the activation of the collagenase MMP-1 by MMP-3 [23].

The primary aim of this project is to study the lack of MMP-3 and the effect of excessive MMP-3 on collagen degradation in skin. For this purpose, we incubate skin explants from MMP-3 knock-out [24] and MMP-3 transgenic mice [25]. Under the applied culture conditions, collagen catabolism dominates and the synthesis of new interstitial collagens is minimal [23]. MMP-1 is not expressed in mice but MMP-13 is thought to be the main collagenase and functional homologue of human MMP-1 [26]. Therefore, we study the interplay between MMP-3 and MMP-13. To enhance the extracellular detection of MMP-13, its endocytosis is blocked by adding the receptor-associated protein (RAP). Finally, the responsiveness of MMP-3-deficient or MMP-3-overexpressing murine skin to an inflammatory stimulus provided by TNF-α is studied. The broad-spectrum MMP inhibitor GM6001 is included in some experiments to elucidate the origin of collagen-degrading enzymes.

2. Results

2.1. MMP-3 Expression in the Skin of the Four Murine Genotypes

First, we quantified the MMP-3 mRNA (Figure 1A) and protein (Figure 1B) in the native skin of the four groups of mice. MMP-3 mRNA and protein were undetectable in the skin of MMP-3-deficient mice (KO). MMP-3 mRNA (p = 0.016) and protein levels were higher in the skin of transgenic (TG) animals compared with their wild-type (WTT) controls. WT and WTT contained similar MMP-3 mRNA and protein (approximately 0.2–0.3 ng) levels.

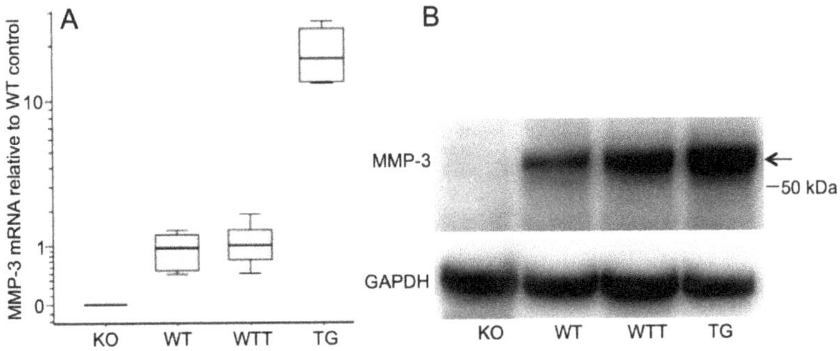

Figure 1. MMP-3 expression in native mouse skin. (**A**) MMP-3 mRNA levels determined by qPCR analysis. Boxes represent the 25th–75th percentile, whiskers the 5th–95th percentile and the horizontal lines within the boxes indicate the median values. (**B**) Representative western blot of six independent experiments of MMP-3 protein expression. Loading was normalized to the total protein (30 µg/lane) of the tissue extracts. GAPDH indicates the loading/transfer of proteins to the PVDF membrane. The arrow indicates the latent MMP-3 at 56 kDa. The position of the 50 kDa molecular weight marker is indicated to the right. KO (BL10), $n = 5$; WT (BL10), $n = 5$; WTT (FVB), $n = 3$; TG (FVB), $n = 4$.

2.2. Collagen Degradation in Incubated Skin Explants of the Four Murine Genotypes

We measured hydroxyproline in the media as an indicator of cleaved collagen molecules in the skin. MMP-3-deficient (KO) skin explants (BL10 genetic background) released a total amount of 90 µg hydroxyproline into the media over the eight-day incubation period, WT (BL10) skin 188 µg, WTT (FVB) skin 304 µg and MMP-3-overexpressing TG (FVB) skin 285 µg hydroxyproline (Table 1). Notably, the accumulated hydroxyproline release was higher in FVB WTT (304 µg) versus BL10 WT (188 µg) mice. Regression analysis was carried out to assess the rate of collagen degradation over the eight-day culture period and showed significant differences in the total hydroxyproline released between KO and WT ($p < 0.001$) skin (Table 1 and Figure 2A) but not between WTT and TG ($p = 0.47$) skin (Table 1 and Figure 2B).

Table 1. Hydroxyproline (µg) in conditioned media over eight days of culture (mean ± SEM).

Day	KO ($n = 10$)	WT ($n = 10$)	WTT ($n = 9$)	TG ($n = 10$)
0	0	0	0	0
2	3.9 ± 0.2	4.3 ± 0.5	6.8 ± 1.7	5.0 ± 1.0
4	29.9 ± 3.5	49.3 ± 4.7	94.3 ± 12.0	74.4 ± 10.5
6	32.4 ± 3.9	74.9 ± 8.4	79.0 ± 5.7	92.4 ± 11.2
8	24.1 ± 3.6	59.0 ± 5.5	124.4 ± 16.7	113.2 ± 11.8
Accumulated	90.3	187.5	304.5	285.0
Regression analysis				
Slope	3.16	9.49	16.9	17.1
p	<0.001		0.47	

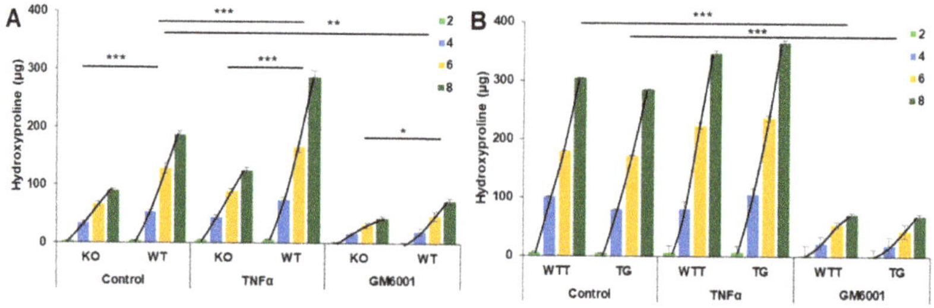

Figure 2. The effect of the MMP-3 genotype (control), TNF-α treatment and GM6001 treatment on collagen degradation in skin measured by the release of hydroxyproline-containing peptides into media from KO and WT (**A**), and WTT and TG (**B**) skin explants incubated over eight days. The timely progression of hydroxyproline release was assessed by regression analysis of the slopes. Mean ± SEM (bars). * $p < 0.05$, ** $p < 0.01$, *** $p < 0.001$. The number of animals (n) used for theses analyses is given in Table 2. TNF-α (10 ng/mL). GM6001, broad-spectrum MMP inhibitor (10 µM). KO, MMP-3 knock-out; WT, wild-type control to KO; WTT, wild-type to transgenic mice (TG); TG, MMP-3 overexpression in skin. Light green bars, day 2; blue bars, day 4; yellow bars, day 6; dark green bars, day 8.

Functional MMP-3 activity was demonstrated by casein zymography in a conditioned medium from skin explants of WT mice but not from skin explants of KO mice (Figure 3A). Tissue extracts of incubated skin explants produced the two characteristic and specific $\frac{3}{4}$ and $\frac{1}{4}$ type I collagen fragments by the action of tissue collagenases. TIMP-1 blocked this endogenous collagenolytic activity. Furthermore, incubation with the organomercurial APMA partly compensated for the lack of MMP-3 (Figure 3B). Consequently, we next wanted to identify the collagenase(s) responsible for the observed collagenolytic activity.

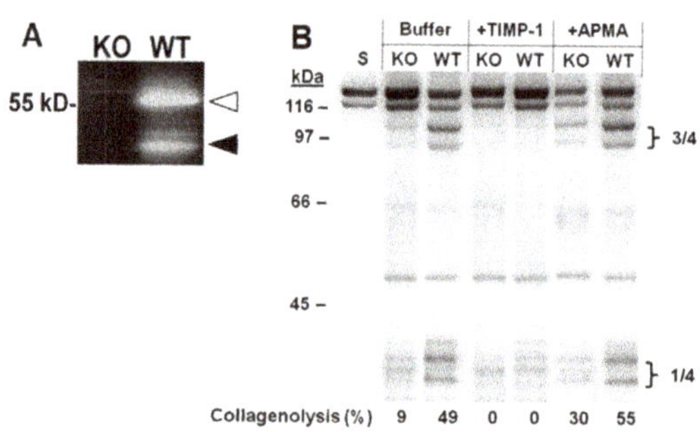

Figure 3. MMP-3 (**A**) and collagenase (**B**) activities. (**A**) Casein zymography of conditioned media collected after eight days of incubation of KO and WT skin explants. Addition of the selective MMP-3 inhibitor UK370106 (1 µM) during incubation of the casein containing SDS-PAGE gels abolished the lysis bands at the arrowheads observed in the control WT media. Open arrowhead, latent MMP-3; solid arrowhead, active MMP-3. (**B**) Type I collagenolytic activity of tissue extracts of skin explants incubated for eight days. Collagenase activity of the extracts was determined in the absence (buffer) or in the presence of rhTIMP-1 (200 nM) or APMA (1 mM). KO, MMP-3 knock-out; WT, wild-type control to KO.

2.3. *MMP-13 in MMP-3-Deficient and MMP-3-Overexpressing Conditions*

Only true tissue collagenases are capable of cleaving native type I collagen. In rodents, these are MMP-1a [27], MMP-8 and MMP-13 [26]. MMP-3 indirectly contributes to collagenolysis via the activation of collagenases [28]. Here, neither MMP-1a nor MMP-8 was detected by western blot analysis. The main collagenase responsible for tissue homeostasis in rodents is also claimed to be

MMP-13 [26]. Therefore, we focused on MMP-13 by determining the mRNA and protein levels in native and incubated skin.

In native skin, MMP-13 mRNA levels were similar in the KO, WT and WTT groups, while MMP-13 transcripts were undetectable in TG skin (Figure 4A). After eight days of incubation, MMP-13 mRNA levels differed among the groups with more ($p = 0.03$) MMP-13 mRNA present in KO skin compared with WTT skin (Figure 4B).

The MMP-13 protein in conditioned media was analyzed by western blot analysis (Figure 4C). After four days of culture, bands corresponding to latent (57 kDa) and active (48 kDa) MMP-13 were fainter in KO compared with WT. The catalytic domain of MMP-13, represented by the band at approximately 29 kDa, was prominent in WT but not in KO. Overall, the expression of MMP-13 was higher in WT compared to KO. In WTT, only faint bands of latent MMP-13 and of the catalytic domain fragment at 29 kDa were found. In contrast, TG-conditioned media presented prominent bands at the positions of latent and active MMP-13. Because MMP-13 detection can be jeopardized by rapid intracellular uptake via the LRP-1 receptor [29,30], the LRP-1 receptor antagonist RAP was used to inhibit MMP-13 endocytosis [31]. The presence of RAP during incubation yielded similar MMP-13 bands but fewer intermediate MMP-13 forms and more of the active MMP-13 compared to the controls (no RAP). After six and eight days of culture, no or minimal amounts of MMP-13 were detected in conditioned media.

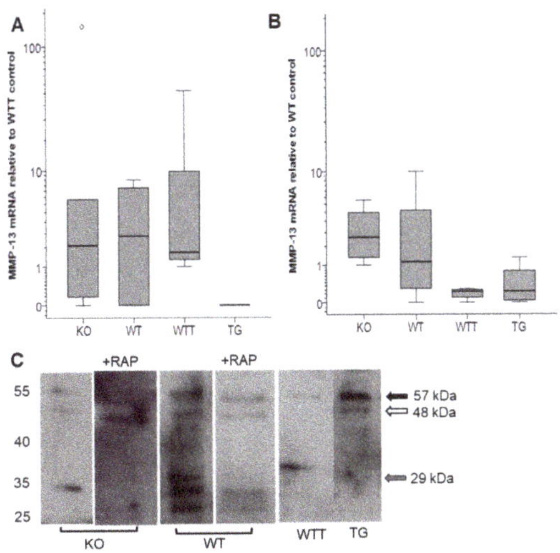

Figure 4. MMP-13 mRNA (**A**,**B**) and protein expression (**C**) in MMP-3 deficiency and overexpression. (**A**,**B**) MMP-13 mRNA expression in (**A**) native skin and (**B**) in incubated skin after eight days of culture. Boxes represent the 25th–75th percentile, whiskers the 5th–95th percentile and the horizontal lines within the boxes indicate the median values. (**C**) MMP-13 protein in conditioned media after four days of the incubation of skin explants analyzed by western blot. Molecular weights in kDa are indicated to the left. The black arrow to the right indicates the position of latent MMP-13 (57 kDa), the white arrow indicates the active MMP-13 (48 kDa) and the grey arrow to the right indicates the fragment containing the catalytic domain (29 kDa). The band below the 29 kDa band represents the C-terminal domain of MMP-13 [30]. The displayed western blot is representative of four animals in each group. The loading of samples was calculated relative to a pool of day four-conditioned media from all control samples (50 μL of each sample were mixed). The reference pool was run together with the samples on 10% SDS-PAGE gels that were stained with Coomassie blue. The density of the common band at approximately 70 kDa was used for normalization (Supplementary Figure S1). KO, MMP-3 knock-out; WT, wild-type control to KO; WTT, wild-type to transgenic mice (TG); TG, MMP-3 overexpression in the skin. +RAP, addition of receptor-associated protein (RAP).

2.4. Effect of Pro-Inflammatory TNF-α on MMP-3 Tissue Levels, MMP-13 Secretion and Collagen Degradation

We next sought to investigate MMP-3 expression under an inflammatory stimulus provided by TNF-α. Expectedly, TNF-α addition increased the presence of latent and active MMP-3 in WT skin but not in KO skin (Figure 5).

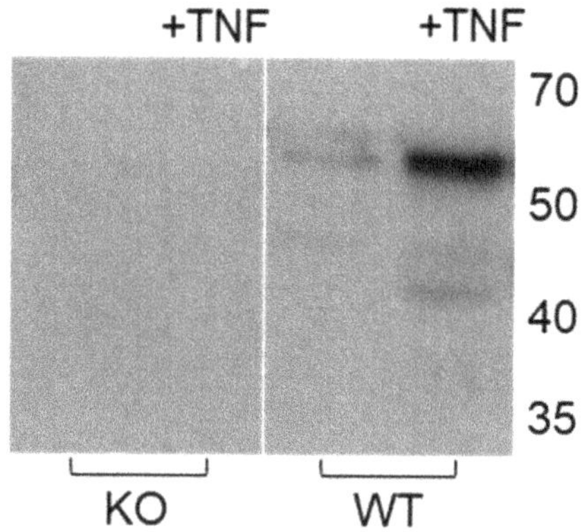

Figure 5. Effect of TNF-α on MMP-3 tissue levels in KO and WT skin after eight days of incubation in the absence or presence of TNF-α (+TNF) analyzed by western blot. Loading was normalized to the total protein (10 μg/lane) of the tissue extracts. Molecular weights in kDa are indicated to the right. Positions of protein markers in kDa are to the right. KO (n = 4), MMP-3 knock-out skin; WT (n = 4), wild-type control to KO skin.

Moreover, TNF-α increased the secretion and activation of latent MMP-13 in the conditioned media of WT and KO after four days of culture with more fragments and less latent MMP-13 in WT and more latent MMP-13 in KO (Figure 6A). In contrast to the effect of TNF-α on skin from these animals with BL10 background, only mild effects were observed in the mice with FVB background (WTT and TG). TNF-α treatment increased the fragmentation of MMP-13 from incubated WTT skin explants and active MMP-13 at 48 kDa from incubated TG skin explants (Figure 6B).

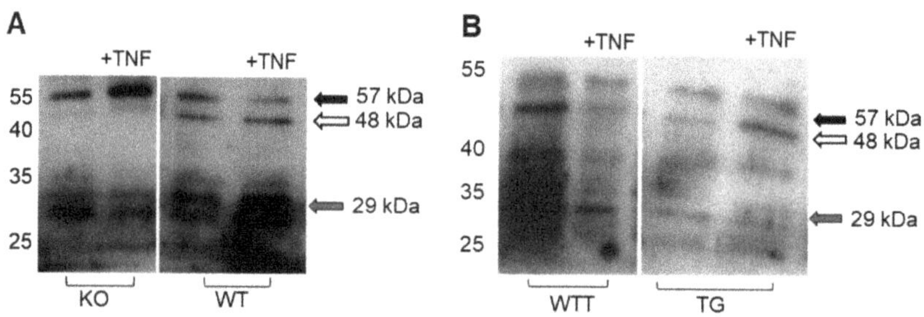

Figure 6. Effect of TNF-α on MMP-13 levels in conditioned media of KO and WT (**A**), and WTT and TG (**B**) skin explants incubated without or with 10 ng/mL TNF-α (+TNF) for four days and analyzed by western blot. The black arrow indicates the position of latent MMP-13 (57 kDa), the white arrow indicates the active MMP-13 (48 kDa) and the grey arrow the fragment containing the catalytic domain (29 kDa). Loading of samples was calculated relative to a pool of day four-conditioned media from all (n = 17) control samples (50 μl of each sample were mixed). The reference pool was run together with the samples on 10% SDS-PAGE gels that were stained with Coomassie blue. The density of the common band at approximately 70 kDa was used for normalization (Supplementary Figure S1). The positions of protein markers in kDa are indicated to the left. KO (n = 5), MMP-3 knock-out; WT (n = 5), wild-type control to KO; WTT (n = 3), wild-type to transgenic (TG); TG (n = 4), MMP-3 overexpression in skin.

The incubation of WT skin explants with TNF-α was translated into increased ($p = 0.0001$) collagen degradation compared with the control-treated WT skin explants over the eight-day experimental period (Figure 2A). In contrast, the TNF-α addition to KO skin explants did not increase ($p = 0.06$) collagen degradation further compared with the control. No significant differences in collagen degradation were found between TNF-α-treated WTT and TG skin explants (Figure 2B).

2.5. Effect of Pro-Inflammatory TNF-α on MMP-2 Secretion in MMP-3-Deficient and -Overexpressing Conditions

Apart from MMP-3 [32], MMP-2 plays an auxiliary role in activating latent MMP-13 with the conversion of the intermediate form into the fully activated form of MMP-13 [33]. In the day eight-conditioned media, no significant differences were found among the four groups with or without TNF-α addition, although the FVB (WTT and TG) mice tended to secrete less MMP-2 compared with the BL10 (KO and WT) mice.

3. Discussion

Pro-inflammatory MMP-3 and TNF-α are associated with many diseases that are characterized by massive tissue destruction and collagen degradation. Previously, we showed that TNF-α promoted collagenolysis in human skin via the activation of MMP-1, possibly mediated by MMP-3 [23]. Furthermore, earlier studies indicated that MMP-3 increases collagenase activity in cultured murine bone and cartilage [34,35]. To investigate if MMP-3 per se has a role in collagen catabolism, we incubated skin from MMP-3-deficient (KO) and MMP-3-overexpressing mice (TG) and their respective wild-type counterparts (WT and WTT). We found massively reduced collagen degradation and fragmentation in MMP-3-deficient skin, which could possibly be ascribed to the reduced activation of the collagenase MMP-13.

The observed effect on collagen degradation could entirely be ascribed to the action of MMP-3 because addition of the non-selective MMP inhibitor GM6001 to the WT skin cultures did not decrease collagen loss more than that of the KO group. It should be emphasized that our sample size was small and therefore the risk of making a type II error is apparent.

We isolated proteinases from the incubated murine KO skin that produced the two $\frac{1}{4}$ and $\frac{3}{4}$ fragments of native type I collagen triple helix. The extracted proteinases could be activated by APMA and completely blocked by TIMP-1. This implied that one or more collagenases degraded the interstitial collagens of the skin and that the lack of collagenase activation appeared to be an underlying mechanism for reduced collagenolysis in MMP-3 deficiency. Corroborating these data are the findings by van Meurs et al. [35], who observed the specific single cleavage site in type II collagen in cultured cartilage from wild-type mice but not from MMP-3-deficient mice.

MMP-13 is thought to be the most powerful collagenase under normal and pathologic conditions [36]. The two other rodent collagenases, MMP-1a and MMP-8, were not expressed in significant amounts in our skin explant model. Consequently, we focused on the presence and activation state of MMP-13. We could show that MMP-3-deficient skin expressed similar amounts of MMP-13 mRNA compared to controls. The detection of the secreted MMP-13 protein is difficult due to its high affinity binding to LRP [29,30]. Hence, we added the LRP-1 receptor agonist RAP [31] to block MMP-13 endocytosis. Despite this manipulation, similar amounts of MMP-13 were present in the conditioned media of KO and WT samples treated with RAP. Nevertheless, differences in the MMP-13 levels were unlikely the cause for differential collagenolysis seen in MMP-3-deficient conditions. On the other hand, activated MMP-13 molecular species were reduced in KO skin in comparison with the wild-type counterpart. Apart from MMP-3, trypsin, plasmin, cathepsin B, MMP-2, MMP-14 and possibly MMP-8 can activate latent MMP-13 [32,33]. Here, MMP-2 was most likely not involved. Despite of multiple possibilities for MMP-13 activation, MMP-3 seemed to be the main activator of MMP-13 in our ex vivo skin model and the absence of the activation function of MMP-3 was not compensated for by other proteinases.

Besides being capable of activating proMMP-13, MMP-3 can activate growth factors and cytokines, notably proIL-1β. IL-1β in turn influences MMP-3 and MMP-13 expression and secretion [37,38].

Having established that collagen loss is severely reduced under MMP-3-deficient conditions and that the activation, rather than the concentration, of MMP-13 decreases collagen in skin, we next investigated the effect of pro-inflammatory TNF-α on MMP-13 expression and collagen loss. TNF-α is known to induce the expression of MMP-3 [23,39], MMP-9 [18,40] and MMP-13 [41]. TNF-α induced MMP-3 in WT but not in KO skin explants. Moreover, TNF-α augmented the activation of MMP-13 in the WT skin explants, presumably via MMP-3 induction. The increased MMP-13 activation by TNF-α would explain the increased collagen degradation and fragmentation in WT skin explants not found in the KO group. One can speculate about the impact of immune cells under in vivo conditions. Lauridsen et al. [42] reported higher levels of MMP-3 when pericytes were activated with TNF-α and co-cultured with neutrophils. Our results may well explain the pronounced collagen degradation and tissue destruction found in highly inflammatory and chronic progressing diseases.

The dependence of collagen degradation on the genetic background of the mice (BL10 versus FVB) was intriguing. Collagenolysis was not increased in the skin of the FVB mice overexpressing MMP-3. If we assume that the MMP-13 activity is the rate-limiting factor for collagenolysis in our model, then the content of MMP-13 is critical. MMP-13 levels were seemingly higher in MMP-3-overexpressed skin. This may indicate that the activity of MMP-3 was not increased in the MMP-3 transgenic group. Many proteinases activate proMMP-3 [28], for example, mast cell-derived proteinases [43].

Wild-type skin from BL10 mice responded with a massive increase of collagen degradation, MMP secretion and activation when TNF-α was added to cultures, whereas FVB murine skin did not respond to the TNF-α challenge. A similar finding was reported by Martin et al. [44] using a breast cancer model in C57Bl/6 and FVB/N mice. MMP-mediated effects on tumor angiogenesis were only found in C57BL/6 mice, with no effect in the FVB/N strain. Therefore, the choice of experimental animals with the appropriate genetic background is pivotal with respect to the interpretation of results and translation to the human situation.

4. Materials and Methods

4.1. Animals

MMP-3-deficient mice and the corresponding wild-type mice were donated by Dr. Gary Rosenberg, University of New Mexico, Health Science Center, Albuquerque, NM, USA [24]. The animals were crossed back to a homogeneous background of BL10 (C57BL10.RIIIH2r-Mmp3tm1Mol/Ztm) mice for 10 generations. Skin from four-month-old or six-month-old homozygous BL10 MMP-3 knock-out (KO) and BL10 wild-type (WT) mice was used in this study.

Mice overexpressing MMP-3 under the keratin-5 promoter were first described by Dr. Lisa McCawley [25]. This overexpression is confined mainly to the keratinocytes of the skin. The sperm of MMP-3-overexpressing mice was used to inseminate mice of the FVB genotype. The offspring were mated to mice with a homozygous background. Backcrossing was performed for 10 generations. Skin from three-month-old FVB MMP-3-overexpressing, transgenic (TG) and FVB wild-type (WTT) mice were used.

The animals were kept in Makrolon® Type 2L cages at 21 °C, with a relative humidity of 55% ± 5% and a light cycle of 10 h/14 h. Sterilized commercial softwood granulate bedding was used (Lignocel, Altromin, Lage, Germany). Autoclaved commercial pellets (Altromin 1314, Altromin, Lage, Germany) and autoclaved water were provided ad libitum. The microbiological status was examined as recommended by FELASA (Federation of European Laboratory Animal Science Associations) and the absence of listed microorganisms was confirmed [45]. We did not test for the presence of *Rodentibacter sp., Helicobacter spp., Staphylococcus aureus* or *Klebsiella oxytoca*.

The experiments were approved by the local Institutional Animal Care and Research Advisory Committee (522-27-11/02-00/118, 10 September 2013).

4.2. Experimental Model and Treatment Groups

After decapitation, the dorsal skin from the male mice of all genotypes was shaved, disinfected and excised under sterile conditions. The skin samples (approximately 5 cm × 5 cm) were prepared within 36 h after harvesting. The skin explants were excised with a sterile 8 mm or a sterile 4 mm trephine and incubated submerged in a serum-free culture medium in 24-well culture plates (Nunc, Roskilde, Denmark) at 37 °C in a humidified atmosphere of 5% CO_2/air in a keratinocyte growth medium (KGM)-2 containing 6 mmol/L glucose, 50 ng/mL amphotericin-B, 100 μg/mL penicillin, 100 U/mL streptomycin and supplemented with 1.4 mM $CaCl_2$ [23].

For each animal genotype, i.e., KO, WT, WTT or TG (Table 2), 8 mm skin explants were incubated in 1.0 mL medium in the absence (control) or presence of rmTNF-α (Sigma-Aldrich, St. Louis, MO, USA) or the broad-spectrum hydroxamate MMP inhibitor GM6001 (Merck-Millipore, Molsheim, France).

Table 2. Experimental groups.

Group	Control	TNF-α [1]	GM6001 [2]
KO	$n = 10$	$n = 10$	$n = 5$
WT	$n = 10$	$n = 10$	$n = 5$
WTT	$n = 9$	$n = 9$	$n = 4$
TG	$n = 10$	$n = 10$	$n = 5$

[1] 10 ng/mL. [2] 10 μM.

Because MMP-13 is rapidly endocytosed intracellularly, extracellular detection of MMP-13 can be difficult. The internalization of MMP-13 is mediated by LRP-1 [29,30]. RAP is a ligand-binding antagonist for LRP-1 and inhibits proteinase endocytosis [31]. Hence, we used RAP in a second series of experiments to facilitate extracellular MMP-13 detection by blocking rapid internalization. Skin explants (4 mm) were incubated in 0.25 mL medium in the absence (control) or presence of the indicated treatments in Table 3.

Table 3. Experimental setting for the RAP (receptor-associated protein) experiment.

Group	Control		TNF-α [1]		GM6001 [2]	
	0	+RAP [3]	0	+RAP [3]	0	+RAP [3]
KO	$n = 4$	$n = 4$	$n = 4$	$n = 4$	$n = 4$	$n = 4$
WT	$n = 4$	$n = 4$	$n = 4$	$n = 4$	$n = 4$	$n = 4$
WTT	$n = 4$	$n = 4$	$n = 4$	$n = 4$	$n = 4$	$n = 4$
TG	$n = 4$	$n = 4$	$n = 4$	$n = 4$	$n = 4$	$n = 4$

[1] 10 ng/mL. [2] 10 μM. [3] 250 nM.

Conditioned media were harvested and replaced by a fresh medium every second day. After eight days of incubation, the experiments were terminated, media were harvested, and skin explants were kept at −80 °C until analysis for MMP-3 and MMP-13 mRNA, and MMP-3 protein. Conditioned media were stored at −20 °C until analysis for hydroxyproline and MMP-13 protein.

4.3. Analyses

4.3.1. Hydroxyproline Assay

Hydroxyproline levels in the conditioned media from days two, four, six and eight were measured colorimetrically [23].

4.3.2. MMP-3 and MMP-13 mRNA Determined by qPCR

RNA was isolated from skin explants (approximately 30 mg) using TRIzol reagent (Thermo Fisher Scientific, Schwerte, Germany). Total RNA (approximately 100 ng) were reverse-transcribed into cDNA and subsequently analyzed for 40 cycles in qPCR. The qPCR was conducted on a 7500 Fast Real Time PCR system (Thermo Fisher Scientific, Schwerte, Germany) with TaqMan probes for murine MMP-3 (Mm00440295_m1) and murine MMP-13 (Mm00439491_m1). Murine β-actin (Mm02619580_g1) and murine peptidylprolyl isomerase A (Mm02342430_g1) were used as housekeeping genes. Each reaction was run in triplicate and the relative quantity of MMP-3 or MMP-13 was determined using the ΔΔCT method [46].

4.3.3. Casein and Gelatin Zymography

Casein zymography (β-casein, Sigma-Aldrich, St. Louis, MO, USA) was performed as described elsewhere [23]. Mark12™ unstained protein standard (LC5677, Thermo Fisher Scientific, Schwerte, Germany) was run in a parallel lane. After electrophoresis, the gels were incubated in 50 mM Tris-HCl (pH 7.5) containing 10 mM $CaCl_2$, 1 μM $ZnCl_2$ and 0.1% Triton X-100 with or without the selective MMP-3 inhibitor UK370106 (R&D Systems, Minneapolis, MN, USA) for 72 h at 37 °C.

Gelatin zymography was used to analyze the MMP-2 content in 4× concentrated (Amicon Ultra-0.5 filter, Merck-Millipore, Darmstadt, Germany) conditioned media [23]. rmMMP-2 (924-MP, R&D Systems, Minneapolis, MN, USA) was run at 2 ng in parallel to estimate the MMP-2 content of the samples by the image analysis of digitized gels [47].

4.3.4. Collagenase Activity

Tissue extracts of skin explants that had been incubated for eight days were prepared together with 1 mM Pefabloc SC (Roche Diagnostics GmbH, Mannheim, Germany) and concentrated 4x by Amicon Ultra-0.5 [23]. The collagenolytic activity was determined by incubating the extracts with native (trypsin-resistant) type I collagen from bovine skin for 240 h at 22 °C, as described elsewhere [23]. rhTIMP-1 (R&D Systems, Minneapolis, MN, USA) was added to one set of reaction vials to a final concentration of 200 nM and APMA (Sigma-Aldrich, St. Louis, MO, USA) to another set of reaction vials to a final concentration of 1 mM. TIMP-1 or APMA were present during the entire 240 h incubation period.

4.3.5. Western Blot Analyses

Native and incubated skin explants were homogenized (Ultra-Turrax® T25 Basic, IKA Werke GmbH, Staufen, Germany) in RIPA (radioimmunoprecipitation assay) buffer or modified RIPA buffer at pH 7.4 (5 μL/mg tissue) containing 0.1 M Tris-HCl, 0.15 M NaCl, 1% Triton X-100, 0.1% SDS (sodium dodecyl sulfate) [48] and EDTA (ethylendiamintetraacetate)-free proteinase inhibitor cocktail and 1 μM pepstatin A. Homogenates were then centrifuged at 12,000× g for 10 min, and the supernatants were stored at −20 °C until analysis. Protein concentration was determined using the Pierce BCA (bicinchoninic acid) Protein Assay kit (Thermo Fisher Scientific, Schwerte, Germany).

Tissue extracts, conditioned media, molecular weight markers (RPN800E, Sigma-Aldrich, St. Louis, MO, USA) and rmMMP-3 (RPA101Mu01, Cloud-Clone Corporation, Katy, TX, USA) were electrophoresed on 10% SDS-PAGE gel under reducing conditions and electrotransferred for seven min using iBlot apparatus (Invitrogen, Carlsbad, CA, USA) onto a polyvinylidene fluoride (PVDF) membrane (Immobilon®, Merck-Millipore, Darmstadt, Germany). Membranes were blocked in 5% fat-free milk (Bio-Rad Laboratories, Hercules, CA, USA) for 1 h at room temperature and incubated with primary antibodies for 18 h at 4 °C (rabbit polyclonal anti-mouse MMP-1a antibody (1:200 dilution; 250750, Abbiotec, San Diego, CA, USA), anti-mouse MMP-3 peptide antibody [49,50], rabbit polyclonal anti-MMP-8 (1:1000 dilution; ABT38, Merck-Millipore, Darmstadt, Germany), rabbit polyclonal anti-MMP-13 antibody (1:250 dilution; ab39012, Abcam, Cambridge, UK), mouse monoclonal anti-β-actin antibody (1:40,000 dilution; A5441, Sigma-Aldrich, St. Louis, MO, USA), rabbit polyclonal anti-β-actin antibody (1:1000 dilution; #4967, Cell Signaling Technology, Leiden, The Netherlands) or rabbit monoclonal anti-GAPDH antibody (1:1000 dilution; #2118, Cell Signaling Technology, Leiden, The Netherlands)). Then, membranes were incubated for 1 h at room temperature with IgG-horseradish peroxidase secondary antibodies (Jackson ImmunoResearch, Ely, UK), and a specifically bound antibody was detected using an immunodetection kit (Amersham ECL Prime Western Blotting Detection Reagent, GE Healthcare Life Sciences, Amersham, UK), and then exposed to X-ray film (Amersham Hyperfilm ECL, GE Healthcare Life Sciences, Amersham, UK).

For the analysis of MMP-13 in conditioned media by western blot, the loading of samples was normalized from Coomassie blue-stained 10% SDS-PAGE gels because the BCA assay was not sensitive enough for these samples. On each gel, the samples were run with a reference sample composed of conditioned media from all control-treated samples (50 μL of each sample were mixed) in each specific experiment. The density of a common band at approximately 70 kDa was determined by ImageJ™ and used for the normalization of protein loading (Supplementary Figure S1).

4.4. Statistics

The Shapiro–Wilk test was used to assess for normality. A Student's t-test was used for the statistical analysis of results with normal distribution, and otherwise the Wilcoxon Rank Sum test and the Mann–Whitney U test using GraphPad Prism 8.0.2 software (GraphPad Software Inc., San Diego, CA, USA) were applied. For the comparison of hydroxyproline release over time, the slopes for two independent samples were compared after regression analysis. The Bonferroni correction was used in the cases of multiple comparisons. $p < 0.05$ was considered significant.

5. Conclusions

Collagen degradation in murine skin is MMP-3 and MMP-13 dependent, with MMP-3 being a potent activator of MMP-13. The pro-inflammatory cytokine TNF-α enhanced collagenolysis via the up-regulation of MMP-3 and increased the activation of MMP-13. Co-targeting MMP-3 in inflammatory diseases seems an appropriate measure to enhance the anti-inflammatory impact of clinically used drugs.

Supplementary Materials: Supplementary materials can be found at http://www.mdpi.com/1422-0067/20/20/5234/s1.

Author Contributions: Conceptualization, U.M., S.K. and M.S.Å.; methodology, B.L., K.M., B.S., A.R. and G.B.; software, U.Z.; validation, U.M., S.K. and M.S.Å.; formal analysis, U.M. and U.Z.; investigation, B.L., K.M., B.S., G.B., A.R., D.W. and M.D.; resources, D.W., M.D., K.Y. and S.K.; data curation, U.M., U.Z. and G.B.; writing—original draft preparation, U.M., B.L., K.M., B.S., G.B., D.W., A.R., G.B. and M.S.Å.; writing—review and editing, U.M., K.M., A.R., G.B., L.J.M. and M.S.Å.; visualization, U.M. and U.Z.; supervision, U.M., K.M., S.K., G.B. and M.S.Å.; project administration, U.M.; funding acquisition, U.M., K.M. and K.Y.

Funding: The research leading to these results has received funding from the European Research Council under the European Community's Seventh Framework Programme (FP7/2007-2013)/ERC grant agreement NO. 243195 (to U.M.). K.Y. is supported by Versus Arthritis Career Development Fellowship (21447). B.L.'s work was supported by the DFG grant MA 4172/13-1 (to K.M.).

Acknowledgments: Antonio Jimenez-Caliani, Cristina Barria, Petra Berger, Katrischa Hennekens and Nilima Dinesh at the University of Bremen are acknowledged for technical support.

Conflicts of Interest: The authors declare no conflict of interest.

Abbreviations

APMA	Aminophenylmercuric acetate
GAPDH	Glyceraldehyde-3-phosphate dehydrogenase
IL-1β	Interleukin-1β
LRP	Lipoprotein receptor-related protein
MMP	Matrix metalloproteinase
PVDF	Polyvinylidene fluoride
RAP	Receptor-associated protein
TIMP	Tissue inhibitor of metalloproteinases
TNF-α	Tumor necrosis factor-α

References

1. Gross, J.; Lapiere, C.M. Collagenolytic activity in amphibian tissues: A tissue culture assay. *Proc. Natl. Acad Sci. USA* **1962**, *48*, 1014–1022. [CrossRef] [PubMed]
2. Rohani, M.G.; Parks, W.C. Matrix remodeling by MMPs during wound repair. *Matrix Biol.* **2015**, *44–46*, 113–121. [CrossRef] [PubMed]
3. Piperi, C.; Papavassiliou, A.G. Molecular mechanisms regulating matrix metalloproteinases. *Curr. Top. Med. Chem.* **2012**, *12*, 1095–1112. [CrossRef] [PubMed]
4. Parks, W.C.; Wilson, C.L.; Lopez-Boado, Y.S. Matrix metalloproteinases as modulators of inflammation and innate immunity. *Nat. Rev. Immunol.* **2004**, *4*, 617–629. [CrossRef] [PubMed]
5. Witty, J.P.; Wright, J.H.; Matrisian, L.M. Matrix metalloproteinases are expressed during ductal and alveolar mammary morphogenesis, and misregulation of stromelysin-1 in transgenic mice induces unscheduled alveolar development. *Mol. Biol. Cell* **1995**, *6*, 1287–1303. [CrossRef] [PubMed]
6. Saarialho-Kere, U.K.; Pentland, A.P.; Birkedal-Hansen, H.; Parks, W.C.; Welgus, H.G. Distinct populations of basal keratinocytes express stromelysin-1 and stromelysin-2 in chronic wounds. *J. Clin. Invest.* **1994**, *94*, 79–88. [CrossRef]
7. Haro, H.; Crawford, H.C.; Fingleton, B.; MacDougall, J.R.; Shinomiya, K.; Spengler, D.M.; Matrisian, L.M. Matrix metalloproteinase-3-dependent generation of a macrophage chemoattractant in a model of herniated disc resorption. *J. Clin. Invest.* **2000**, *105*, 133–141. [CrossRef]
8. Nerusu, K.C.; Warner, R.L.; Bhagavathula, N.; McClintock, S.D.; Johnson, K.J.; Varani, J. Matrix metalloproteinase-3 (stromelysin-1) in acute inflammatory tissue injury. *Exp. Mol. Pathol.* **2007**, *83*, 169–176. [CrossRef]
9. Lochter, A.; Galosy, S.; Muschler, J.; Freedman, N.; Werb, Z.; Bissell, M.J. Matrix metalloproteinase stromelysin-1 triggers a cascade of molecular alterations that leads to stable epithelial-to-mesenchymal conversion and a premalignant phenotype in mammary epithelial cells. *J. Cell Biol.* **1997**, *139*, 1861–1872. [CrossRef]
10. Brenneisen, P.; Sies, H.; Scharffetter-Kochanek, K. Ultraviolet-B irradiation and matrix metalloproteinases: From induction via signaling to initial events. *Ann. N Y Acad Sci.* **2002**, *973*, 31–43. [CrossRef]
11. Kurz, B.; Lemke, A.K.; Fay, J.; Pufe, T.; Grodzinsky, A.J.; Schünke, M. Pathomechanisms of cartilage destruction by mechanical injury. *Ann. Anat.* **2005**, *187*, 473–485. [CrossRef] [PubMed]
12. Raffetto, J.D.; Khalil, R.A. Matrix metalloproteinases and their inhibitors in vascular remodeling and vascular disease. *Biochem. Pharmacol.* **2008**, *75*, 346–359. [CrossRef] [PubMed]
13. Yoo, J.; Rodriguez Perez, C.E.; Nie, W.; Sinnett-Smith, J.; Rozengurt, E. Protein kinase D1 mediates synergistic MMP-3 expression induced by TNF-α and bradykinin in human colonic myofibroblasts. *Biochem. Biophys Res. Commun.* **2011**, *413*, 30–35. [CrossRef] [PubMed]
14. Wang, M.; Qin, X.; Mudgett, J.S.; Ferguson, T.A.; Senior, R.M.; Welgus, H.G. Matrix metalloproteinase deficiencies affect contact hypersensitivity: Stromelysin-1 deficiency prevents the response and gelatinase B deficiency prolongs the response. *Proc. Natl. Acad Sci. USA* **1999**, *96*, 6885–6889. [CrossRef]
15. Warner, R.L.; Beltran, L.; Younkin, E.M.; Lewis, C.S.; Weiss, S.J.; Varani, J.; Johnson, K.J. Role of stromelysin 1 and gelatinase B in experimental acute lung injury. *Am. J. Respir Cell Mol. Biol.* **2001**, *24*, 537–544. [CrossRef]
16. Bullard, K.M.; Lund, L.; Mudgett, J.S.; Mellin, T.N.; Hunt, T.K.; Murphy, B.; Ronan, J.; Werb, Z.; Banda, M.J. Impaired wound contraction in stromelysin-1-deficient mice. *Ann. Surg.* **1999**, *230*, 260–265. [CrossRef]
17. Black, R.A.; Rauch, C.T.; Kozlosky, C.J.; Peschon, J.J.; Slack, J.L.; Wolfson, M.F.; Castner, B.J.; Stocking, K.L.; Reddy, P.; Srinivasan, S.; et al. A metalloproteinase disintegrin that releases tumour-necrosis factor-alpha from cells. *Nature* **1997**, *385*, 729–733. [CrossRef]
18. Steenport, M.; Khan, K.M.; Du, B.; Barnhard, S.E.; Dannenberg, A.J.; Falcone, D.J. Matrix metalloproteinase (MMP)-1 and MMP-3 induce macrophage MMP-9: Evidence for the role of TNF-alpha and cyclooxygenase-2. *J. Immunol.* **2009**, *183*, 8119–8127. [CrossRef]
19. Lee, E.J.; Moon, P.G.; Baek, M.C.; Kim, H.S. Comparison of the effects of matrix metalloproteinase inhibitors on TNF-alpha release from activated microglia and TNF-alpha converting enzyme activity. *Biomol. Ther.* **2014**, *22*, 414–419. [CrossRef]

20. Gearing, A.J.; Beckett, P.; Christodoulou, M.; Churchill, M.; Clements, J.; Davidson, A.H.; Drummond, A.H.; Galloway, W.A.; Gilbert, R.; Gordon, J.L.; et al. Processing of tumour necrosis factor-alpha precursor by metalloproteinases. *Nature* **1994**, *370*, 555–557. [CrossRef]
21. Voigt, H.; Lemke, A.K.; Mentlein, R.; Schunke, M.; Kurz, B. Tumor necrosis factor alpha-dependent aggrecan cleavage and release of glycosaminoglycans in the meniscus is mediated by nitrous oxide-independent aggrecanase activity in vitro. *Arthritis Res. Ther.* **2009**, *11*, R141. [CrossRef] [PubMed]
22. Ahn, S.J.; Rhim, E.M.; Kim, J.Y.; Kim, K.H.; Lee, H.W.; Kim, E.C.; Park, S.H. Tumor necrosis factor-alpha induces matrix metalloproteinases-3, -10, and -13 in human periodontal ligament cells. *J. Periodontol.* **2014**, *85*, 490–497. [CrossRef] [PubMed]
23. Ågren, M.S.; Schnabel, R.; Christensen, L.H.; Mirastschijski, U. Tumor necrosis factor-alpha-accelerated degradation of type I collagen in human skin is associated with elevated matrix metalloproteinase (MMP)-1 and MMP-3 ex vivo. *Eur. J. Cell Biol.* **2015**, *94*, 12–21. [CrossRef] [PubMed]
24. Walker, E.J.; Rosenberg, G.A. TIMP-3 and MMP-3 contribute to delayed inflammation and hippocampal neuronal death following global ischemia. *Exp. Neurol.* **2009**, *216*, 122–131. [CrossRef]
25. McCawley, L.J.; Wright, J.; LaFleur, B.J.; Crawford, H.C.; Matrisian, L.M. Keratinocyte expression of MMP3 enhances differentiation and prevents tumor establishment. *Am. J. Pathol.* **2008**, *173*, 1528–1539. [CrossRef]
26. Henriet, P.; Rousseau, G.G.; Eeckhout, Y. Cloning and sequencing of mouse collagenase cDNA. Divergence of mouse and rat collagenases from the other mammalian collagenases. *FEBS letters* **1992**, *310*, 175–178. [CrossRef]
27. Balbin, M.; Fueyo, A.; Knauper, V.; Lopez, J.M.; Alvarez, J.; Sanchez, L.M.; Quesada, V.; Bordallo, J.; Murphy, G.; Lopez-Otin, C. Identification and enzymatic characterization of two diverging murine counterparts of human interstitial collagenase (MMP-1) expressed at sites of embryo implantation. *J. Biol. Chem.* **2001**, *276*, 10253–10262. [CrossRef]
28. Nagase, H. Chapter 158: Matrix Metalloproteinase-3/Stromelysin-1. In *Handbook of Proteolytic Enzymes*, 3rd ed.; Rawlings, N.D., Salvesen, G., Eds.; Academic Press: London, UK, 2013; Volume 1, pp. 763–774.
29. Barmina, O.Y.; Walling, H.W.; Fiacco, G.J.; Freije, J.M.; Lopez-Otin, C.; Jeffrey, J.J.; Partridge, N.C. Collagenase-3 binds to a specific receptor and requires the low density lipoprotein receptor-related protein for internalization. *J. Biol. Chem.* **1999**, *274*, 30087–30093. [CrossRef]
30. Yamamoto, K.; Okano, H.; Miyagawa, W.; Visse, R.; Shitomi, Y.; Santamaria, S.; Dudhia, J.; Troeberg, L.; Strickland, D.K.; Hirohata, S.; et al. MMP-13 is constitutively produced in human chondrocytes and co-endocytosed with ADAMTS-5 and TIMP-3 by the endocytic receptor LRP1. *Matrix Biol.* **2016**, *56*, 57–73. [CrossRef]
31. Yamamoto, K.; Troeberg, L.; Scilabra, S.D.; Pelosi, M.; Murphy, C.L.; Strickland, D.K.; Nagase, H. LRP-1-mediated endocytosis regulates extracellular activity of ADAMTS-5 in articular cartilage. *FASEB J.* **2013**, *27*, 511–521. [CrossRef]
32. Knäuper, V.; Lopez-Otin, C.; Smith, B.; Knight, G.; Murphy, G. Biochemical characterization of human collagenase-3. *J. Biol. Chem.* **1996**, *271*, 1544–1550. [CrossRef] [PubMed]
33. Knäuper, V.; Will, H.; López-Otin, C.; Smith, B.; Atkinson, S.J.; Stanton, H.; Hembry, R.M.; Murphy, G. Cellular mechanisms for human procollagenase-3 (MMP-13) activation. Evidence that MT1-MMP (MMP-14) and gelatinase A (MMP-2) are able to generate active enzyme. *J. Biol. Chem.* **1996**, *271*, 17124–17131. [CrossRef] [PubMed]
34. Eeckhout, Y.; Vaes, G. Further studies on the activation of procollagenase, the latent precursor of bone collagenase. Effects of lysosomal cathepsin B, plasmin and kallikrein, and spontaneous activation. *Biochem. J.* **1977**, *166*, 21–31. [CrossRef]
35. Van Meurs, J.; van Lent, P.; Stoop, R.; Holthuysen, A.; Singer, I.; Bayne, E.; Mudgett, J.; Poole, R.; Billinghurst, C.; van der Kraan, P.; et al. Cleavage of aggrecan at the Asn341-Phe342 site coincides with the initiation of collagen damage in murine antigen-induced arthritis: A pivotal role for stromelysin 1 in matrix metalloproteinase activity. *Arthritis Rheum* **1999**, *42*, 2074–2084. [CrossRef]
36. Henriet, P.; Eeckhout, Y. Chapter 154: Matrix Metallopeptidase-13/Collagenase-3. In *Handbook of Proteolytic Enzymes*, 3rd ed.; Rawlings, N.D., Salvesen, G., Eds.; Academic Press: London, UK, 2013; Volume 1, pp. 734–744.

37. Gosset, M.; Pigenet, A.; Salvat, C.; Berenbaum, F.; Jacques, C. Inhibition of matrix metalloproteinase-3 and -13 synthesis induced by IL-1beta in chondrocytes from mice lacking microsomal prostaglandin E synthase-1. *J. Immunol.* **2010**, *185*, 6244–6252. [CrossRef] [PubMed]
38. Kusano, K.; Miyaura, C.; Inada, M.; Tamura, T.; Ito, A.; Nagase, H.; Kamoi, K.; Suda, T. Regulation of matrix metalloproteinases (MMP-2, -3, -9, and -13) by interleukin-1 and interleukin-6 in mouse calvaria: Association of MMP induction with bone resorption. *Endocrinology* **1998**, *139*, 1338–1345. [CrossRef] [PubMed]
39. Arnott, C.H.; Scott, K.A.; Moore, R.J.; Hewer, A.; Phillips, D.H.; Parker, P.; Balkwill, F.R.; Owens, D.M. Tumour necrosis factor-alpha mediates tumour promotion via a PKC alpha- and AP-1-dependent pathway. *Oncogene* **2002**, *21*, 4728–4738. [CrossRef] [PubMed]
40. Han, Y.P.; Tuan, T.L.; Hughes, M.; Wu, H.; Garner, W.L. Transforming growth factor-beta - and tumor necrosis factor-alpha -mediated induction and proteolytic activation of MMP-9 in human skin. *J. Biol. Chem.* **2001**, *276*, 22341–22350. [CrossRef] [PubMed]
41. Borden, P.; Solymar, D.; Sucharczuk, A.; Lindman, B.; Cannon, P.; Heller, R.A. Cytokine control of interstitial collagenase and collagenase-3 gene expression in human chondrocytes. *J. Biol. Chem.* **1996**, *271*, 23577–23581. [CrossRef]
42. Lauridsen, H.M.; Pellowe, A.S.; Ramanathan, A.; Liu, R.; Miller-Jensen, K.; McNiff, J.M.; Pober, J.S.; Gonzalez, A.L. Tumor necrosis factor-alpha and IL-17A activation induces pericyte-mediated basement membrane remodeling in human neutrophilic dermatoses. *Am. J. Pathol.* **2017**, *187*, 1893–1906. [CrossRef]
43. Lees, M.; Taylor, D.J.; Woolley, D.E. Mast cell proteinases activate precursor forms of collagenase and stromelysin, but not of gelatinases A and B. *Eur. J. Biochem.* **1994**, *223*, 171–177. [CrossRef] [PubMed]
44. Martin, M.D.; Carter, K.J.; Jean-Philippe, S.R.; Chang, M.; Mobashery, S.; Thiolloy, S.; Lynch, C.C.; Matrisian, L.M.; Fingleton, B. Effect of ablation or inhibition of stromal matrix metalloproteinase-9 on lung metastasis in a breast cancer model is dependent on genetic background. *Cancer Res.* **2008**, *68*, 6251–6259. [CrossRef] [PubMed]
45. Mähler, M.; Berard, M.; Feinstein, R.; Gallagher, A.; Illgen-Wilcke, B.; Pritchett-Corning, K.; Raspa, M. FELASA recommendations for the health monitoring of mouse, rat, hamster, guinea pig and rabbit colonies in breeding and experimental units. FELASA working group on revision of guidelines for health monitoring of rodents and rabbits. *Lab. Anim.* **2014**, *48*, 178–192. [CrossRef]
46. Livak, K.J.; Schmittgen, T.D. Analysis of relative gene expression data using real-time quantitative PCR and the 2(-Delta Delta C(T)) method. *Methods* **2001**, *25*, 402–408. [CrossRef] [PubMed]
47. Henriksen, N.A.; Sørensen, L.T.; Jorgensen, L.N.; Ågren, M.S. Circulating levels of matrix metalloproteinases and tissue inhibitor of metalloproteinases in patients with incisional hernia. *Wound Repair Regen* **2013**, *21*, 661–666. [CrossRef]
48. Hinke, S.A.; Navedo, M.F.; Ulman, A.; Whiting, J.L.; Nygren, P.J.; Tian, G.; Jimenez-Caliani, A.J.; Langeberg, L.K.; Cirulli, V.; Tengholm, A.; et al. Anchored phosphatases modulate glucose homeostasis. *EMBO J.* **2012**, *31*, 3991–4004. [CrossRef]
49. Mirastschijski, U.; Dinesh, N.; Baskaran, S.; Wedekind, D.; Gavrilovic, J.; Murray, M.Y.; Bevan, D.; Kelm, S. Novel specific human and mouse stromelysin-1 (MMP-3) and stromelysin-2 (MMP-10) antibodies for biochemical and immunohistochemical analyses. *Wound Repair Regen* **2019**, *27*, 309–323. [CrossRef]
50. Ardestani, A.; Paroni, F.; Azizi, Z.; Kaur, S.; Khobragade, V.; Yuan, T.; Frogne, T.; Tao, W.; Oberholzer, J.; Pattou, F.; et al. MST1 is a key regulator of beta cell apoptosis and dysfunction in diabetes. *Nat. Med.* **2014**, *20*, 385–397. [CrossRef]

International Journal of
Molecular Sciences

MDPI

Article

Stimulation of Peritoneal Mesothelial Cells to Secrete Matrix Metalloproteinase-9 (MMP-9) by TNF-α: A Role in the Invasion of Gastric Carcinoma Cells

Teruaki Oku [1,†], Kentaro Shimada [1,†], Hiroki Kenmotsu [1], Yusuke Ando [1], Chisato Kurisaka [1], Rikio Sano [1], Makoto Tsuiji [1], Shinya Hasegawa [2], Tetsuya Fukui [2] and Tsutomu Tsuji [1,*]

1 Department of Microbiology, Hoshi University School of Pharmacy and Pharmaceutical Sciences, Tokyo 142-8501, Japan; oku@hoshi.ac.jp (T.O.); m842@hoshi.ac.jp (K.S.); s061084@hoshi.ac.jp (H.K.); p17andoh@hoshi.ac.jp (Y.A.); d1403@hoshi.ac.jp (C.K.); m220@hoshi.ac.jp (R.S.); m-tsuiji@hoshi.ac.jp (M.T.)

2 Department of Health Chemistry, Hoshi University School of Pharmacy and Pharmaceutical Sciences, Tokyo 142-8501, Japan; s-hasegawa@hoshi.ac.jp (S.H.); fukui@hoshi.ac.jp (T.F.)

* Correspondence: tsuji@hoshi.ac.jp; Tel.: +81-3-5498-5753

† These authors contributed equally to this work.

Received: 12 October 2018; Accepted: 6 December 2018; Published: 9 December 2018

Abstract: It has recently been recognized that inflammatory cytokines, such as tumor necrosis factor-α (TNF-α), upregulate the secretion of matrix metalloproteinase-9 (MMP-9) from cancer cells and thereby promote peritoneal dissemination. In this study, we found that TNF-α also stimulated peritoneal mesothelial cells to secrete MMP-9 as assessed by zymography. MMP-9 gene expression in mesothelial cells induced by TNF-α was confirmed by quantitative RT-PCR analysis. We then utilized the reconstituted artificial mesothelium, which was composed of a monolayer of mesothelial cells cultured on a Matrigel layer in a Boyden chamber system, to examine the effects of TNF-α on carcinoma cell invasion. The transmigration of MKN1 human gastric carcinoma cells through the reconstituted mesothelium was promoted by TNF-α in a dose-dependent manner. The increased MKN1 cell migration was partially inhibited by the anti-α3 integrin antibody, indicating that the invasion process involves an integrin-dependent mechanism. Finally, we observed that the invasion of MMP-9-knockdown MKN1 cells into Matrigel membranes was potentiated by the exogenous addition of purified proMMP-9. These results suggest that TNF-α-induced MMP-9 secretion from mesothelial cells plays an important role in the metastatic dissemination of gastric cancer.

Keywords: TNF-α; matrix metalloproteinase; peritoneal mesothelial cell; gastric cancer; metastatic dissemination

1. Introduction

Metastatic peritoneal dissemination causes a poor prognosis for patients with advanced gastric cancer. Cancer metastasis depends on multiple interactions between cancer cells and various factors derived from the host microenvironment, including cytokines, motility factors, cell adhesion molecules, and matrix-degrading enzymes. On the other hand, many recent studies have suggested the relevance of inflammation to cancer progression. In particular, tumor necrosis factor-α (TNF-α) and other inflammatory cytokines have been shown to participate in the initiation and progression of cancer [1–5]. TNF-α is a multifunctional cytokine that modulates various aspects of cancer cell phenotypes, such as cell proliferation, migration, invasion, and metastatic potential, in addition to causing the death of cancer cells. TNF-α, either alone or in combination with transforming growth factor-β, has also been shown to induce so-called epithelial-mesenchymal transition (EMT), which includes changes in both morphological and invasive phenotypes [6,7]. It has been well established that TNF-α induces the production of matrix metalloproteinases (MMPs) by carcinoma cells originating from various tissues, and these MMPs facilitate cancer cell invasion and metastasis [8–13]. A recent clinical study of patients with gastric cancer indicated that TNF-α expression was significantly different between patients with and those without peritoneal metastasis and was one of the risk factors for peritoneal metastasis of gastric cancer [14].

We have studied the interaction of gastric carcinoma cells with the extracellular matrix (ECM) deposited by peritoneal mesothelial cells and found that MKN1 human gastric carcinoma cells secreted MMP-9 accompanied by enhanced invasion through Matrigel-reconstituted basement membranes when these cells adhered to ECM containing laminin-332, a major component of submesothelial basement membranes [15]. Moreover, the phenotype of MKN1 cells was found to be more invasive in the presence of TNF-α derived from tumor-associated macrophage-like cells in association with enhanced MMP-9 secretion [16]. In the present study, we co-cultured MKN1 cells with mesothelial cells isolated from murine peritoneum and examined the effects of treatment with TNF-α on MMP-9 production by these cells. The TNF-α treatment of these cells resulted in the induction of MMP-9 secretion not only from MKN1 carcinoma cells but also from mesothelial cells. This result prompted us to characterize the enhanced MMP-9 production by mesothelial cells upon TNF-α treatment. We also present evidence showing that MMP-9 secreted from mesothelial cells plays a significant role in the invasion of gastric carcinoma cells. For this purpose, we used a reconstituted mesothelium consisting of peritoneal mesothelial cells and Matrigel basement membranes in a Boyden chamber system.

2. Results

2.1. Mesothelial Cells Secreted MMP-9 in Response to Stimulation with TNF-α

We first examined the effects of TNF-α on MMP-9 secretion in the co-cultures of MKN1 and peritoneal mesothelial cells. Zymographic analysis of the conditioned medium of the co-culture of these cells revealed that MMP-9 secretions from both MKN1 and murine peritoneal mesothelial cells were markedly increased in the presence of TNF-α (10 ng/mL) (Figure 1). Since murine MMP-9 has a slightly higher molecular mass (105 kDa for proMMP-9) than its human counterpart (92 kDa) due to the presence of an extra amino acid sequence in murine MMP-9 [17], the two molecules can be distinguished from each other by zymography. By contrast, the MMP-2 level was not significantly changed in the conditioned medium after TNF-α treatment. The upregulation of MMP-9 secretion from both types of cells by TNF-α treatment (10 ng/mL) was confirmed when TNF-α was added to the separate cultures of MKN1 and mesothelial cells (Figure 2A). We observed low-level secretion of MMP-2 from both cell types in the absence of TNF-α, and the secretion was not substantially influenced by TNF-α treatment in either case. We then examined the dose-response of the potentiation of MMP-9 secretion by TNF-α. In the zymographic analysis, we detected MMP-9 activity in the conditioned medium of mesothelial cell culture in the presence of TNF-α (1 ng/mL) for 24 or 48 h, and the MMP-9 secretion was dose-dependently increased (up to 100 ng/mL) (Figure 2B). In the kinetic analysis, we detected MMP-9 activity after the culture with TNF-α for 12 h, and MMP-9 secretion was increased at 24 and 48 h (Figure 2C). These results clearly indicated that TNF-α treatment induced MMP-9 secretion from mesothelial cells, in addition to enhancing MMP-9 secretion from MKN1 gastric carcinoma cells.

We next conducted RT-qPCR to measure mRNA levels for MMP-9 in mesothelial cells after TNF-α treatment. The results shown in Figure 3A indicated that the mRNA level in mesothelial cells was elevated at 6 h after the stimulation by TNF-α (10 ng/mL) and reached the maximum level at 12 h, where it remained until 24 h after the stimulation. The increase in MMP-9 mRNA expression was found to be dose-dependent within the range of 1–100 ng/mL of TNF-α (Figure 3B), which was in good agreement with the results of the zymographic analysis (Figure 2B). However, the mRNA for MMP-2 exhibited no significant change after TNF-α treatment.

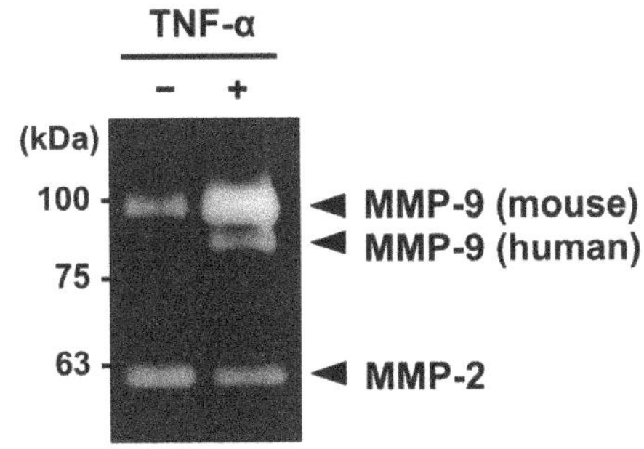

Figure 1. Zymographic analysis of MMPs in the conditioned medium of the co-culture of MKN1 and mesothelial cells after TNF-α treatment. A suspension of MKN1 cells (5 × 10^4 cells/0.2 mL) in ASF104 medium was added to a monolayer of murine peritoneal mesothelial cells in a 96-well culture plate and the cells were cultured with/without TNF-α (10 ng/mL) for 48 h. The culture supernatant was analyzed by gelatin zymography.

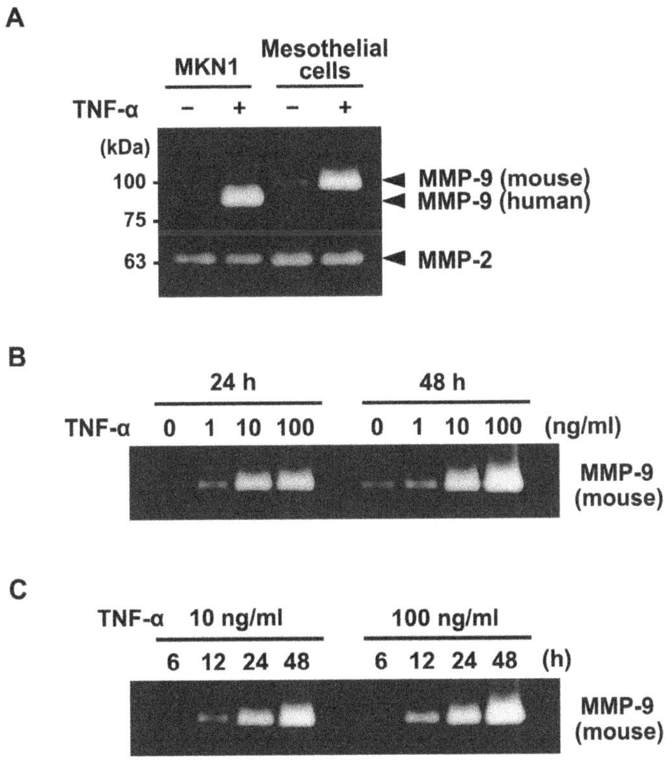

Figure 2. Effects of TNF-α on the secretion of MMP-9 from peritoneal mesothelial cells as assessed by zymography. (**A**) Monolayers of peritoneal mesothelial cells and MKN1 cells were separately cultured in ASF104 serum-free medium with/without TNF-α (10 ng/mL) for 48 h. (**B**) Mesothelial cells were cultured with TNF-α (0, 1, 10 or 100 ng/mL) for 24 or 48 h. (**C**) Mesothelial cells were cultured with TNF-α (10 or 100 ng/mL) for 6, 12, 24 or 48 h.

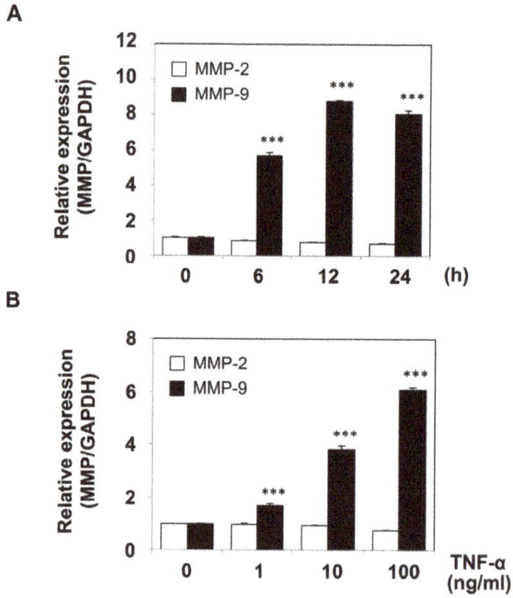

Figure 3. RT-qPCR analysis of mRNA for MMP-2 and MMP-9 in mesothelial cells after treatment with TNF-α. (**A**) Mesothelial cells were cultured in ASF104 serum-free medium containing TNF-α (10 ng/mL) at 37 °C for 0, 6, 12 or 24 h. Total RNA was then isolated, and RT-qPCR for the genes for MMP-2 (*open bar*) and MMP-9 (*closed bar*) was carried out as described in Section 4. (**B**) Mesothelial cells were cultured in ASF104 medium containing TNF-α (0, 1, 10 or 100 ng/mL) for 6 h. The gene expressions of MMP-2 (*open bar*) and MMP-9 (*closed bar*) were analyzed by the relative standard curve method using Gapdh as an internal control. Experiments were performed in triplicate, and the data are presented as the mean ± SEM. Statistical data analysis was conducted using the Student's *t*-test. *** $p < 0.005$ vs. control.

2.2. TNF-α Potentiates MKN1 Cell Invasion through the Reconstituted Mesothelium

Because the above experiments indicated that mesothelial cells secreted MMP-9 in response to TNF-α treatment, we designed an artificial, reconstituted mesothelium where a monolayer of mesothelial cells was cultured on a Matrigel layer in a Boyden chamber system (Figure 4A) and examined the effects of TNF-α on carcinoma cell invasion. Mesothelial cells isolated from the murine peritoneum grew as a monolayer with polygonal morphology after 4–5 days (Figure 4B). The transmigration of MKN1 cells through the reconstituted mesothelium was promoted by TNF-α in a dose-dependent manner (Figure 4C).

We previously found that the interaction between α3β1 integrin on cancer cells and laminin in the mesothelium played an important role in the cancer cell adhesion and invasion [15,18]. Next, we examined the effects of the anti-α3 integrin antibody on the transmigration of MKN1 cells through the reconstituted mesothelium. The cell invasion potentiated by TNF-α was significantly inhibited by the anti-α3 integrin antibody (Figure 5A), suggesting the importance of an α3β1 integrin-dependent process in the invasion. The adhesion of MKN1 cells to a monolayer of mesothelial cells was also increased after the TNF-α treatment of mesothelial cells and was partially inhibited by the anti-α3 integrin antibody (Figure 5B). Mochizuki et al. [19] reported that the treatment of mesothelial cells with TNF-α induced their morphological change followed by an increase in the areas of intercellular gaps. This process may cause exposure of the submesothelial extracellular matrix (ECM) in the intercellular gaps. Because laminin-332, a counter-ligand for α3β1 integrin, is a major component of submesothelial ECM, TNF-α treatment might facilitate the adhesion of MKN1 cells to the mesothelium via α3β1 integrin/laminin-332 interaction. In RT-qPCR analysis, we observed a slight increase in expression of the γ2 subunit of laminin-332 after TNF-α treatment of mesothelial cells (Figure S1), and this might also have caused the increased adhesion of MKN1 cells.

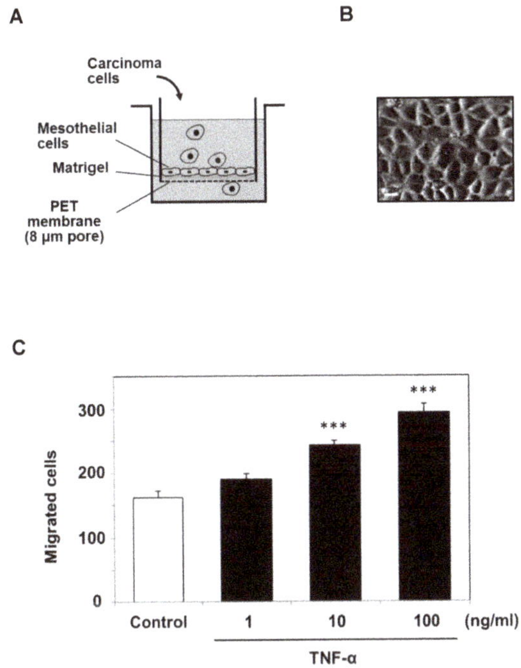

Figure 4. Cell invasion assay using a reconstituted artificial mesothelium in a Boyden chamber (Transwell) system. (**A**) The inner chamber with a membrane (8.0 µm pore) was composed of a monolayer of peritoneal mesothelial cells on a Matrigel layer and was utilized to examine the migration of MKN1 cells. The outer chamber was filled with ASF104 medium supplemented with HT1080 serum-free conditioned medium as a chemoattractant. (**B**) Microscopic observation of a monolayer of mesothelial cells (scale bar = 20 µm). (**C**) After mesothelial cells were treated with TNF-α (1, 10 or 100 ng/mL) and washed with ASF104 medium, MKN1 cells (1×10^5 cells/0.2 mL) were placed in the inner chamber and incubated at 37 °C for 16 h. The cells migrating into the outer chamber through the membrane were counted under a microscope after staining with Diff-Quik. Experiments were performed in triplicate, and the data are presented as the mean ± SEM. Statistical data analysis was conducted using the Student's *t*-test. *** $p < 0.005$ vs. the control.

2.3. The Exogenous Addition of MMP-9 Promotes MKN1 Cell Invasion

Because our experiments demonstrated that TNF-α potentiated MMP-9 secretion from both mesothelial cells and cancer cells, we next addressed the effect of the exogenous addition of MMP-9 on the invasive behavior of cancer cells. When MMP-9 purified from human THP-1 leukemic cells (Figure S2) was added to the culture of MKN1 cells, their invasion into Matrigel basement membranes was increased by approximately 29% (Figure 6A). To exclude the contribution of secretion of MMP-9 and MMP-2 from cancer cells, we established MMP-knockdown cells by the RNA interference procedure (Figure S3). The silencing of MMP-9 expression in MKN1 cells resulted in marked impairment of the invasion (~90% inhibition), whereas MMP-2 silencing exhibited only a moderate effect (~30% inhibition) (Figure 6A,B). The addition of MMP-9 to the culture potentiated the invasion of MMP-9-knockdown cells in a dose-dependent manner (Figure 6B). These results suggest that the invasion of cancer cells is influenced by MMP-9 secreted from TNF-α-stimulated mesothelial cells.

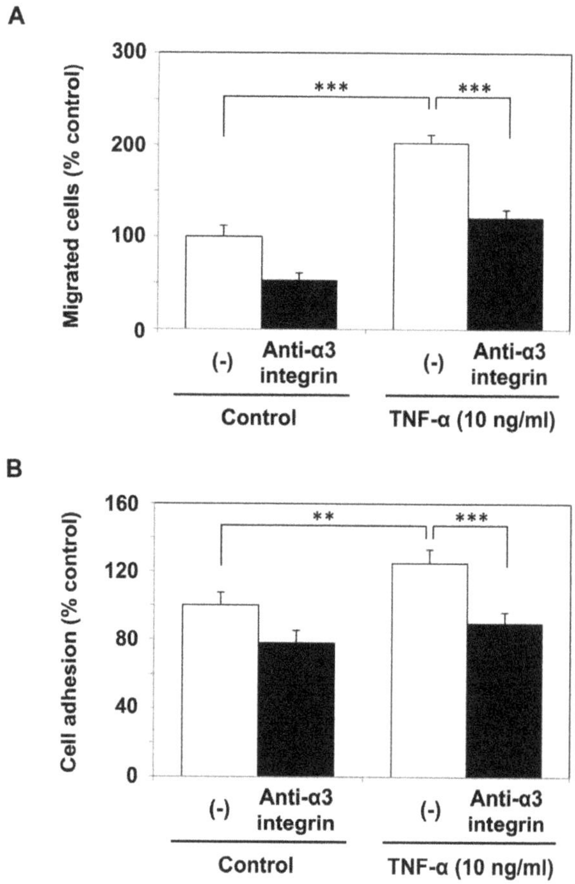

Figure 5. Invasion and adhesion of MKN1 cells and effects of the anti-α3 integrin antibody. A monolayer of mesothelial cells was stimulated with TNF-α (10 ng/mL) for 6 h. (**A**) MKN1 cells (1×10^5 cells/0.2 mL) in ASF104 serum-free medium were added to the inner chamber of the reconstituted mesothelium and incubated at 37 °C for 16 h. The cells that had migrated into the outer chamber through the membrane were counted under a microscope. (**B**) Fluorescently labeled MKN1 cells were added to the monolayer of mesothelial cells in a 96-well culture plate, and incubated at 37 °C for 40 min. After non-adherent cells were removed by washing, adherent cells were lysed with 1% Nonidet P-40 and measured with a fluorescence spectrophotometer (Ex = 490 nm, Em = 520 nm). For the inhibition experiments, MKN1 cells were treated with the anti-α3 integrin antibody (SM-T1, 10 μg/mL) at 0 °C for 30 min. Experiments were performed in triplicate, and the data are presented as the mean ± SEM. Statistical data analysis was conducted using the Student's *t*-test. ** $p < 0.01$, *** $p < 0.005$.

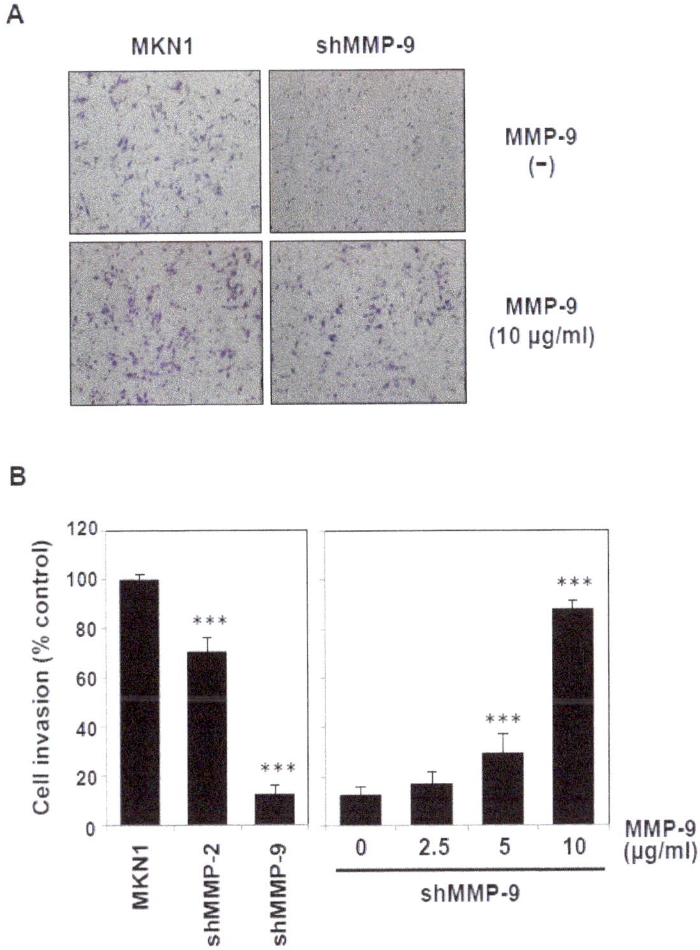

Figure 6. Potentiation of the invasion of MMP-9-knockdown MKN1 cells by exogenous MMP-9. MMP-2- or MMP-9-knockdown MKN1 cells were prepared by RNA interference as described in the Materials and Methods section. (**A**) The MMP-9-knockdown MKN1 cells (*right*) and parent cells (*left*) were assayed for in vitro invasion using Matrigel basement membranes in the presence or absence of MMP-9 purified from THP-1 cells. After culturing for 16 h, the cells that had migrated into the lower chamber were microscopically observed (40×). (**B**) MMP-2- or MMP-9-knockdown MKN1 cells were subjected to an invasion assay, and the migrated cells were counted under a microscope (*Left*). The invasion assay was performed in the presence of purified MMP-9 (2.5–10 μg/mL) (*right*). Experiments were performed in triplicate, and the data are presented as the mean ± SEM. Statistical data analysis was conducted using the Student's *t*-test. *** $p < 0.005$.

3. Discussion

It has recently been recognized that inflammatory cytokines such as TNF-α play crucial roles in metastatic dissemination. For example, TNF-α treatment of human gastric cancer cells was shown to increase MMP-9 mRNA and protein expression levels [20]. In the present study, we demonstrated that peritoneal mesothelial cells also secreted MMP-9 in response to TNF-α (Figures 1 and 2) and that the enhanced secretion of MMP-9 was associated with upregulation of MMP-9 transcription (Figure 3). In addition, we suggested the importance of the MMP-9 secreted by host mesothelial cells in gastric carcinoma invasion by utilizing a reconstituted mesothelium consisting of a monolayer of peritoneal mesothelial cells layered on a Matrigel basement membrane (Figure 4). The increase in cell invasion was shown to partly involve an integrin-dependent process (Figure 5). In addition to matrix-degrading enzymes secreted by the cancer cells themselves, MMP-9 derived from the host microenvironments, including peritoneal mesothelial cells, appeared to contribute to the cancer cell invasion into submesothelial matrices (Figure 6). Although mesothelial cells were thought to play a structural role in maintaining the smooth surface of the peritoneum and possess rather static properties, recent studies have shown that mesothelial cells produce various growth factors (e.g., vascular endothelial growth factor, basic fibroblast growth factor, epidermal growth factor, hepatocyte growth factor, and transforming growth factor-β) and ECM components (e.g., cytokeratin, vimentin and type I and III collagen) [21]. Therefore, peritoneal mesothelial cells are considered to be potentially active cells that dynamically respond to environmental changes. Chen and co-workers reported that TNF-α induced MMP-9 synthesis in human pleural mesothelial cells associated with *Mycobacterium tuberculosis* infection and that this process might be implicated in the pathogenesis of tuberculous pleuritis [22]. TNF-α is a strong activator for NF-κB, and MMP-9 expression is known to be upregulated via activation of NF-κB signaling in various types of cells [23,24]. The signaling pathway involved in the enhanced MMP-9 expression in mesothelial cells should be clarified in future studies.

The origin of TNF-α involved in the stimulation of mesothelial cells is unclear at present. Considering that micrometastasis in the early stage of peritoneal dissemination takes place mainly in the omentum, where many small lymphoid tissues called milky spots are present [25,26], macrophages and lymphoid cells accumulated in milky spots are likely to secrete TNF-α. Another candidate would be the macrophages infiltrating into tumors, which are collectively called tumor-associated macrophages (TAMs) and are known to produce various cytokines depending on the stage and types of tumors [16,27,28].

A recent clinical study on patients with gastric cancer indicated a positive correlation between TNF-α expression and peritoneal metastasis [14]. The results showed that the TNF-α expressions were significantly different between the metastatic group and non-metastatic group, and the logistic regression analysis indicated that TNF-α expression was a risk factor for peritoneal metastasis of gastric cancer. TNF-α has been shown to deteriorate malignant phenotypes of cancer cells, including cell proliferation, migration, epithelial-mesenchymal transition, and productivity of cytokines and growth factors. In addition to these effects on cancer cells, it seems important that the phenotypic changes of mesothelial cells induced by TNF-α, especially the enhancement of MMP-9 secretion, may cause peritoneal metastasis, and thus a poor prognosis for patients with gastric cancer. The relationship between the level of MMP-9 in the gastric carcinoma environment and the malignant invasive phenotypes of gastric carcinoma, including lymph node metastasis, has been extensively studied, and the data suggest that MMP-9 plays important roles in the progression of gastric cancer [29–31]. Therefore, it would be feasible that MMP-9 and its endogenous inhibitors (tissue inhibitors of metalloproteinases, TIMPs) are preventive and therapeutic targets for cancer metastasis.

We have previously studied the interaction of gastric carcinoma cells with the mesothelium and found that the integrin-dependent adhesion of carcinoma cells to the submesothelial matrix promotes MMP-9 secretion from cancer cells [15]. TNF-α treatment of mesothelial cells has been shown to induce morphological changes characterized by cell rounding and disruption of the cell–cell contacts, leading to the expansion of intercellular gaps [19]. It is likely that the adhesion of cancer cells to the exposed submesothelial matrix is stabilized, which might promote further upregulation of MMP-9 secretion and successive invasion of cancer cells. In conclusion, the present study strongly suggests that TNF-α-induced MMP-9 secretions from mesothelial cells play an important role in the metastatic dissemination of gastric cancer.

4. Materials and Methods

4.1. Reagents

RPMI1640 medium, trypsin, Triton X-100, Brij 35, TriReagent and hexadimethrine bromide were purchased from Sigma-Aldrich (St. Louis, MO, USA). Nonidet P-40 was from Nacalai Tesque (Kyoto, Japan). Bovine serum albumin (BSA), gelatin, dimethyl sulfoxide (DMSO), phorbol 12-myristate 13-acetate (PMA) and polyethylene glycol (PEG) 6000 were purchased from Wako Pure Chemical Industries (Osaka, Japan). Polyethylenimine "Max" was purchased from Polysciences Inc. (Warrington, PA, USA). Fetal calf serum (FCS) was obtained from Biosera (Boussens, France). ASF104 serum-free medium was supplied by Ajinomoto (Tokyo, Japan). Matrigel was purchased from BD Biosciences (San Diego, CA, USA). The PrimeScript RT Reagent Kit and KAPA SYBR FAST qPCR Kit Master Mix (2×) ABI Prism were obtained from Takara Bio Inc. (Shiga, Japan) and KAPA Biosystems (Boston, MA, USA), respectively. ViraPower Lentiviral Packaging Mix and puromycin dihydrochloride were supplied by Life Technologies (Carlsbad, CA, USA). Oligonucleotides were supplied by FASMAC (Kanagawa, Japan). Recombinant murine TNF-α was a product of Peprotech (Rocky Hill, NJ, USA). A fluorescent dye, 3′-*O*-acetyl-2′, 7′-*bis*(carboxyethyl)-4 or 5-carboxyfluorescein, diacetoxymethyl ester (BCECF-AM), was obtained from Dojindo Laboratories (Kumamoto, Japan). Gelatin Sepharose 4B was a product of GE Healthcare (Piscataway, NJ, USA).

4.2. Cells

The MKN1 human gastric carcinoma, HT1080 human fibrosarcoma, and THP-1 human acute monocytic leukemia cell lines were supplied by the RIKEN Cell Bank (Tsukuba, Japan) and cultured in RPMI 1640 medium supplemented with 10% FCS at 37 °C under 5% CO_2. Mesothelial cells were isolated under sterile conditions essentially as described by Nakashio et al. [32]. Briefly, the parietal peritoneum and diaphragm excised from ddY mice were washed with phosphate-buffered saline (PBS) and incubated with 0.25% trypsin in PBS at 37 °C for 30 min with gentle mixing every 5 min. After an equal volume of RPMI 1640/10% FCS was added to the cell suspension, the mixture was centrifuged at 2000 rpm for 10 min. The pelleted cells were suspended and incubated with ammonium-chloride-potassium (ACK) lysis buffer (150 mM NH_4Cl, 10 mM $KHCO_3$, 0.1 mM EDTA, pH 7.4) to remove red blood cells at room temperature for 1 min. After 10 volumes of RPMI 1640 medium/10% FCS was added, the mesothelial cells were recovered by centrifugation at 2000 rpm for 10 min. The pelleted cells were resuspended in RPMI 1640 medium/10% FCS and passed through nylon mesh (40 μm), and an aliquot (2×10^5 cells/0.1 mL) was placed in a 96-well culture plate that had been coated with 10% Matrigel at 4 °C for 16 h. The cells were cultured at 37 °C for 4–5 days until the growth of a monolayer of polygonal mesothelial cells. This procedure was conducted in accordance with the Guide for the Care and Use of Laboratory Animals of Hoshi University School of Pharmacy and Pharmaceutical Sciences (accredited by the Ministry of Education, Culture, Sports, Science and Technology, Japan).

4.3. Gelatin Zymography

MMP activity was assayed by gelatin zymography essentially as described previously [33]. Briefly, MKN1 and/or mesothelial cells were treated with TNF-α (0, 1, 10, 100 ng/mL) in ASF104 serum-free medium for 0–48 h, and the culture supernatants were mixed with sample buffer for SDS-polyacrylamide gel electrophoresis (Laemmli's buffer without reducing agent). Specimens were electrophoresed on a polyacrylamide gel (7.5%) containing 0.15% gelatin. The gel was washed three times with 1% Triton X-100 for 20 min each, and three times with water for 20 min each. Then gelatinolytic activity was detected by incubating the gel for 16 h at 37 °C in 50 mM Tris-HCl, 150 mM NaCl, 5 mM $CaCl_2$, and 1 μM $ZnCl_2$ followed by staining with Coomassie brilliant blue (CBB) R-250 (Merck, Darmstadt, Germany).

4.4. Quantitative RT-PCR (RT-qPCR)

Total RNA was isolated from mesothelial cells using TriReagent, and cDNA was synthesized with a PrimeScript RT reagent kit using the total RNA as a template according to the manufacturer's protocols. The qPCR reactions were conducted with an Applied Biosystems StepOne system (Life Technologies) using a KAPA SYBR FAST qPCR Kit Master Mix (2×) ABI Prism under the following conditions: Denaturation at 95 °C for 20 s followed by 40 cycles of 95 °C for 5 s, 55 °C for 30 s and 72 °C for 30 s. Melting curve analysis and agarose gel electrophoresis of the PCR products were performed to verify the specific amplification. All samples were analyzed in triplicate and quantified by the relative standard curve method using the *Gapdh* housekeeping gene as an internal control. The sequences of the primers used were as follows: *Mmp2*, 5′-ATC GCT CAG ATC CGT GGT G-3′ (forward) and 5′-GGA GCT CAG GCC AGA ATG T-3′ (reverse); *Mmp9*, 5′-ATG TAC CCG CTG TAT AGC TAC C-3′ (forward) and 5′-ATA GTG GGA CAC ATA GTG GGA G-3′ (reverse); *Gapdh*, 5′-TGA AGC AGG CAT CTG AGG G-3′ (forward) and 5′-CGA AGG TGG AAG AGT GGG AG-3′ (reverse).

4.5. Cell Invasion Assay Using Reconstituted Mesothelium

Reconstituted artificial mesothelium consisting of a monolayer of mesothelial cells on a Matrigel layer in a Boyden chamber (Transwell) system was utilized for cell invasion assay. Murine mesothelial cells were isolated as described above and cultured in a 10% Matrigel-coated inner chamber with a polyethylene terephthalate (PET) membrane (pore size, 8 μm; Falcon 353097; BD Biosciences) combined with an outer chamber (Falcon 353504) for 4–5 days until the cells grew as a confluent monolayer. An MKN1 gastric carcinoma cell suspension (1×10^5 cells/0.2 mL) was placed in the inner (upper) chamber, and the outer (lower) chamber was filled with ASF104 serum-free medium supplemented with HT1080 serum-free conditioned medium. The addition of HT1080 conditioned medium (20%) increased the chemotactic activity, although the conditioned medium contained a moderate level of MMP-9 secreted by HT1080 cells. After the chambers were incubated at 37 °C for 16 h, the cells that had migrated into the lower chamber through the membrane were stained with Diff-Quik (International Reagents, Kobe, Japan) and counted under a microscope.

4.6. Cell Adhesion Assay

MKN1 cells were labeled with the fluorescent dye BCECF-AM (3 μM) at 37 °C for 40 min [18]. The labeled cells (1×10^5 cells/0.1 mL in 1% BSA containing RPMI 1640 medium) were added to the monolayer of mesothelial cells in a 96-well culture plate as described above. After the plate was incubated at 37 °C for 40 min, non-adherent cells were removed by gentle washing three times with PBS. Adherent cells were lysed with 0.1 mL of 1% Nonidet P-40, and the lysates were diluted with 9 volumes of PBS. The fluorescence intensities were measured with a fluorescence spectrophotometer (Ex = 490 nm, Em = 520 nm). The percent adhesion was calculated as follows: (The fluorescence intensity of adherent cells/fluorescence intensity of total cells) × 100.

4.7. Flow Cytometry

The expression of α3 integrin in MKN1 cells was measured by a flow cytometer (BD FACSVerse; BD Biosciences, San Diego, CA, USA), using monoclonal antibodies against human integrin α3 subunit (SM-T1) [34] and FITC-labeled anti-mouse IgG (Kirkegaard & Perry Laboratories, Gaithersburg, MD, USA).

4.8. MMP-Knockdown MKN1 Cells

Knockdown of MMP-2 and MMP-9 was conducted by using lentivirus-mediated RNA interference. The short hairpin RNAs (shRNAs) of target sequences against MMP-2 (185–203: AGG AGA GCT GCA TCC TGT T) [35] and MMP-9 (343–361: AAG TGG TAC CAC CAT AAC A) [36] were cloned into the pSIH-H1 vector (System Biosciences, Palo Alto, CA, USA) according to the manufacturer's protocol.

Lentiviral particles were produced in HEK293FT cells after the co-transfection of pSIH-H1 vector and ViraPower (three packaging plasmids) using polyethylenimine "MAX" [37] for 48 h and concentrated by the PEG precipitation method. Briefly, the culture medium containing viral particles was mixed with a one-third volume of 4 × PEG solution (32% PEG, 400 mM NaCl, 40 mM HEPES-NaOH, pH 7.4) and incubated at 4 °C overnight. After centrifugation (3000× *g*, 30 min), the obtained precipitate was dissolved in serum-free RPMI 1640 medium. The virus solution was mixed with hexadimethrine bromide (8 μg/mL at final concentration), and the mixture was added to pre-cultured MKN1 cells (1×10^5 cells/2 mL in a 6-well plate). The plate was then centrifuged (1800× *g*, 90 min) at 32 °C and incubated at 37 °C for 48 h. The cells were successively cultured for 24 h after the medium was replaced with RPMI 1640 medium/10% FCS, followed by selection with puromycin (2.5 μg/mL).

4.9. Purification of MMP-9

MMP-9 was purified from the conditioned medium of human monocytic leukemia THP-1 cells based on the method of Morodomi et al. [38]. THP-1 cells were cultured in ASF104 medium containing PMA (5 nM) at 37 °C for 3–4 days. The conditioned medium was mixed with gelatin Sepharose 4B which had been equilibrated with TNC buffer (50 mM Tris-HCl, 0.15 M NaCl, 10 mM $CaCl_2$, 0.05% Brij 35, pH 7.5) and incubated at 4 °C for 16 h with gentle agitation. After gelatin Sepharose 4B was washed three times with TNC buffer, MMP-9 was eluted with DMSO (5%)-containing TNC buffer. The eluate was then applied to a PD-10 desalting column (GE Healthcare, Piscataway, NJ, USA) to replace the buffer with ASF104 medium. The purity of MMP-9 was estimated to be >95%, as analyzed by SDS-polyacrylamide gel electrophoresis followed by CBB and silver staining. The amino acid sequence of the purified protein was analyzed by peptide mass fingerprinting. The tryptic fragments obtained from the specimen were applied to MALDI/TOFMS (AXIMA-QIT; Shimadzu/Kratos, Kyoto, Japan) as described previously [39]. Major mass peaks detected by the MS analysis and the corresponding amino acid sequences to human MMP-9 were as follows: m/z 1345.74 for ^{25}QSTLVLFPGDLR^{36}, m/z 1184.62 for ^{43}QLAEEYLYR^{51}, m/z 1701.90 for ^{77}QLSLPETGELDSATLK^{92}, m/z 1084.55 for ^{107}FQTFEGDLK^{115}, m/z 1680.91 for ^{144}AFALWSAVTPLTFTR^{158}, m/z 2350.10 for ^{163}DADIVIQFGVAEHGDGYPFDGK^{184}, m/z 3175.37 for ^{185}DGLLAHAFPPGPGIQGDAHFDDDELWSLGK^{214}, m/z 2012.89 for ^{222}FGNADGAACHFPFIFEGR^{239}, m/z 2073.87 for ^{250}SDGLPWCSTTANYDTDDR^{267}, m/z 999.47 for ^{268}FGFCPSER^{275}, m/z 997.50 for ^{325} LFGFCPTR^{332}, m/z 1385.79 for ^{547}GSRPQGPFLIADK^{559}, m/z 739.30 for ^{560}WPALPR^{565}, m/z 873.45 for ^{579}LFFFSGR^{585}, m/z 1532.83 for ^{586}QVWVYTGASVLGPR^{599}, m/z 1440.80 for ^{604}LGLGADVAQVTGALR^{618}, m/z 823.45 for ^{624}MLLFSGR^{630} and m/z 1921.91 for ^{653}MFPGVPLDTHDVFQYR^{668} (UniProt P14780). Based on these results, the purified protein was confirmed to be MMP-9.

Supplementary Materials: Supplementary materials can be found at http://www.mdpi.com/1422-0067/19/12/3961/s1. Figure S1: RT-qPCR analysis of mRNA for the γ2 subunit of laminin-332 in mesothelial cells; Figure S2: Purification of MMP-9 from THP-1 cells; Figure S3: Zymographic analysis of MMPs in the conditioned medium of MMP-2- or MMP-9-knockdown MKN1 cells.

Author Contributions: Conceptualization, T.O., K.S. and T.T.; Formal analysis, T.O., K.S., H.K., Y.A., C.K., R.S. and S.H.; Investigation, T.O., K.S., H.K., Y.A., C.K., R.S. and S.H.; Methodology, T.O., M.T., S.H. and T.F.; Project administration, T.T.; Resources, M.T., S.H. and T.F.; Supervision, T.T.; Writing—original draft, T.O.; Writing—review & editing, T.T.

Funding: This work was supported in part by Grants-in-Aid for Scientific Research from the Ministry of Education, Culture, Sports, Science and Technology of Japan (23590092).

Acknowledgments: We thank Yoko Sawaki, Ayako Takai, Misaki Ide, and Hiromi Akiyama (Hoshi University School of Pharmacy and Pharmaceutical Sciences) for their technical assistance.

Conflicts of Interest: The authors disclose no potential conflicts of interest related to this work.

Abbreviations

BCECF-AM	3′-*O*-acetyl-2′,7′-*bis*(carboxyethyl)-4 or 5-carboxyfluorescein, diacetoxymethyl ester
CBB	Coomassie brilliant blue
ECM	extracellular matrix
FCS	fetal calf serum
MALDI/TOFMS	matrix-assisted laser desorption ionization/time-of-flight mass spectrometer
MMP	matrix metalloproteinase
PBS	phosphate-buffered saline
PMA	phorbol 12-myristate 13-acetate
RT-PCR	reverse transcription-polymerase chain reaction
TNF-α	tumor necrosis factor-α

References

1. Waters, J.P.; Pober, J.S.; Bradley, J.R. Tumour necrosis factor and cancer. *J. Pathol.* **2013**, *230*, 241–248. [CrossRef] [PubMed]
2. Landskron, G.; De la Fuente, M.; Thuwajit, P.; Thuwajit, C.; Hermoso, M.A. Chronic inflammation and cytokines in the tumor microenvironment. *J. Immunol. Res.* **2014**, *2014*, 149185. [CrossRef] [PubMed]
3. Balkwill, F. Tumour necrosis factor and cancer. *Nat. Rev. Cancer* **2009**, *9*, 361–371. [CrossRef] [PubMed]
4. Li, Q.; Withoff, S.; Verma, I.M. Inflammation-associated cancer: NF-κB is the lynchpin. *Trends Immunol.* **2005**, *26*, 318–325. [CrossRef] [PubMed]
5. Wolczyk, D.; Zaremba-Czogalla, M.; Hryniewicz-Jankowska, A.; Tabola, R.; Grabowski, K.; Sikorski, A.F.; Augoff, K. TNF-α promotes breast cancer cell migration and enhances the concentration of membrane-associated proteases in lipid rafts. *Cell. Oncol.* **2016**, *39*, 353–363. [CrossRef] [PubMed]
6. Yamauchi, Y.; Kohyama, T.; Takizawa, H.; Kamitani, S.; Desaki, M.; Takami, K.; Kawasaki, S.; Kato, J.; Nagase, T. Tumor necrosis factor-α enhances both epithelial-mesenchymal transition and cell contraction induced in A549 human alveolar epithelial cells by transforming growth factor-β1. *Exp. Lung Res.* **2010**, *36*, 12–24. [CrossRef] [PubMed]
7. Tsubota, Y.; Ogawa, T.; Oyanagi, J.; Nagashima, Y.; Miyazaki, K. Expression of laminin γ2 chain monomer enhances invasive growth of human carcinoma cells in vivo. *Int. J. Cancer* **2010**, *127*, 2031–2041. [CrossRef] [PubMed]
8. Chuang, M.J.; Sun, K.H.; Tang, S.J.; Deng, M.W.; Wu, Y.H.; Sung, J.S.; Cha, T.L.; Sun, G.H. Tumor-derived tumor necrosis factor-α promotes progression and epithelial-mesenchymal transition in renal cell carcinoma cells. *Cancer Sci.* **2008**, *99*, 905–913. [CrossRef]
9. Ho, M.Y.; Tang, S.; Chuang, M.J.; Cha, T.L.; Li, J.Y.; Sun, G.H.; Sun, K.H. TNF-α induces epithelial-mesenchymal transition of renal cell carcinoma cells via a GSK3β-dependent mechanism. *Mol. Cancer Res.* **2012**, *10*, 1109–1119. [CrossRef]
10. Lee, S.J.; Park, S.S.; Cho, Y.H.; Park, K.; Kim, E.J.; Jung, K.H.; Kim, S.K.; Kim, W.J.; Moon, S.K. Activation of matrix metalloproteinase-9 by TNF-α in human urinary bladder cancer HT1376 cells: The role of MAP kinase signaling pathways. *Oncol. Rep.* **2008**, *19*, 1007–1013. [CrossRef]
11. Itatsu, K.; Sasaki, M.; Yamaguchi, J.; Ohira, S.; Ishikawa, A.; Ikeda, H.; Sato, Y.; Harada, K.; Zen, Y.; Sato, H.; et al. Cyclooxygenase-2 is involved in the up-regulation of matrix metalloproteinase-9 in cholangiocarcinoma induced by tumor necrosis factor-α. *Am. J. Pathol.* **2009**, *174*, 829–841. [CrossRef] [PubMed]
12. Jayasooriya, R.G.; Lee, Y.G.; Kang, C.H.; Lee, K.T.; Choi, Y.H.; Park, S.Y.; Hwang, J.K.; Kim, G.Y. Piceatannol inhibits MMP-9-dependent invasion of tumor necrosis factor-α-stimulated DU145 cells by suppressing the Akt-mediated nuclear factor-κB pathway. *Oncol. Lett.* **2013**, *5*, 341–347. [CrossRef] [PubMed]
13. Roomi, M.W.; Kalinovsky, T.; Niedzwiecki, A.; Rath, M. Modulation of MMP-2 and -9 secretion by cytokines, inducers and inhibitors in human melanoma A-2058 cells. *Oncol. Rep.* **2017**, *37*, 3681–3687. [CrossRef] [PubMed]
14. Guo, L.; Ou, J.L.; Zhang, T.; Ma, L.; Qu, L.F. Effect of expressions of tumor necrosis factor α and interleukin 1B on peritoneal metastasis of gastric cancer. *Tumour Biol.* **2015**, *36*, 8853–8860. [CrossRef] [PubMed]

15. Saito, Y.; Sekine, W.; Sano, R.; Komatsu, S.; Mizuno, H.; Katabami, K.; Shimada, K.; Oku, T.; Tsuji, T. Potentiation of cell invasion and matrix metalloproteinase production by α3β1 integrin-mediated adhesion of gastric carcinoma cells to laminin-5. *Clin. Exp. Metastasis* **2010**, *27*, 197–205. [CrossRef] [PubMed]
16. Kamoshida, G.; Matsuda, A.; Miura, R.; Takashima, Y.; Katsura, A.; Tsuji, T. Potentiation of tumor cell invasion by co-culture with monocytes accompanying enhanced production of matrix metalloproteinase and fibronectin. *Clin. Exp. Metastasis* **2013**, *30*, 289–297. [CrossRef] [PubMed]
17. Yabluchanskiy, A.; Ma, Y.; Iyer, R.P.; Hall, M.E.; Lindsey, M.L. Matrix metalloproteinase-9: Many shades of function in cardiovascular disease. *Physiology* **2013**, *28*, 391–403. [CrossRef]
18. Takatsuki, H.; Komatsu, S.; Sano, R.; Takada, Y.; Tsuji, T. Adhesion of gastric carcinoma cells to peritoneum mediated by α3β1 integrin (VLA-3). *Cancer Res.* **2004**, *64*, 6065–6070. [CrossRef]
19. Mochizuki, Y.; Nakanishi, H.; Kodera, Y.; Ito, S.; Yamamura, Y.; Kato, T.; Hibi, K.; Akiyama, S.; Nakao, A.; Tatematsu, M. TNF-α promotes progression of peritoneal metastasis as demonstrated using a green fluorescence protein (GFP)-tagged human gastric cancer cell line. *Clin. Exp. Metastasis* **2004**, *21*, 39–47. [CrossRef]
20. Kim, S.; Choi, M.G.; Lee, H.S.; Lee, S.K.; Kim, S.H.; Kim, W.W.; Hur, S.M.; Kim, J.H.; Choe, J.H.; Nam, S.J.; et al. Silibinin suppresses TNF-alpha-induced MMP-9 expression in gastric cancer cells through inhibition of the MAPK pathway. *Molecules* **2009**, *14*, 4300–4311. [CrossRef]
21. Kawanishi, K.; Nitta, K.; Yamato, M.; Okano, T. Therapeutic applications of mesothelial cell sheets. *Ther. Apher. Dial.* **2015**, *19*, 1–7. [CrossRef]
22. Chen, W.L.; Sheu, J.R.; Chen, R.J.; Hsiao, S.H.; Hsiao, C.J.; Chou, Y.C.; Chung, C.L.; Hsiao, G. *Mycobacterium tuberculosis* upregulates TNF-α expression via TLR2/ERK signaling and induces MMP-1 and MMP-9 production in human pleural mesothelial cells. *PLoS ONE* **2015**, *10*, e0137979. [CrossRef] [PubMed]
23. Hozumi, A.; Nishimura, Y.; Nishiuma, T.; Kotani, Y.; Yokoyama, M. Induction of MMP-9 in normal human bronchial epithelial cells by TNF-α via NF-κB-mediated pathway. *Am. J. Physiol. Lung Cell. Mol. Physiol.* **2001**, *281*, L1444–L1452. [CrossRef] [PubMed]
24. Chen, Y.J.; Chang, L.S. NFκB- and AP-1-mediated DNA looping regulates matrix metalloproteinase-9 transcription in TNF-α-treated human leukemia U937 cells. *Biochim. Biophys. Acta* **2015**, *1849*, 1248–1259. [CrossRef] [PubMed]
25. Liu, J.; Geng, X.; Li, Y. Milky spots: Omental functional units and hotbeds for peritoneal cancer metastasis. *Tumour Biol.* **2016**, *37*, 5715–5726. [CrossRef] [PubMed]
26. Kanda, M.; Kodera, Y. Molecular mechanisms of peritoneal dissemination in gastric cancer. *World J. Gastroenterol.* **2016**, *22*, 6829–6840. [CrossRef] [PubMed]
27. Torisu, H.; Ono, M.; Kiryu, H.; Furue, M.; Ohmoto, Y.; Nakayama, J.; Nishioka, Y.; Sone, S.; Kuwano, M. Macrophage infiltration correlates with tumor stage and angiogenesis in human malignant melanoma: Possible involvement of TNFα and IL-1α. *Int. J. Cancer* **2000**, *85*, 182–188. [CrossRef]
28. Hagemann, T.; Robinson, S.C.; Schulz, M.; Trümper, L.; Balkwill, F.R.; Binder, C. Enhanced invasiveness of breast cancer cell lines upon co-cultivation with macrophages is due to TNF-α dependent up-regulation of matrix metalloproteases. *Carcinogenesis* **2004**, *25*, 1543–1549. [CrossRef]
29. Yao, Z.; Yuan, T.; Wang, H.; Yao, S.; Zhao, Y.; Liu, Y.; Jin, S.; Chu, J.; Xu, Y.; Zhou, W.; et al. MMP-2 together with MMP-9 overexpression correlated with lymph node metastasis and poor prognosis in early gastric carcinoma. *Tumour Biol.* **2017**, *39*, 1010428317700411. [CrossRef] [PubMed]
30. Wroblewski, L.E.; Pritchard, D.M.; Carter, S.; Varro, A. Gastrin-stimulated gastric epithelial cell invasion: The role and mechanism of increased matrix metalloproteinase 9 expression. *Biochem. J.* **2002**, *365 Pt 3*, 873–879. [CrossRef]
31. Shan, Y.Q.; Ying, R.C.; Zhou, C.H.; Zhu, A.K.; Ye, J.; Zhu, W.; Ju, T.F.; Jin, H.C. MMP-9 is increased in the pathogenesis of gastric cancer by the mediation of HER2. *Cancer Gene Ther.* **2015**, *22*, 101–107. [CrossRef] [PubMed]
32. Nakashio, T.; Narita, T.; Akiyama, S.; Kasai, Y.; Kondo, K.; Ito, K.; Takagi, H.; Kannagi, R. Adhesion molecules and TGF-β1 are involved in the peritoneal dissemination of NUGC-4 human gastric cancer cells. *Int. J. Cancer* **1997**, *70*, 612–618. [CrossRef]
33. Tsuji, T.; Kawada, Y.; Kai-Murozono, M.; Komatsu, S.; Han, S.A.; Takeuchi, K.; Mizushima, H.; Miyazaki, K.; Irimura, T. Regulation of melanoma cell migration and invasion by laminin-5 and α3β1 integrin (VLA-3). *Clin. Exp. Metastasis* **2002**, *19*, 127–134. [CrossRef]

34. Takeuchi, K.; Tsuji, T.; Hakomori, S.; Irimura, T. Intercellular adhesion induced by anti-α3 integrin (VLA-3) antibodies. *Exp. Cell Res.* **1994**, *211*, 133–141. [CrossRef] [PubMed]
35. Ries, C.; Egea, V.; Karow, M.; Kolb, H.; Jochum, M.; Neth, P. MMP-2, MT1-MMP, and TIMP-2 are essential for the invasive capacity of human mesenchymal stem cells: Differential regulation by inflammatory cytokines. *Blood* **2007**, *109*, 4055–4063. [CrossRef] [PubMed]
36. Gondi, C.S.; Dinh, D.H.; Gujrati, M.; Rao, J.S. Simultaneous downregulation of uPAR and MMP-9 induces overexpression of the FADD-associated protein RIP and activates caspase 9-mediated apoptosis in gliomas. *Int. J. Oncol.* **2008**, *33*, 783–790. [PubMed]
37. Oku, T.; Ando, Y.; Ogura, M.; Tsuji, T. Development of splice variant-specific monoclonal antibodies against human α3 Integrin. *Monoclon. Antib. Immunodiagn. Immunother.* **2016**, *35*, 12–17. [CrossRef]
38. Morodomi, T.; Ogata, Y.; Sasaguri, Y.; Morimatsu, M.; Nagase, H. Purification and characterization of matrix metalloproteinase 9 from U937 monocytic leukaemia and HT1080 fibrosarcoma cells. *Biochem. J.* **1992**, *285 (Pt 2)*, 603–611. [CrossRef]
39. Itoh, S.; Hamada, E.; Kamoshida, G.; Takeshita, K.; Oku, T.; Tsuji, T. Staphylococcal superantigen-like protein 5 (SSL5) inhibits matrix metalloproteinase-9 from human neutrophils. *Infect. Immun.* **2010**, *78*, 3298–3305. [CrossRef]

International Journal of
Molecular Sciences

MDPI

Article

ADAMTS-9 in Mouse Cartilage Has Aggrecanase Activity That Is Distinct from ADAMTS-4 and ADAMTS-5

Fraser M. Rogerson [1,2,3], Karena Last [1,2], Suzanne B. Golub [1,2], Stephanie J. Gauci [2], Heather Stanton [1,2], Katrina M. Bell [2] and Amanda J. Fosang [1,2,*]

1 University of Melbourne Department of Paediatrics, Royal Children's Hospital, Parkville, Victoria 3052, Australia; fraser.rogerson@rmit.edu.au (F.M.R.); karena.last@mcri.edu.au (K.L.); sue.golub@mcri.edu.au (S.B.G.); heather.stanton@mcri.edu.au (H.S.)
2 Murdoch Children's Research Institute, Royal Children's Hospital, Parkville, Victoria 3052, Australia; steph_pascoe@hotmail.com (S.J.G.); katrina.bell@mcri.edu.au (K.M.B.)
3 Royal Melbourne Institute of Technology, 124 La Trobe Street, Melbourne, Victoria 3000, Australia
* Correspondence: amanda.fosang@mcri.edu.au

Received: 19 December 2018; Accepted: 24 January 2019; Published: 29 January 2019

Abstract: A disintegrin and metalloproteinase with thrombospondin motifs (ADAMTS)-4 and ADAMTS-5 are the principal aggrecanases in mice and humans; however, mice lacking the catalytic domain of both enzymes (TS-4/5Δcat) have no skeletal phenotype, suggesting there is an alternative aggrecanase for modulating normal growth and development in these mice. We previously identified aggrecanase activity that (a) cleaved at E↓G rather than E↓A bonds in the aggrecan core protein, and (b) was upregulated by retinoic acid but not IL-1α. The present study aimed to identify the alternative aggrecanase. Femoral head cartilage explants from TS-4/5Δcat mice were stimulated with IL-1α or retinoic acid and total RNA was analysed by microarray. In addition to ADAMTS-5 and matrix metalloproteinase (MMP)-13, which are not candidates for the novel aggrecanase, the microarray analyses identified MMP-11, calpain-5 and ADAMTS-9 as candidate aggrecanases upregulated by retinoic acid. When calpain-5 and MMP-11 failed to meet subsequent criteria, ADAMTS-9 emerged as the most likely candidate for the novel aggrecanase. Immunohistochemistry revealed ADAMTS-9 expression throughout the mouse growth plate and strong expression, particularly in the proliferative zone of the TS-4/5-Δcat mice. In conclusion, ADAMTS-9 has a novel specificity for aggrecan, cleaving primarily at E↓G rather than E↓A bonds in mouse cartilage. ADAMTS-9 might have more important roles in normal skeletal development compared with ADAMTS-4 and ADAMTS-5, which have key roles in joint pathology.

Keywords: aggrecan; aggrecanase; ADAMTS; cartilage; arthritis

1. Introduction

The weight-bearing properties of articular cartilage are conferred by the large and highly charged glycosaminoglycan side chains that are present on the proteoglycan aggrecan. In arthritic disease, the resilience of cartilage is compromised when large aggrecan aggregates are degraded by aggrecanases that are members of the ADAMTS (a disintegrin and metalloproteinase with thrombospondin motifs) family of zinc-dependent enzymes within the metzincin family of metalloproteinases. The principal aggrecanases thought to be responsible for aggrecan degradation are ADAMTS-4 [1] and ADAMTS-5 [2].

These enzymes were originally identified by their ability to cleave the aggrecan core protein at the TEGE $^{373\downarrow374}$ ALGS bond in the aggrecan interglobular domain (IGD) [3–5] and at four additional sites within the chondroitin sulphate (CS)-rich region [6,7] (Figure 1) in response to stimulation with pro-catabolic agents, such as interleukin-1-alpha (IL-1α), IL-1β and retinoic acid. In contrast, ADAMTS-1 [8], -8 [9], -9 [10,11], -16 [11] and -18 [11] were thought to have only weak activity against aggrecan, in vitro and in vivo.

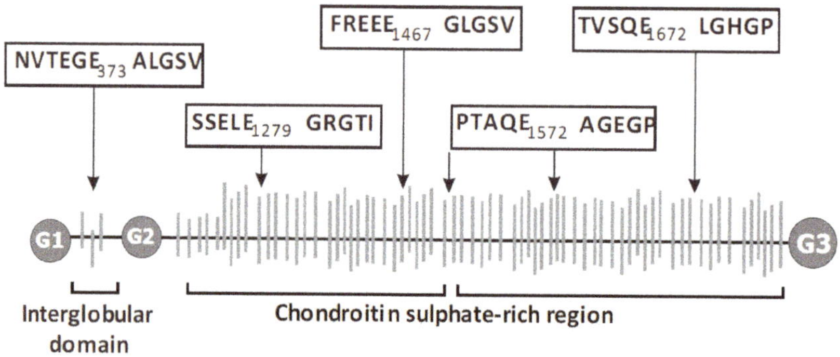

Figure 1. Aggrecanase cleavage sites and neoepitopes in the mouse aggrecan core protein. The domain structure of the mouse aggrecan core protein, with globular G1, G2 and G3 domains and intervening interglobular and chondroitin sulphate-rich regions are shown. The amino acid sequences flanking the conserved cleavage sites are shown in boxes with amino acid numbers at the P1 position. Modified from [12].

In the course of our studies, we identified a new aggrecanase activity that was present in cartilage explants from mice deficient in both ADAMTS-4 and ADAMTS-5 catalytic activity (TS-4/5Δcat mice) [13]. We found that retinoic acid treatment of TS-4/5Δcat cartilage explants did not stimulate aggrecanase activity at the classical TEGE $^{373\downarrow374}$ ALGS or TAQE $^{1572\downarrow1573}$ AGEG sites, but instead at the FREEE $^{1467\downarrow1468}$ GLGS and SSELE $^{1279\downarrow1280}$ GRGTI sites located in the chondroitin sulphate-rich region of the core protein (Figure 1). Thus, retinoic acid stimulates the expression of a novel aggrecanase activity in TS-4/5Δcat cartilage. This unexpected finding reveals that the substrate specificity of this unknown enzyme is distinct from ADAMTS-4 and -5. In addition, and in contrast to ADAMTS-4 and ADAMTS-5, the activity of the novel aggrecanase is stimulated by retinoic acid, but not by interleukin-1α (IL-1α) treatment [13].

In order to identify this novel aggrecanase, we cultured mouse femoral head cartilage explants from wildtype and TS-4/5Δcat mice. We used discrimination between the known effects of IL-1α and retinoic acid on the up-regulation of aggrecanase activity to screen for candidate extracellular proteinases whose mRNA expression levels were specifically increased by retinoic acid, but not by IL-1α treatment. Our results reveal that ADAMTS-9 is a major aggrecanase in TS-4/5Δcat mice, raising the possibility that ADAMTS-9 might be active in cartilage, in vivo, in the presence of inhibitors designed to block ADAMTS-4 and ADAMTS-5 activity.

2. Results

2.1. Retinoic Acid Regulates the Expression of a Novel Aggrecanase Activity via a Transcriptional Mechanism

We previously identified a novel aggrecanase activity that cleaves aggrecan in the CS-rich region in cartilage explants from mice deficient in both ADAMTS-4 and ADAMTS-5 [13]. Since this novel activity is stimulated by treatment with retinoic acid, but not IL-1α, we reasoned that the differential in susceptibility to retinoic acid and IL-1α could be used to identify aggrecanase candidates by microarray. We first confirmed that retinoic acid stimulates aggrecanolysis in TS-4/5Δcat cartilage via a transcriptional mechanism by culturing TS-4/5Δcat cartilage explants in the presence or absence of the transcriptional inhibitor, actinomycin D. Western blot analysis of the conditioned media showed that retinoic acid treatment stimulated the release of the aggrecan FREEE ^{1467} neoepitope from TS-4/5Δcat cartilage into conditioned medium (Figure 2a, lane 3), compared with the untreated control (Figure 2a, lane 1). However, the retinoic acid-induced increase in FREEE ^{1467} neoepitope was blocked in the presence of actinomycin D (Figure 2a, lane 4), confirming that the effects of retinoic acid in this system were indeed transcription-dependent. The chondrocytes remained viable in culture for a further 6 days in the presence or absence of retinoic acid, confirming that inhibition of aggrecan cleavage was not due to actinomycin-induced cell death.

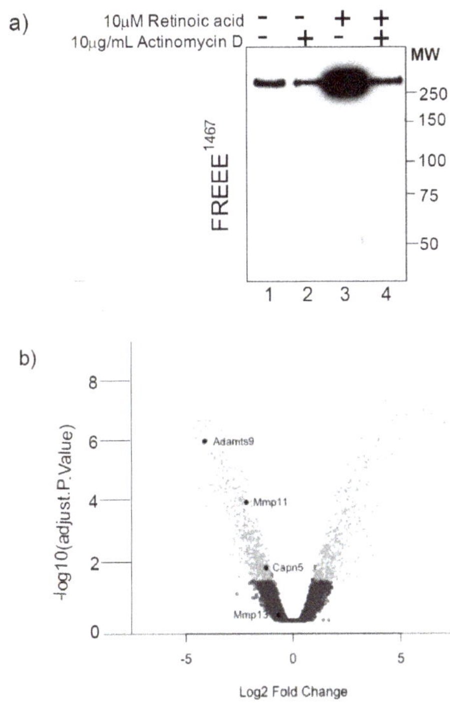

Figure 2. Identifying candidate aggrecanases in TS4/5Δcat mice. (**a**) Aggrecanase cleavage in TS-4/5Δcat femoral head cartilage explants, cultured untreated (lane 1) or in the presence of actinomycin D (lane 2), retinoic acid (lane 3) or both actinomycin D and retinoic acid together (lane 4) was analysed by FREEE ^{1467} neo-epitope Western blotting. (**b**) The results of the microarray studies, represented as a volcano plot, identify candidate proteinases for the novel aggrecanase.

2.2. Microarray and qPCR Studies Reveal ADAMTS-9 as a Strong Candidate for the Novel Aggrecanase

Based on our conclusion that retinoic acid exerts transcriptional control over aggrecanolysis in TS-4/5Δcat chondrocytes, we next used microarrays to shortlist genes encoding aggrecan-degrading proteinases, based on their up-regulation in response to retinoic acid but not IL-1α treatment. We focused on enzyme families known to cleave aggrecan, including the a disintegrin and metalloproteinase (ADAM), ADAMTS, matrix metalloproteinase (MMP), cathepsin and calpain families. The results presented as a volcano plot in Figure 2b show that only three genes matched these criteria: *Adamts-9*, *Mmp-11* and *Calpain-5* (Figure 2b). There was no increase greater than 1.5-fold in the expression of any ADAM or cathepsin gene in IL-1α-treated compared with retinoic acid-treated cultures.

To validate the microarray expression data, we used qPCR to measure the expression of key genes in cultured chondrocytes from TS-4/5Δcat mice. The results shown in Figure 3 confirm the up-regulation of mRNA expression for *Adamts-5* (Figure 3a), *Mmp-13* (Figure 3b) and *Mmp-11* (Figure 3d) by both IL-1α and retinoic acid, as reported previously, thus excluding these enzymes as the novel aggrecanase. The expression of *Calpain-5* was only weakly increased by IL-1α and retinoic acid (Figure 3c) and was therefore also considered not to be an aggrecanase. Overall, the mRNA data suggested that ADAMTS-9 was the strongest and most likely candidate for the novel aggrecanase. Figure 3e shows that while *Adamts-9* mRNA expression is increased by treatment with retinoic acid, there is no increase in *Adamts-9* gene expression induced by IL-1α. Similarly, retinoic acid, but not IL-1α, increased the expression of ADAMTS-9 protein detected as a band of approximately 200 kDa by Western blot (Figure 3f).

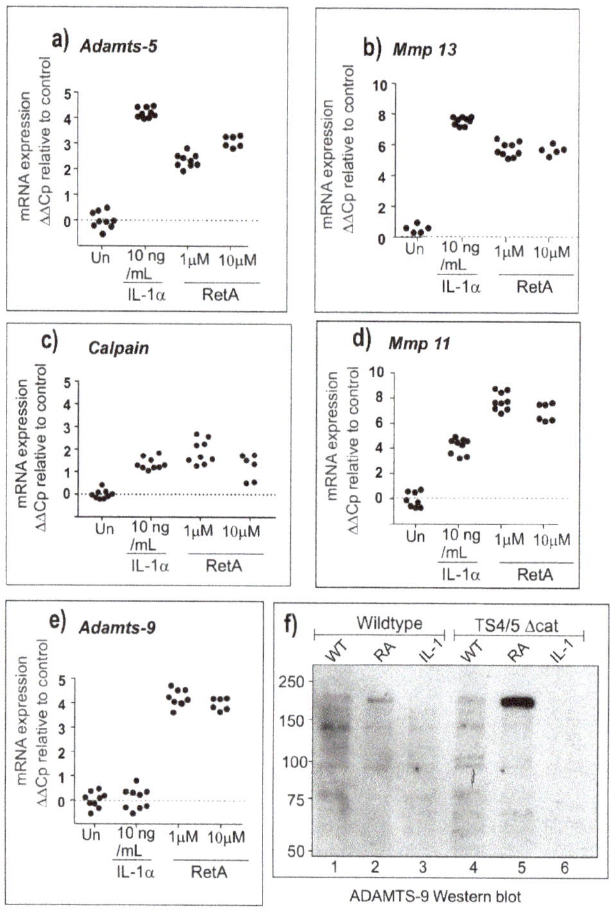

Figure 3. q-PCR and Western blot analyses of candidate aggrecanases. Candidate aggrecanases expressed in cultured cartilage explants were analysed by qPCR for genes including *Adamts-5* (**a**), *Mmp-13* (**b**), *Calpain* (**c**), *Mmp-11* (**d**), and *Adamts-9* (**e**). Protein extracts from Wildtype and TS4/TS5Δcat cartilage extracts were also analysed for ADAMTS-9 expression by Western blotting (**f**).

2.3. The Novel Aggrecanase is Inhibited by TIMP-3

Tissue inhibitor of metalloproteinase (TIMP)-3 is a potent endogenous inhibitor of ADAMTS and MMP family members [14]. We therefore used sensitivity to inhibition with TIMP-3 to further confirm the identity of the novel aggrecanase and found that when femoral head explants from TS-4/5Δcat mice were incubated with retinoic acid in the presence or absence of 500 nM TIMP-3, aggrecan cleavage at the FREEE $^{1467\downarrow1468}$ GLGSV site was inhibited (Figure 4a, lane 5). An additional FREEE ^{1467} band migrating at approximately 50 kD and most likely the SSELE $^{1279\_1467}$ FREEE fragment previously identified in human cartilage [15] was also generated in the presence of retinoic acid (Figure 4a, lanes 2 and 3) but blocked by high concentrations of TIMP-3 (Figure 4a, lane 5). As a control for the potency of TIMP-3 inhibition, cleavage at the MMP-sensitive DIPEN $^{342\downarrow343}$ FFGVG site was also examined. TIMP-3 blocked cleavage at DIPEN $^{342\downarrow343}$ FFGVG at the same high concentration (Figure 4b, lanes 4 and 5).

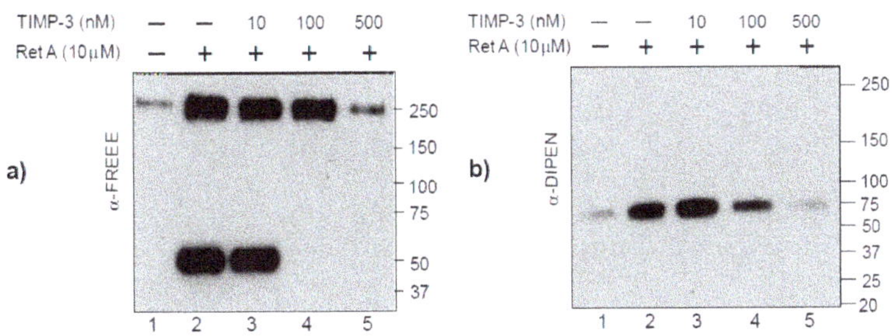

Figure 4. The novel aggrecanase is inhibited by tissue inhibitor of metalloproteinase (TIMP)-3. Western blot analyses show that TIMP-3 inhibits retinoic acid-induced aggrecanolysis at FREEE $^{1467\downarrow1468}$ GLGSV (**a**) and DIPEN $^{341\downarrow342}$ FFGVG (**b**) in TS-4/5Δcat cartilage explants.

2.4. ADAMTS9 Expression and Aggrecanase Activity in the Growth Plate

Finally, we used immunohistochemistry to compare sites of ADAMTS-9 expression and aggrecanase activity in sections of the 3-week mouse growth plate. For both genotypes, FREEE ^{1467} neoepitope was present in selected regions of the growth plate including the proliferative zone (Figure 5a,b,e,f; P) and a thin layer of cells surrounding the secondary centre of ossification (Figure 5a,b,e,f; right-pointing arrow). The staining surrounding the secondary centre was more punctate in sections from TS-4/5Δcat mice (Figure 5e,f) than from wildtype mice (Figure 5a,b). FREEE ^{1467} neoepitope was also present at the metaphyseal border in wildtype mice (Figure 5a,b; left-pointing arrow) but was absent at this site in TS-4/5Δcat mice (Figure 5e,f). The different patterns of FREEE ^{1467} immunostaining at the metaphyseal border in wildtype mice, compared with TS4/5Δcat mice, suggest that FREEE ^{1467} immunoreactivity in the growth plate is a product of ADAMTS-4 and/or ADAMTS-5 activity. We therefore conclude that ADAMTS-9 is not essential for aggrecanase activity at the metaphyseal border.

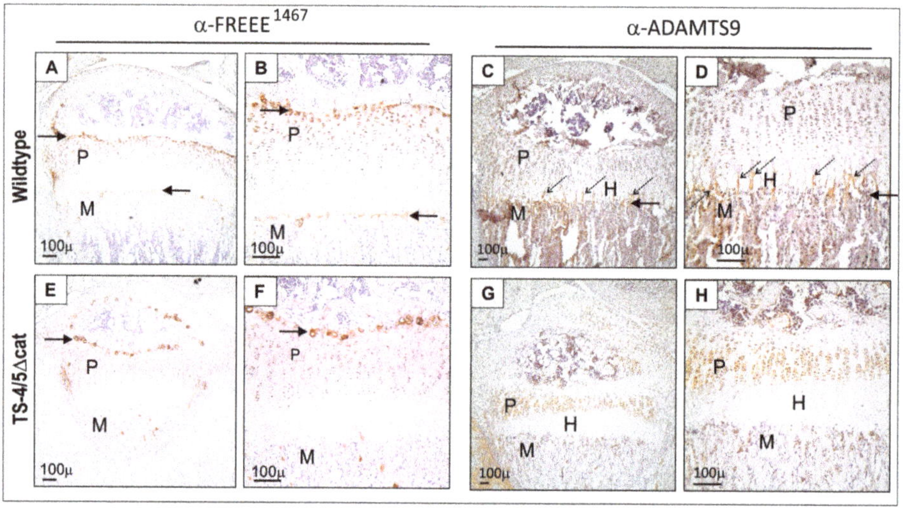

Figure 5. ADAMTS-9 protein and FREEE [1467] neoepitope are present in the 3-week-old mouse joint. M, metaphysis; P, proliferative zone of the growth plate; left-pointing arrow, metaphyseal border; right-pointing arrow, boundary of the secondary centre of ossification; diagonal dotted arrows, columns of calcified cartilage; H, hypertrophic zone of the growth plate. Panels (**a**,**b**,**e**,**f**) show staining for the FREEE [1467] neoepitope. Panels (**c**,**d**,**g**,**h**) show staining for ADAMTS-9 protein. Panels (**a**–**d**) show cartilage from wildtype mice. Panels (**e**–**h**) show cartilage from TS-4/5Δcat mice.

Unlike the limited staining for the aggrecan FREEE [1467] neoepitope seen in wildtype growth plates (Figure 5 a,b,e,f), ADAMTS-9 immunostaining was strong throughout all zones of the wildtype growth plate, including the metaphysis (M), proliferative zone (P), hypertrophic zones (H) and in bony spicules (*diagonal dotted arrows*) protruding into the calcified region of the hypertrophic zone (Figure 5c,d). In contrast, there was no immunostaining for ADAMTS-9 in the hypertrophic/pre-hypertrophic zones (H) of the growth plate in TS4/5Δcat mice (Figure 5g,h). One interpretation of this unintuitive staining pattern is that ADAMTS-9 is not expressed in the hypertrophic zone (Figure 5g,h). However, a more likely interpretation is that the physical entrapment of residual, undegraded and highly charged aggrecan in the hypertrophic zone blocks diffusion of immunoreagents into this region of the sections, producing a false negative result.

3. Discussion

ADAMTS-9 is a multi-domain proteinase with key roles in the remodelling of extracellular matrices during growth and development [16–21]. Versican is a major substrate for ADAMTS-9, and in this study we show that the related proteoglycan, aggrecan, which is highly expressed in joint cartilage, is also a substrate for ADAMTS-9. Our results suggest that unlike ADAMTS-4 and ADAMTS-5, which are widely recognised as destructive aggrecanases in joint disease [22,23], the aggrecanolytic activity of ADAMTS-9 might be more important for normal joint development [18] than for pathology.

Previous studies in wildtype mice have reported the mRNA expression of ADAMTS-9 in chondrocytes and cartilage explants [20,21]; however, in the presence of ADAMTS-4 and ADAMTS-5 expression, it has not been possible to show direct evidence of ADAMTS-9 activity. Here, we have used TS-4/5Δcat mice to show that ADAMTS-9 is a bona fide aggrecanase with activity, not at the classical TEGE $^{373\downarrow374}$ ALGS site in the IGD and TAQE $^{1572\downarrow1573}$ AGEG site in the C-terminal domains, but instead ADAMTS-9 cleaves at FREEE $^{1467\downarrow1468}$ GLGSV and SELE $^{1279\downarrow1280}$ GRGT sites located centrally within the aggrecan core protein. These results extend previous studies, suggesting that ADAMTS-9 might also cleave at TAQE $^{1771\downarrow1772}$ AGEG near the C-terminal G3 domain [10].

Our results are consistent with previous studies showing *Adamts-9* mRNA expression in neonatal growth plates [21] and in the proliferative and hypertrophic zones of the 7-week-old mouse growth plate [21]. In addition, we have used immunoreactivity against the FREEE ^{1467} neoepitope to show that TS-4/5Δcat mice express aggrecanase activity, and we propose that this activity is likely to be the product of ADAMTS-9 (Figure 5).

The distribution of FREEE ^{1467} neoepitope in the TS-4/5Δcat epiphysis suggests that ADAMTS-9 has a role in the remodelling of the secondary centre of ossification and the proliferative zone of the growth plate. However, whereas the absence of FREEE ^{1467} neoepitope at the metaphyseal border in TS4/5-Δcat mice suggests that ADAMTS-9 might not degrade aggrecan at this site, a more likely interpretation (and complication) of this staining pattern is that an accumulation of undegraded aggrecan in calcified regions of the hypertrophic zone impedes antibody penetration and compromises immunoreactivity.

Beyond the G2 domain, the aggrecan core protein is heavily decorated with glycosaminoglycan side chains. When ADAMTS-9 cleaves at FREEE $^{1467\downarrow1468}$ GLGSV, the C-terminal third of aggrecan is lost from the cartilage, while the remaining N-terminal two thirds, immobilised in the cartilage by G1-domain binding to hyaluronan, provide the weight-bearing properties. Accordingly, cleavage at FREEE $^{1467\downarrow1468}$ GLGSV is less deleterious for weight-bearing than cleavage in the IGD, which releases the entire CS-containing portion of aggrecan.

There are very few studies on the roles of regional aggrecanolysis. In vitro testing of cartilage explants showed that aggrecan loss mediated by IGD cleavage correlates with a loss of mechanical properties, whereas aggrecan loss due to cleavage at sites in the CS-2 domain does not appear to affect biomechanics [24]. In vivo studies to specifically explore the role of C-terminal domain cleavage have not been done, but it is widely held that C-terminal processing of aggrecan is part of normal homeostasis [25,26]. Furthermore, the proportion of C-terminally shortened aggrecan increases with age [27,28], suggesting that loss of the C-terminus is part of normal tissue maturation. Thus, ADAMTS-9 might have a role in maturation and normal cartilage homeostasis.

Alternatively, CS-2 domain cleavage might indeed have a role in cartilage pathology. In vitro evidence suggests that there is a hierarchy for aggrecanase cleavage and that cleavage in the IGD occurs after cleavage in the CS-2 domain [29–31]. If CS-2 domain cleavage is a prerequisite for IGD cleavage, then a CS-2 aggrecanase such as ADAMTS-9 might initiate an aggrecan-degrading cascade. Aggrecan fragments resulting from CS-2 domain cleavage are elevated in human synovial fluids following knee joint injury [32], raising the possibility that ADAMTS-9 contributes to the production of G3-containing fragments from injured cartilage. Furthermore, the G3 domain binds complement factors C1q and C3, thus activating the classical and alternative complement pathways [33], suggesting a possible role for ADAMTS-9 in complement regulation in inflamed joints.

4. Materials and Methods

4.1. Mice

The TS-4/5Δcat mice deficient in both ADAMTS-4 and ADAMTS-5 activity have been described previously [13]. The Δ-cat mice were bred from the TS-4 Δ-cat and TS-5 Δ-cat strains [13,22], which have in-frame deletions of exon 4 of the *Adamts-4* gene (TS-4 Δ-cat) and exon 3 of the *Adamts-5* gene (TS-5 Δcat), encoding the respective catalytic domains [22]. Mutant ADAMTS-5 mRNA lacking exon 3 is translated and includes exons encoding the ADAMTS-5 C-terminal domain [34]. The TS-4/5Δcat mice are healthy with no abnormalities in skeletal growth or development. The Murdoch Children's Research Institute Animal Ethics Committee approved all animal procedures under projects A605, A710, A795 and A868. Each experiment was done at least twice with a minimum of two biological replicates, except for the microarray which was done once with three mice per treatment.

4.2. Preparation of Primary Chondrocytes

Chondrocytes were isolated from the distal femoral and proximal tibial epiphyses of 6-day-old mice. The dissected epiphyses were digested for 1 h with 0.25% trypsin/EDTA to remove loose fibrous tissue, washed with PBS, then incubated overnight at 37 °C/5%CO_2 with 300 U/mL collagenase II (Worthington Biochemical Corp, Lakewood, NJ, USA) in DMEM and 5% fetal bovine serum (FBS). The isolated cells were washed extensively with medium and seeded at 5×10^5 cells/well in 24-well plates in DMEM, 10% FBS, 100 U/mL penicillin, 100 U/mL streptomycin, 2 mM L-glutamine and 20 mM HEPES. After 3 days in culture, the cells were washed, placed in serum-free DMEM and stimulated with either 10 ng/mL IL-1α (Peprotech, Rocky Hill, NJ, USA) or 10 µM retinoic acid (Sigma-Aldrich, St Louis, MO, USA) for a further 3 days. These concentrations of IL-1α and retinoic acid were previously confirmed by dose response to elicit maximal gene expression.

4.3. Culture of Femoral Head Cartilage Explants

Mouse femoral head cartilage explants were cultured as described previously [12,13]. Explants were incubated with 10 µM retinoic acid ± 10µg/mL actinomycin D (Life Technologies, Carlsbad, CA, USA) for 24 h. Conditioned medium was harvested, and the explants were washed and incubated with fresh retinoic acid for a further 6 days to confirm that actinomycin D had no adverse effect on chondrocyte activity or viability. To examine the inhibition of aggrecanase activity by tissue inhibitor of metalloproteinase-3 (TIMP-3), femoral head explants were co-cultured with 10 µM retinoic acid and 10–500 nM recombinant human TIMP-3 (a gift from Professor Hideaki Nagase, UK) and then analysed for aggrecanase products by Western blotting.

4.4. Western Blotting

To detect aggrecan neoepitopes, conditioned media from mouse femoral head cultures were dialysed against proteinase inhibitors and deglycosylated as described previously [34]. Deglycosylated samples were resolved by reducing SDS-PAGE on 4–20% gradient 7TGX gels (Bio-Rad, Hercules, CA, USA) and then transferred to polyvinylidene difluoride membranes. Aggrecan degradation was analysed using polyclonal antibodies [35] that recognise the FREEE ^{1467} neoepitope [36], created by aggrecanase cleavage in the CS-rich region, and the DIPEN ^{341} neoepitope [37], created by MMP cleavage in the IGD. To detect ADAMTS-9 protein, conditioned medium from primary chondrocyte cell cultures was blotted and detected with anti-ADAMTS-9 antibody from Abcam (Cambridge, UK), used at 1 µg/mL and detected with anti-rabbit-HRP (1:2000) from Dako (Jena, Germany). The volume of medium loaded was determined empirically for each neoepitope examined and results were normalised to the values for total aggrecan as described previously [12,13]. The anti-DIPEN and anti-FREEE antibodies were used at a concentration of 5 µg/mL.

4.5. Microarray Analysis

Femoral head cartilage explants harvested from 3-week-old TS-4/5Δcat mice were cultured for 3 days in serum-free DMEM containing 100 U/mL penicillin, 100 U/mL streptomycin, 2 mM L-glutamine and 20 mM HEPES. The explants were either (i) unstimulated, (ii) stimulated with 10 ng/mL recombinant human IL-1 or (iii) stimulated with 10 µM retinoic acid. These concentrations of IL-1α and retinoic acid stimulated maximal aggrecan loss from cartilage explants as described previously [13]. After treatment, the explants were snap frozen and stored at −80 °C. Three separate mice (unpooled) were included for each treatment. Total RNA from frozen, crushed cartilage samples was isolated using the RNeasy kit (Qiagen, Hilden, Germany) and the RNA quality was assessed by BioAnalyser. The microarray analyses were done under contract at the Australian Genome Research Facility using the Illumina Sentrix Mouse 1.1 chip. The data were analysed using the R limma package from Bioconductor, utilising the neqc function (quantile normalisation) with non-detected probes removed and a factor included for batch and biological replicates in the linear mode [38]. The microarray data is available at Gene Expression Omnibus (GSE110754).

4.6. Quantitative Real-Time Polymerase Chain Reaction (qPCR) Analysis of mRNA Expression

To validate the microarray data identifying the putative aggrecanases, mRNA expression of the candidates was analysed by real-time quantitative PCR (qPCR) with probes and primers from the Universal Probes Library (UPL) (Roche, Basel, Switzerland). The ΔΔCp values derived from the Cp values determined by the Roche Light Cycler 480 were calculated for each gene and normalised against the geometric mean of the genes *Rpl10* and *Rpl26* [39], two reference genes specific for chondrocytes. The primer sequences used for mouse *Mmp13, Mmp11, Adamts5, Adamts9* and *calpain 5* genes are shown in Table 1.

Table 1. Primer sequences for qPCR analyses.

Gene	Sequence
Mmp-13 F	TCAAGGCTATGCACACTGGT
Mmp-13 R	CACTATGGTCTTTTCAATGCCTAA
Mmp-11 F	CAGGCCAAAAGGTACACAGC
Mmp-11 R	CCTTTGAGGTTCCGTGTCTC
Adamts-5 F	ATGCAGCCATCCTGTTCAC
Adamts-5 R	CATTCCCAGGGTGTCACAT
Adamts-9 F	ACAGCCATCAGAGAGTGCAA
Adamts-9 R	TCCTACACAGTACTTCCCACCAT
Calpain 5 F	CTAGCCTCCGCTCCAGTG
Calpain 5 R	AAGAAGGGGAGGCACCTG

4.7. Immunohistochemistry

Knee joints from 3-week-old mice were fixed in 4% (*w*/*v*) paraformaldehyde for 24 h and then decalcified in 7% EDTA/PBS, pH 7.4, for 3 weeks. Five-micron sections of paraffin-embedded epiphyses were treated with 3% (*v*/*v*) H_2O_2 to block endogenous peroxidase activity. Antigen retrieval included treatment with 10 mM citrate, pH 6.0, with 0.05% (*v*/*v*) Tween 20 for 30 min at 60 °C and then cooling to room temperature prior to incubation with 0.2% (*w*/*v*) hyaluronidase (Sigma) for 1 h at 37 °C. Non-specific antibody binding was reduced by treatment with 1.5% (*v*/*v*) normal goat serum. Sections were incubated with either rabbit anti-FREEE ^{1467} primary antibody (0.5 µg/mL), rabbit anti-ADAMTS-9 antibody (Abcam, Cambridge, UK; cat No 28279, 1:2000 dilution) or no primary antibody overnight at 4 °C. The following day, sections were incubated with HRP-labelled anti-rabbit IgG (Dako, Jena, Germany), visualised with diaminobenzidine and counterstained with haematoxylin.

4.8. Declarations Ethics Approval

The animal procedures used in this study were approved by the Murdoch Children's Research Institute Animal Ethics Committee under ethics project numbers A605 (20 May 2008), A710 (18 April 2012) A795 (16 April 2015), A868 (17 May 2018).

4.9. Availability of Data and Materials

The data sets generated and analysed during this study are available in the GEO repository at accession number GSE110754.

5. Conclusions

The results of this study are significant in the context of continuing efforts to identify the physiological aggrecanases and to develop aggrecanase inhibitors for the clinical management of degenerative joint diseases. Overall, we show that in the absence of ADAMTS-4 and -5 activity,

ADAMTS-9 is expressed and might therefore have a role in normal aggrecan remodelling during skeletal growth and development. While our findings suggest that ADAMTS-9 has important roles in skeletal growth, it is not clear what role(s), if any, ADAMTS-9 might have in pathological aggrecanolysis. In future work, it will be important to determine whether aggrecanase inhibitors targeting ADAMTS-4 and/or ADAMTS-5 might also affect ADAMTS-9 activity in adult tissues or influence endochondral ossification in developing and juvenile skeletal tissues.

Author Contributions: Conceptualisation: A.J.F.; Methodology: F.M.R.; Validation: F.M.R., K.L., S.B.G., S.J.G., A.J.F.; Formal analysis: K.M.B.; Investigation: F.M.R., K.L., S.B.G., S.J.G.; Writing—original draft preparation: F.M.R., A.J.F.; Writing—Review and editing: H.S., A.J.F.; Supervision: A.J.F.; Funding acquisition: F.M.R., A.J.F.

Funding: This work was supported by grants from the National Health and Medical research Council Australia (App ID 491203) and by the Victorian Government's Operational Infrastructure Support Program.

Acknowledgments: The authors thank Johnson & Johnson Pharmaceutical Research & Development, L.L.C. for supplying the *Adamts-4-* and *Adamts-5*-floxed mice; the Australian Genome Research Facility, Parkville, Victoria, who did the microarrays and Prof. Hideaki Nagase (Kennedy Institute of Rheumatology, UK) for providing TIMP-3.

Conflicts of Interest: The authors declare no conflict of interest.

References

1. Tortorella, M.D.; Burn, T.C.; Pratta, M.A.; Abbaszade, I.; Hollis, J.M.; Liu, R.; Rosenfeld, S.A.; Copeland, R.A.; Decicco, C.P.; Wynn, R.; et al. Purification and Cloning of Aggrecanase-1: A Member of the ADAMTS Family of Proteins. *Science* **1999**, *284*, 1664–1666. [CrossRef] [PubMed]
2. Abbaszade, I.; Liu, R.Q.; Yang, F.; Rosenfeld, S.A.; Ross, O.H.; Link, J.R.; Ellis, D.M.; Tortorella, M.D.; Pratta, M.A.; Hollis, J.M.; et al. Cloning and Characterization of ADAMTS11, an Aggrecanase from the ADAMTS Family. *J. Biol. Chem.* **1999**, *274*, 23443–23450. [CrossRef]
3. Sandy, J.D.; Neame, P.J.; Boynton, R.E.; Flannery, C.R. Catabolism of aggrecan in cartilage explants. Identification of a major cleavage site within the interglobular domain. *J. Biol. Chem.* **1991**, *266*, 8683–8685. [PubMed]
4. Sandy, J.D.; Flannery, C.R.; Neame, P.J.; Lohmander, L.S. The structure of aggrecan fragments in human synovial fluid. Evidence for the involvement in osteoarthritis of a novel proteinase which cleaves the Glu 373-Ala 374 bond of the interglobular domain. *J. Clin. Investig.* **1992**, *89*, 1512–1516. [CrossRef] [PubMed]
5. Lohmander, L.S.; Neame, P.J.; Sandy, J.D. The structure of aggrecan fragments in human synovial fluid: Evidence that aggrecanase mediates cartilage degradation in inflammatory joint disease, joint injury, and osteoarthritis. *Arthritis Rheum.* **1993**, *36*, 1214–1222. [CrossRef] [PubMed]
6. Ilic, M.Z.; Handley, C.J.; Robinson, H.C.; Mok, M.T. Mechanism of catabolism of aggrecan by articular cartilage. *Arch. Biochem. Biophys.* **1992**, *294*, 115–122. [CrossRef]
7. Loulakis, P.; Shrikhande, A.; Davis, G.; Maniglia, C.A. N-terminal sequence of proteoglycan fragments isolated from medium of interleukin-1-treated articular-cartilage cultures. Putative site(s) of enzymic cleavage. *Biochem. J.* **1992**, *284 Pt 2*, 589–593. [CrossRef]
8. Kuno, K.; Okada, Y.; Kawashima, H.; Nakamura, H.; Miyasaka, M.; Ohno, H.; Matsushima, K. ADAMTS-1 cleaves a cartilage proteoglycan, aggrecan. *FEBS Lett.* **2000**, *478*, 241–245. [CrossRef]
9. Collins-Racie, L.A.; Flannery, C.R.; Zeng, W.; Corcoran, C.; Annis-Freeman, B.; Agostino, M.J.; Arai, M.; DiBlasio-Smith, E.; Dorner, A.J.; Georgiadis, K.E.; et al. ADAMTS-8 exhibits aggrecanase activity and is expressed in human articular cartilage. *Matrix Biol.* **2004**, *23*, 219–230. [CrossRef]
10. Somerville, R.P.T.; Longpre, J.-M.; Jungers, K.A.; Engle, J.M.; Ross, M.; Evanko, S.; Wight, T.N.; Leduc, R.; Apte, S.S. Characterization of ADAMTS-9 and ADAMTS-20 as a Distinct ADAMTS Subfamily Related to *Caenorhabditis elegans* GON-1. *J. Biol. Chem.* **2003**, *278*, 9503–9513. [CrossRef]
11. Zeng, W.; Corcoran, C.; Collins-Racie, L.A.; Lavallie, E.R.; Morris, E.A.; Flannery, C.R. Glycosaminoglycan-binding properties and aggrecanase activities of truncated ADAMTSs: Comparative analyses with ADAMTS-5, -9, -16 and -18. *Biochim. Biophys. Acta* **2006**, *1760*, 517–524. [CrossRef] [PubMed]
12. Stanton, H.; Golub, S.B.; Rogerson, F.M.; Last, K.; Little, C.B.; Fosang, A.J. Investigating ADAMTS-mediated aggrecanolysis in mouse cartilage. *Nat. Protoc.* **2011**, *6*, 388–404. [CrossRef] [PubMed]

13. Rogerson, F.M.; Stanton, H.; East, C.J.; Golub, S.B.; Tutolo, L.; Farmer, P.J.; Fosang, A.J. Evidence of a novel aggrecan-degrading activity in cartilage: Studies of mice deficient in both ADAMTS-4 and ADAMTS-5. *Arthritis Rheum.* **2008**, *58*, 1664–1673. [CrossRef] [PubMed]
14. Kashiwagi, M.; Tortorella, M.; Nagase, H. Brew K TIMP-3 is a potent inhibitor of aggrecanase-1 (ADAM-TS4) and aggrecanase 2 (ADAM-TS5). *J. Biol. Chem.* **2001**, *276*, 12501–12504. [CrossRef] [PubMed]
15. Struglics, A.; Lohmander, L.S.; Last, K.; Akikusa, J.; Allen, R.; Fosang, A.J. Aggrecanase cleavage in juvenile idiopathic arthritis patients is minimally detected in the aggrecan interglobular domain but robust at the aggrecan C-terminus. *Arthritis Rheum.* **2012**, *64*, 4151–4161. [CrossRef]
16. McCulloch, D.R.; Nelson, C.M.; Dixon, L.J.; Silver, D.L.; Wylie, J.D.; Lindner, V.; Sasaki, T.; Cooley, M.A.; Argraves, W.S.; Apte, S.S. ADAMTS metalloproteases generate active versican fragments that regulate interdigital web regression. *Dev. Cell* **2009**, *17*, 687–698. [CrossRef]
17. Enomoto, H.; Nelson, C.M.; Somerville, R.P.; Mielke, K.; Dixon, L.J.; Powell, K.; Apte, S.S. Cooperation of two ADAMTS metalloproteases in closure of the mouse palate identifies a requirement for versican proteolysis in regulating palatal mesenchyme proliferation. *Development* **2010**, *137*, 4029–4038. [CrossRef]
18. Dubail, J.; Aramaki-Hattori, N.; Bader, H.L.; Nelson, C.M.; Katebi, N.; Matuska, B.; Olsen, B.R.; Apte, S.S. A new Adamts9 conditional mouse allele identifies its non-redundant role in interdigital web regression. *Genesis* **2014**, *52*, 702–712. [CrossRef]
19. Mead, T.J.; Du, Y.; Nelson, C.M.; Gueye, N.A.; Drazba, J.; Dancevic, C.M.; Vankemmelbeke, M.; Buttle, D.J.; Apte, S.S. ADAMTS9-Regulated Pericellular Matrix Dynamics Governs Focal Adhesion-Dependent Smooth Muscle Differentiation. *Cell Rep.* **2018**, *23*, 485–498. [CrossRef]
20. Demircan, K.; Topcu, V.; Takigawa, T.; Akyol, S.; Yonezawa, T.; Ozturk, G.; Ugurcu, V.; Hasgul, R.; Yigitoglu, M.R.; Akyol, O.; et al. ADAMTS4 and ADAMTS5 knockout mice are protected from versican but not aggrecan or brevican proteolysis during spinal cord injury. *Biomed. Res. Int.* **2014**, *2014*, 693746. [CrossRef]
21. Kumagishi, K.; Nishida, K.; Yamaai, T.; Momota, R.; Miyaki, S.; Hirohata, S.; Naito, I.; Asahara, H.; Ninomiya, Y.; Ohtsuka, A. A disintegrin and metalloproteinase with thrombospondin motifs 9 (ADAMTS9) expression by chondrocytes during endochondral ossification. *Arch. Histol. Cytol.* **2009**, *72*, 175–185. [CrossRef] [PubMed]
22. Stanton, H.; Rogerson, F.M.; East, C.J.; Golub, S.B.; Lawlor, K.E.; Meeker, C.T.; Little, C.B.; Last, K.; Farmer, P.J.; Campbell, I.K.; et al. ADAMTS5 is the major aggrecanase in mouse cartilage in vivo and in vitro. *Nature* **2005**, *434*, 648–652. [CrossRef] [PubMed]
23. Larkin, J.; Lohr, T.A.; Elefante, L.; Shearin, J.; Matico, R.; Su, J.L.; Xue, Y.; Liu, F.; Genell, C.; Miller, R.E.; et al. Translational development of an ADAMTS-5 antibody for osteoarthritis disease modification. *Osteoarthr. Cartil.* **2015**, *23*, 1254–1266. [CrossRef] [PubMed]
24. Patwari, P.; Kurz, B.; Sandy, J.D.; Grodzinsky, A.J. Mannosamine inhibits aggrecanase-mediated changes in the physical properties and biochemical composition of articular cartilage. *Arch. Biochem. Biophys.* **2000**, *374*, 79–85. [CrossRef] [PubMed]
25. Paulsson, M.; Mörgelin, M.; Wiedemann, H.; Beardmore-Gray, M.; Dunham, D.G.; Hardingham, T.E.; Heinegård, D.; Timpl, R.; Engel, J. Extended and globular protein domains in cartilage proteoglycans. *Biochem. J.* **1987**, *245*, 763–772. [CrossRef]
26. Buckwalter, J.A.; Rosenberg, L.C. Electron microscopic studies of cartilage proteoglycans. Direct evidence for the variable length of the chondroitin sulfate-rich region of proteoglycan subunit core protein. *J. Biol. Chem.* **1982**, *257*, 9830–9839.
27. Flannery, C.; Stanescu, V.; Mörgelin, M.; Boynton, R.; Gordy, J.; Sandy, J. Variability in the G3 domain content of bovine aggrecan from cartilage extracts and chondrocyte cultures. *Arch. Biochem. Biophys.* **1992**, *15*, 52–60. [CrossRef]
28. Dudhia, J.; Davidson, C.M.; Wells, T.M.; Vynios, D.H.; Hardingham, T.E.; Bayliss, M.T. Age-related changes in the content of the C-terminal region of aggrecan in human articular cartilage. *Biochem. J.* **1996**, *313*, 933–940. [CrossRef]
29. Sandy, J.D.; Thompson, V.; Doege, K.; Verscharen, C. The intermediates of aggrecanase- dependent cleavage of aggrecan in rat chondrosarcoma cells treated with interleukin-1. *Biochem. J.* **2000**, *351 Pt 1*, 1–166. [CrossRef]

30. Sandy, J.D.; Verscharen, C. Analysis of aggrecan in human knee cartilage and synovial fluid indicates that aggrecanase (ADAMTS) activity is responsible for the catabolic turnover and loss of whole aggrecan whereas other protease activity is required for C-terminal processing in vivo. *Biochem. J.* **2001**, *358 Pt 3*, 615–626. [CrossRef]
31. Ilic, M.Z.; Robinson, H.C.; Handley, C.J. Characterization of Aggrecan Retained and Lost from the Extracellular Matrix of Articular Cartilage. Involvement of carboxyl-terminal processing in the catabolism of aggrecan. *J. Biol. Chem.* **1998**, *273*, 17451–17458. [CrossRef]
32. Struglics, A.; Hansson, M.; Lohmander, L.S. Human aggrecanase generated synovial fluid fragment levels are elevated directly after knee injuries due to proteolysis both in the inter globular and chondroitin sulfate domains. *Osteoarthr. Cartil.* **2011**, *19*, 1047–1057. [CrossRef] [PubMed]
33. Melin Furst, C.; Morgelin, M.; Vadstrup, K.; Heinegard, D.; Aspberg, A.; Blom, A.M. The C-type lectin of the aggrecan G3 domain activates complement. *PLoS ONE* **2013**, *8*, e61407. [CrossRef] [PubMed]
34. Kosasih, H.J.; Last, K.; Rogerson, F.M.; Golub, S.B.; Gauci, S.J.; Russo, V.C.; Stanton, H.; Wilson, R.; Lamande, S.R.; Holden, P.; et al. A Disintegrin and Metalloproteinase with Thrombospondin Motifs-5 (ADAMTS-5) Forms Catalytically Active Oligomers. *J. Biol. Chem.* **2016**, *291*, 3197–3208. [CrossRef]
35. Fosang, A.J.; Last, K.; Stanton, H.; Golub, S.B.; Little, C.B.; Brown, L.; Jackson, D.C. Neoepitope antibodies against MMP-cleaved and aggrecanase-cleaved aggrecan. *Methods Mol. Biol.* **2010**, *622*, 312–347.
36. East, C.J.; Stanton, H.; Golub, S.B.; Rogerson, F.M.; Fosang, A.J. ADAMTS-5 deficiency does not block aggrecanolysis at preferred cleavage sites in the chondroitin sulphate-rich region of aggrecan. *J. Biol. Chem.* **2007**, *282*, 8632–8640. [CrossRef] [PubMed]
37. Fosang, A.J.; Neame, P.J.; Hardingham, T.E.; Murphy, G.; Hamilton, J.A. Cleavage of cartilage proteoglycan between G1 and G2 domains by stromelysins. *J. Biol. Chem.* **1991**, *266*, 15579–15582.
38. Ritchie, M.E.; Phipson, B.; Wu, D.; Hu, Y.; Law, C.W.; Shi, W.; Smyth, G.K. Limma powers differential expression analyses for RNA-sequencing and microarray studies. *Nucleic Acids Res.* **2015**, *43*, e47. [CrossRef]
39. Vandesompele, J.; De Preter, K.; Pattyn, F.; Poppe, B.; Van Roy, N.; De Paepe, A.; Speleman, F. Accurate normalization of real-time quantitative RT-PCR data by geometric averaging of multiple internal control genes. *Genome Biol.* **2002**, *3*, RESEARCH0034. [CrossRef]

International Journal of
Molecular Sciences

MDPI

Article

Differences in Shedding of the Interleukin-11 Receptor by the Proteases ADAM9, ADAM10, ADAM17, Meprin α, Meprin β and MT1-MMP

Martin Sammel [1], Florian Peters [1], Juliane Lokau [2], Franka Scharfenberg [1], Ludwig Werny [1], Stefan Linder [3], Christoph Garbers [2], Stefan Rose-John [1] and Christoph Becker-Pauly [1,*]

1 Institute of Biochemistry, University of Kiel, Otto-Hahn-Platz 9, 24118 Kiel, Germany
2 Institute of Pathology, Otto-von-Guericke University Magdeburg, Leipziger Str. 44, 39120 Magdeburg, Germany
3 Institute for Medical Microbiology, Virology and Hygiene, University Medical Center Eppendorf, 20246 Hamburg, Germany
* Correspondence: cbeckerpauly@biochem.uni-kiel.de

Received: 28 June 2019; Accepted: 25 July 2019; Published: 26 July 2019

Abstract: Interleukin-11 (IL-11) has been associated with inflammatory conditions, bone homeostasis, hematopoiesis, and fertility. So far, these functions have been linked to classical IL-11 signaling via the membrane bound receptor (IL-11R). However, a signaling cascade via the soluble IL-11R (sIL-11R), generated by proteolytic cleavage, can also be induced. This process is called IL-11 trans-signaling. A disintegrin and metalloprotease 10 (ADAM10) and neutrophil elastase were described as ectodomain sheddases of the IL-11R, thereby inducing trans-signaling. Furthermore, previous studies employing approaches for the stimulation and inhibition of endogenous ADAM-proteases indicated that ADAM10, but not ADAM17, can cleave the IL-11R. Herein, we show that several metalloproteases, namely ADAM9, ADAM10, ADAM17, meprin β, and membrane-type 1 matrix metalloprotease/matrix metalloprotease-14 (MT1-MMP/MMP-14) when overexpressed are able to shed the IL-11R. All sIL-11R ectodomains were biologically active and capable of inducing signal transducer and activator of transcription 3 (STAT3) phosphorylation in target cells. The difference observed for ADAM10/17 specificity compared to previous studies can be explained by the different approaches used, such as stimulation of protease activity or making use of cells with genetically deleted enzymes.

Keywords: interleukin; IL-6; IL-11; trans-signaling; metalloproteases; ADAM; MMP; meprin

1. Introduction

IL-11 is a crucial factor in several physiological and pathophysiological signaling pathways [1]. It plays an important role in inflammation, osteogenesis, hematopoiesis, and fertility [1–3]. So far, those functions have only been linked to interleukin-11 (IL-11) signaling via the membrane bound IL-11 receptor (IL-11R). In this so-called classic signaling pathway, IL-11 binds to its membrane bound α-receptor, inducing the dimerization of the ubiquitously expressed β-receptor glycoprotein-130 (gp130) on the same cell. This complex then activates intracellular JAK/STAT, MAPK, PI3K signaling pathways [4–6]. The IL-11R can be released from the cell surface [7] by proteolytic ectodomain-shedding, which is analogous to IL-6R solubilization [8]. Soluble receptors such as sIL-6R and sIL-11R can then bind their ligands, IL-6 or IL-11, respectively, before inducing dimerization of gp130 on distinct cells. This pathway is called trans-signaling. sIL-11R can be found in serum of healthy individuals [7], strongly suggesting a physiological function for IL-11 trans-signaling, as known for sIL-6R [9].

Similar to the IL-6R [10,11], the IL-11R consists of three extracellular domains, an immunoglobulin (Ig)-like domain (D1), two fibronectin type III domains (D2 and D3), a stalk region, which is a 54 amino acid long stretch connecting the extracellular part to the transmembrane region (TM), followed by an intracellular domain (ICD) [11,12]. The extracellular domains D2 and D3 form the cytokine binding module, which is essential for signaling [9].

Ectodomain-shedding of the IL-11R and IL-6R requires proteolytic cleavage of the receptors within their stalk regions. IL-6 trans-signaling can be proteolytically induced by a disintegrin and metalloprotease 10 (ADAM10) and ADAM17, as well as by meprin α and meprin β [13]. IL-11R shedding has been attributed to ADAM10 and the neutrophil-derived serine protease neutrophil elastase (NE), thereby inducing trans-signaling [7].

Cleavage of the IL-6R by its major sheddase ADAM17 takes place between P^{355} and V^{356} [14,15]. For meprin β-mediated cleavage of the IL-6R, the cleavage site at P^{355}/V^{356} is in too close proximity to the cell surface and thus sterically not accessible [13]. Additionally, the meprin β-generated sIL-6R fragment has a molecular weight of about 50 kDa, which is different from the 70 kDa fragment produced by ADAM17. This indicates that the meprin β cleavage site in the IL-6R stalk region is further N-terminal, as compared to the one used by ADAM17 [13]. So far, no precise cleavage site for the IL-11R has been identified. However, experiments with deletion variants of the IL-11R indicated that R^{355} in the IL-11R stalk region is important for ADAM10-mediated shedding [7].

ADAM10 and ADAM17 are ubiquitously expressed membrane bound metalloproteases promoting important biological functions. ADAM10 is an essential factor in Notch signaling and ADAM10 knock-out mice resemble the phenotype of Notch deficient animals [16]. ADAM17 plays an essential role in inflammation and cell proliferation by releasing TNFα and ligands of the EGFR [17]. Structurally, ADAM10 and ADAM17 are closely related and exhibit a characteristic domain composition comprising a N-terminal signal peptide followed by a prodomain, a metalloprotease domain, a disintegrin domain, a cysteine-rich domain, a single transmembrane helix, and a cytoplasmic region [18]. ADAM9 is another member of the ADAM family of proteases. It has the same domain topology, except for an EGF-like domain between the cysteine-rich domain and the transmembrane helix [17].

The metalloproteases meprin α and meprin β are multidomain oligomeric metalloproteases playing critical roles in inflammation and maturation of fibrillar collagen [19–21]. Meprin α is cleaved by furin on the secretory pathway and is thus released as a soluble protease, whereas pro-meprin β is expressed as a membrane bound protein and can be activated by matriptase-2 or the bacterial protease RgpB at the cell surface [22–24]. In its inactive proform, meprin β can also be shed from the cell surface by ADAM10 and ADAM17 and subsequently becomes activated by tryptic proteases [24–27]. Prominent substrates of meprins are CD99 [28,29], procollagen [20], mucine 2 [30], as well as IL-6 [31], and the IL-6R [13], highlighting their role in inflammation and extracellular matrix homeostasis and indicate a partially shared substrate pool with ADAM proteases.

Membrane-type 1 matrix metalloprotease (MT1-MMP) as a member of the MMP-family seems to act as a counterpart to meprins by degrading fibrillary collagens I, II, III. It also cleaves ADAM9 [32].

Meprins, as astacin-proteases, together with ADAMs and MMPs, belong to the metzincin superfamily. As described above, these proteases have a partially overlapping substrate pool [19,33–35]. The IL-6R and IL-11R are cleaved within their respective stalk regions [7,13]. The length and amino acid composition of these regions indicate broad susceptibility for proteolysis of both receptors. Hence, we hypothesized that IL-6R and IL-11R are susceptible for shedding by a larger group of proteases.

In this study we employed genetically-modified HEK293T cells deficient for ADAM10 and ADAM17 to avoid basal shedding of the IL-11R by these well-described sheddases. In this proof-of-concept study we were able to specifically address the capability of certain metalloproteases to cleave the IL-11R and the IL-6R. This would not have been possible in the HEK293T wild-type cell-line, which was often used to study ADAM10/17 activity after stimulation with ionomycin or phorbol myristate acetate (PMA), respectively. These compounds however are not solely influencing ADAM10/17 and we therefore compared in a proof-of-concept approach stimulation of endogenous ADAM proteases in HEK293T wild-type cells with retransfection of ADAM10/17 and other metalloproteases on an ADAM10/17-free background.

2. Results

2.1. The IL-11R Is Cleaved by Different Metalloproteases

The IL-11R can, at least in vitro, signal through its membrane bound as well as its soluble form. ADAM10 and neutrophil elastase have been described as sheddases of the IL-11R [7]. We hypothesized that, due to the length of the stalk region and its amino acid composition, the IL-11R should potentially be prone to proteolysis by additional proteases. To further investigate the production of sIL-11R by proteolysis, we transiently transfected human embryonic kidney 293T cells deficient for ADAM10 and ADAM17 (HEK293T ADAM $10/17^{-/-}$) with cDNAs coding for IL-11R and different metalloproteases. Indeed, upon cotransfection with ADAM9, meprin β, MT1-MMP, ADAM17, and ADAM10, we could detect sIL-11R in the cell culture supernatant, demonstrating that all these proteases are putative sheddases of the IL-11R (Figure 1A). Of note, among the tested proteases only cotransfected meprin α showed no activity towards the IL-11R.

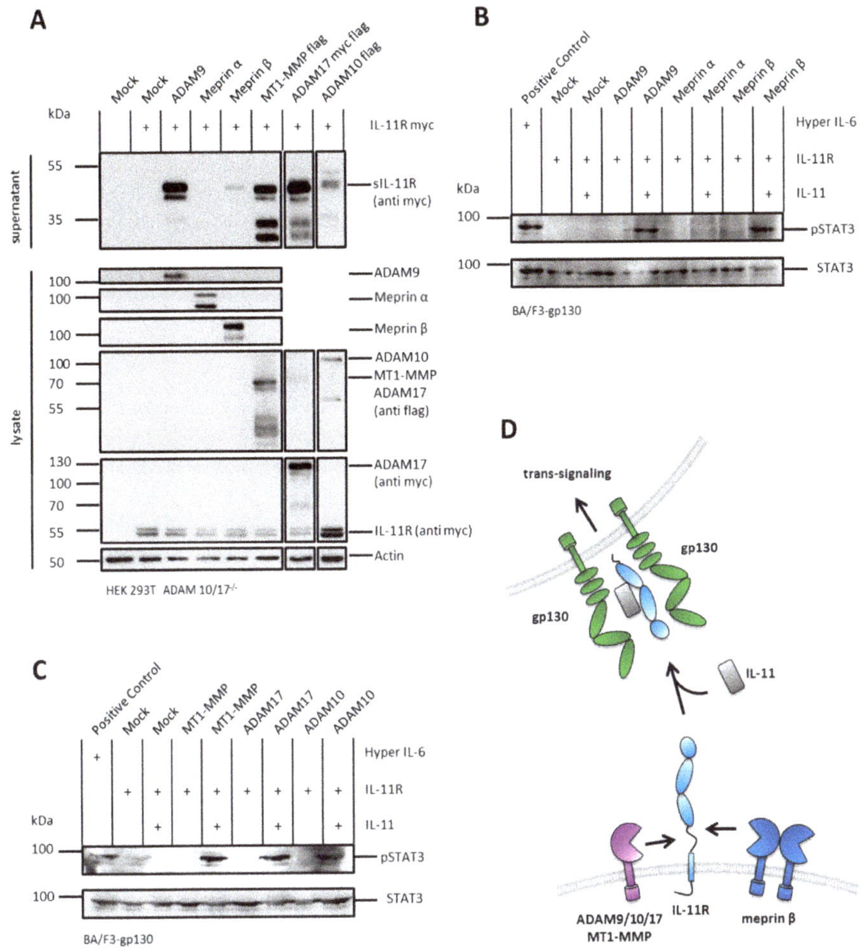

Figure 1. The membrane bound interleukin-11 receptor (IL-11R) is shed by different metalloproteases and induces interleukin-11 (IL-11) trans-signaling. (**A**) Shedding of the IL-11R in HEK293T ADAM $10/17^{-/-}$ cells. Cells were transfected with a N-terminally myc-tagged variant of the IL-11R alone or

with ADAM9, meprin α, meprin β, membrane-type 1 matrix metalloprotease (MT1-MMP), ADAM17, or ADAM10. Supernatants were ultracentrifuged, trichloroacetic acid (TCA)-precipitated, and analyzed by immunoblotting with an antibody raised against the myc-tag. ADAM9, meprin α, and meprin β were detected with specific antibodies in lysate controls. MT1-MMP was flag-tagged, IL-11R was myc-tagged, and ADAM17 flag-/myc-tagged. pcDNA3.1 served as mock and actin as loading control. Cotransfection of IL-11R and ADAM10 was performed in an independent experiment. (**B**) Phosphorylation of STAT3 in Ba/F3-gp130 cells stably transfected with gp130. Cells were treated with supernatants from the experiments in (**A**) and with recombinant IL-11. Phosphorylation of STAT3 was analyzed for cotransfection experiments of IL-11R with ADAM9, meprin α, and meprin β. A fusion protein consisting of soluble IL-6R and IL-6 (Hyper IL-6) served as positive control for pSTAT3. Mock control from (**A**) served as negative control. Phosphorylation was detected with an antibody raised against phosphorylated STAT3. Total STAT3 protein served as loading control. (**C**) Same experimental setting as in (**B**) but for MT1-MMP, ADAM17, and ADAM10 as IL-11R sheddases. (**D**) Schematic overview of IL-11R shedding by ADAM9, ADAM10, ADAM17, MT1-MMP, and meprin β.

2.2. sIL-11R Shed by Different Metalloproteases Is Able to Induce Phosphorylation of STAT3

It has recently been shown that a complex of IL-11 and sIL-11R is able to induce the JAK/STAT-pathway via phosphorylation of STAT3 [7]. We tested sIL-11R generated by different proteases for biological activity on Ba/F3-gp130 cells. Therefore, we treated these cells with the conditioned media resulting from the cotransfection experiments and with recombinant IL-11. In a second step, the cell lysates were stained for phosphorylated STAT3 to validate biological activity of the IL-11R ectodomains.

Indeed, sIL-11R resulting from cotransfection experiments with the proteases ADAM9, meprin β, MT1-MMP, ADAM17, or ADAM10 (Figure 1A) induced phosphorylation of STAT3, thus proving the ability to induce IL-11 trans-signaling (Figure 1B–D).

2.3. ADAM 10 and ADAM 17 Are Both Able to Shed the IL-11R

To examine cleavage of the IL-11R by ADAM proteases, we either stimulated endogenous ADAM proteases in HEK293T wild-type cells transiently transfected with the IL-11R, or we retransfected HEK293T ADAM $10/17^{-/-}$ cells with ADAM10 or ADAM17 together with IL-11R. Upon stimulation with the phorbolester PMA, a strong inducer of ADAM17 activity via activation of Protein Kinase C, shedding of the IL-11R was slightly increased (Figure 2) [36]. Stimulation with ionomycin, which activates ADAM10 via cellular influx of calcium, also induced shedding, albeit to a much higher extent. This implies that ADAM10 is the major sheddase of the IL-11R in HEK293T wild-type cells, as described previously [7]. ADAM17 on the other hand sheds the IL-11R only to a much lower extent. However, upon retransfection of HEK293T ADAM $10/17^{-/-}$ cells with cDNAs coding for the proteases ADAM10 and ADAM17, shedding of the IL-11R was observed (Figure 2). Interestingly, ionomycin also induced IL-11R shedding in the HEK293T ADAM $10/17^{-/-}$ cells without retransfection of proteases, indicating that not only ADAM10/17, but also other sheddases of the IL-11R exist, at least in the cell line used herein.

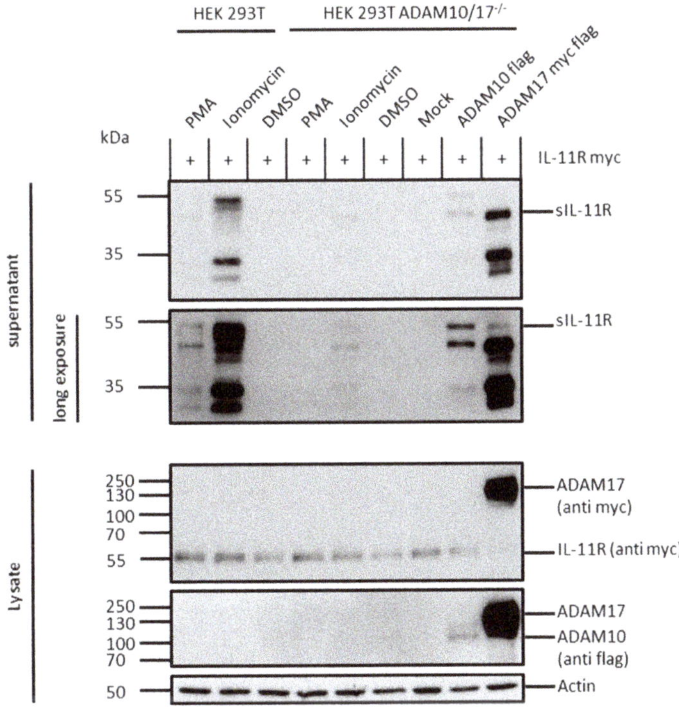

Figure 2. ADAM10 retransfected in HEK293T ADAM 10/17$^{-/-}$ cells is capable of shedding the IL-11R. HEK293T wild-type cells (lanes 1–3) and HEK293T ADAM10/17$^{-/-}$ cells (lanes 4–6) were transfected with a N-terminally myc-tagged variant of the IL-11R and stimulated with phorbol myristate acetate (PMA) or ionomycin. Dimethyl sulfoxide (DMSO) served as control. HEK293T ADAM10/17$^{-/-}$ were transfected with an N-terminally myc-tagged variant of the IL-11R and ADAM10 or ADAM17 (lanes 7–9). For both setups the supernatant was ultracentrifuged, TCA-precipitated, and analyzed by immunoblotting with different antibodies. ADAM10 flag-tagged, IL-11R myc-tagged, and ADAM17 flag-/myc-tagged were detected in lysate controls. pcDNA3.1 served as mock and actin as loading control.

2.4. Arginine 355 Is not Obligatory for Ectodomain Shedding of the IL-11R

The cotransfection experiments described above were repeated with an IL-11R variant lacking amino acids H353-S362 (IL-11RΔ353). The amino acid Arg355 is located within the stalk region, which was previously described to be critical for ADAM10-mediated shedding of the IL-11R, as a mutation of arginine to glutamic acid at this position resulted in significantly reduced proteolysis [7]. However, cotransfection of HEK293T ADAM 10/17$^{-/-}$ cells with ADAM9, meprin β, MT1-MMP, ADAM17, or ADAM10 with IL-11RΔ353 resulted in shedding of this IL-11R variant (Figure 3A). Moreover, sIL-11RΔ353 was able to induce trans-signaling, as measured by pSTAT3 in stimulated Ba/F3-gp130 cells (Figure 3B,C).

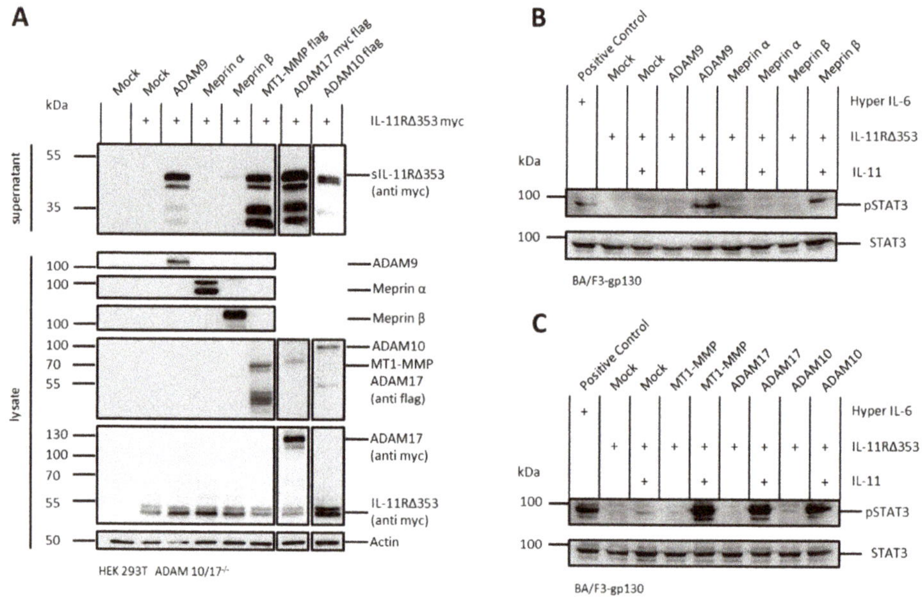

Figure 3. Shedding of the IL-11RΔ353 in HEK293T ADAM10/17$^{-/-}$ cells. (**A**) HEK293T ADAM10/17$^{-/-}$ cells were transfected with a N-terminally myc-tagged variant of the IL-11RΔ353 and ADAM9, meprin α, meprin β, MT1-MMP, ADAM17, or ADAM10. Supernatants were ultracentrifuged, TCA-precipitated, and analyzed by immunoblotting with different antibodies. pcDNA3.1 served as mock and actin as loading control. Cotransfection of IL-11RΔ353 and ADAM10 was performed in an independent experiment. (**B**) Phosphorylation of STAT3 in Ba/F3-gp130 cells stably transfected with gp130. Cells were treated with supernatants from the experiments in (**A**) and with recombinant IL-11. Phosphorylation of STAT3 was analyzed for cotransfection of IL-11RΔ353 with ADAM9, meprin α, and meprin β. A fusion protein consisting of soluble IL-6R and IL-6 (Hyper IL-6) served as positive control. Mock control from (**A**) served as negative control. Phosphorylation was detected with an antibody raised against phosphorylated STAT3. Total STAT3 protein served as loading control. (**C**) Same experimental setting as in (**B**) but for MT1-MMP, ADAM17, and ADAM10 as IL-11RΔ353 sheddases.

2.5. The IL-6R Is Cleaved by Several Metalloproteases

It has been shown that, similarly to the IL-11R, the IL-6R can signal through its membrane bound as well as its soluble form. The sIL-6R can be, comparable to sIL-11R, the result of a proteolytic cleavage event induced by ADAMs or meprins [7,13]. To analyze the production of sIL-6R by proteolysis, we applied the same experimental setting as for the cotransfection with the IL-11R. Upon cotransfection with ADAM9, meprin α, meprin β, MT1-MMP, ADAM17, or ADAM10, we could detect sIL-6R in the cell culture supernatants, thus demonstrating that the IL-6R can be shed by all these proteases (Figure 4A).

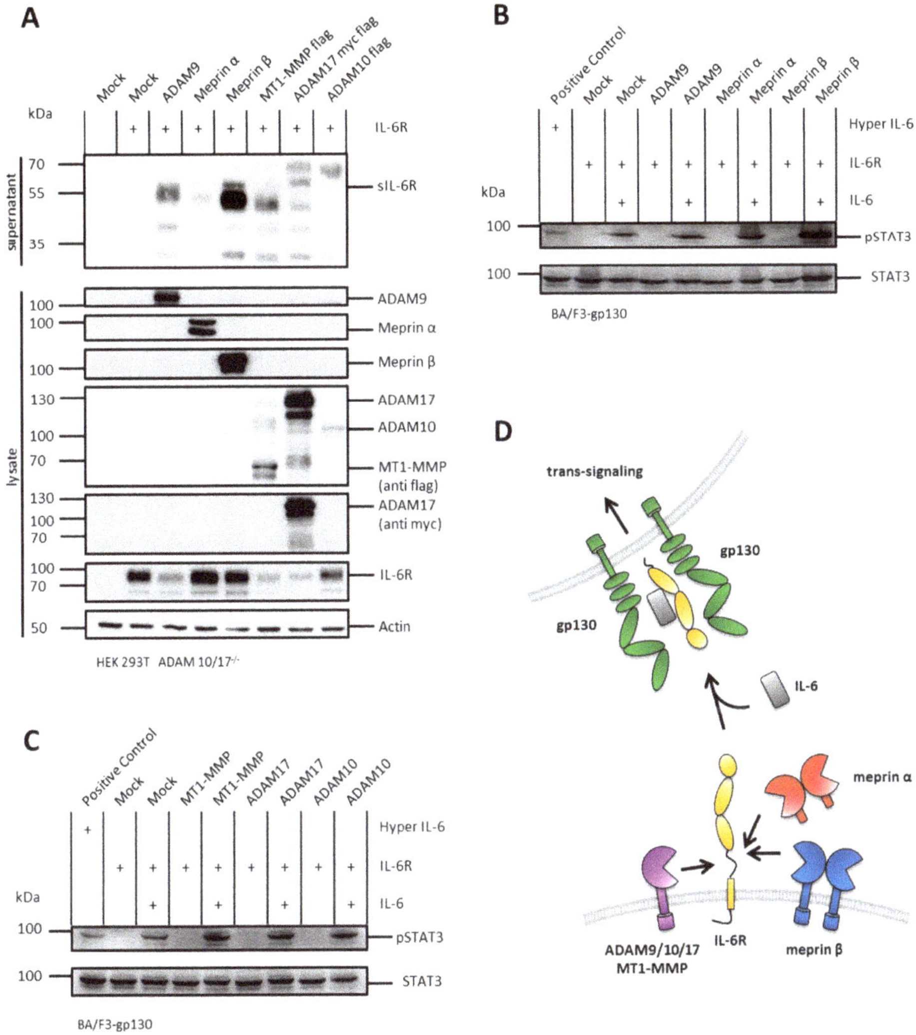

Figure 4. The IL-6R is shed by different metalloproteases and induces IL-6 trans-signaling. (**A**) HEK293T ADAM10/17$^{-/-}$ cells were transfected with IL-6R and/or ADAM9, meprin α, meprin β, MT1-MMP, ADAM17, or ADAM10. Supernatants were ultracentrifuged, TCA-precipitated, and analyzed by immunoblotting with an antibody raised against the D1-domain of the IL-6R. ADAM9, meprin α, and meprin β were detected with specific antibodies in lysate controls. MT1-MMP and ADAM10 are flag-tagged and ADAM17 flag-/myc-tagged and detected respectively in lysate controls. pcDNA3.1 served as mock and actin as loading control. (**B**) Phosphorylation of STAT3 in Ba/F3-gp130 cells stably transfected with gp130. Cells were treated with supernatants from the experiments in (**A**) and with recombinant IL-6. Phosphorylation of STAT3 was analyzed for cotransfection with ADAM9, meprin α, and meprin β. A fusion protein consisting of soluble IL-6R and IL-6 (Hyper IL-6) served as positive control. Mock control from (**A**) served as negative control. Phosphorylation was detected with an antibody raised against phosphorylated STAT3. Total STAT3 protein served as loading control. (**C**) Same experimental setting as in (**B**) but for MT1-MMP, ADAM17, and ADAM10 as IL-6R sheddases. (**D**) Schematic overview of IL-6R shedding by ADAM9, ADAM10, ADAM17, MT1-MMP, meprin α, and meprin β.

Similar to the above described IL-11R shedding and trans-signaling experiments (Figure 1A–D, Figure 3A–C), the proteases ADAM9, meprin α, meprin β, MT1-MMP, ADAM17, and ADAM10 induced phosphorylation of STAT3 by shedding of the IL-6R, thus proving the ability to also induce IL-6 trans-signaling (Figure 4A–D).

2.6. N-Terminal Cleavage of the IL-6R and the IL-11R Apart from the Stalk Regions

The results from the cotransfection experiments showed that at least two different shedding sites within the stalk region of the IL-6R can be assessed, which are in accordance with previous studies [13]. Conclusively, this resulted in a 70 kDa fragment produced by ADAM10 and ADAM17 and a 50 kDa fragment produced by ADAM9, meprin α, meprin β, MT1-MMP, and ADAM17 (Figure 4A). Of note, cotransfection with ADAM9, meprin β, MT1-MMP, ADAM10, or ADAM17 led to additional fragments of about 30 kDa and/or 40 kDa. Immunodetection of the cell culture supernatants was performed with an antibody raised against the D1-domain of the IL-6R. Therefore, these two fragments could be the product of cleavage events within the linker regions between domains D3 and D2, as well as D2 and D1.

Comparable to IL-6R-, also IL-11R cleavage results in multiple fragments. Upon stimulation of HEK293T wild-type cells, as well as cotransfection of HEK293T ADAM $10/17^{-/-}$ cells with different proteases, a characteristic pattern of at least four different receptor bands appeared (Figures 1A, 2 and 3A). This implied that there could be more than one cleavage site within the IL-11R. The two bands between 35 and 55 kDa likely represent shedding within the stalk region resulting in the full-length soluble receptor. The bands at 30 and 35 kDa could be due to cleavage in the linker region between extracellular domains D1 to D3. Analogous to the IL-6R, this band pattern underlines cleavage specificities for the different proteases.

3. Discussion

The pro-inflammatory cytokines IL-6 and IL-11 fulfill several important functions. IL-6 is an important mediator in the innate and acquired immune system, involved in differentiation of T-cells, release of c-reactive protein [9], and acts as acute phase trigger [6]. IL-6 trans-signaling is driven by ectodomain shedding of the IL-6R and has been shown to be a critical event in inflammatory diseases such as rheumatoid arthritis and inflammatory bowel disease [9]. IL-11 is involved in hematopoiesis, especially production of thrombocytes, which led to its approval as the drug 'Oprelvekin' for the treatment of chemotherapy-induced thrombocytopenia [7]. Additionally, IL-11 can activate the hepatic acute phase response [37]. IL-11R knock-out mice underlined an important role in fertility [2] and bone homeostasis [3], as IL-11R knock-out mice show elevated bone density. Mutations within the ectodomain of the IL-11R are associated with craniosynostosis [2,3] and gastric tumorigenesis [38,39]. Moreover, IL-11 is produced by osteoblasts, which further implies a possible role of IL-11 signaling in osteoporosis [40]. For IL-11 trans-signaling, no definite physiological or pathophysiological event has been defined so far, although, sIL-11R can be detected in the plasma of healthy individuals [7]. We could detect both IL-11R shedding and IL-11 trans-signaling in vitro, which strongly suggests an in vivo function for sIL-11R, as demonstrated for sIL-6R.

In this study, we demonstrated that ectodomain-shedding and subsequent induction of trans-signaling of the IL-6R and the IL-11R can be mediated by different metalloproteases. We hypothesize that the length and amino acid composition of both receptors are the reason for their broad susceptibility for proteolysis, which in turn underlines the importance of this trans-signaling event.

So far, possible sheddases of the IL-11R have been identified on the basis of stimulation experiments to induce proteolytic activity of endogenous ADAM10 and ADAM17 in HEK293T wild-type cells [7]. However, stimulation with PMA or ionomycin may additionally induce other proteases besides ADAM10/17 potentially cleaving the IL-11R and IL-6R. To address this issue in a genetically modified system, we employed HEK293T cells deficient for ADAM10 and ADAM17 and retransfected these cells with the proteases and the IL-11R, which allowed us to associate the shedding events precisely with the transfected proteases. Of note, retransfection of the genetically-modified HEK293T cells resulted in some experiments in slightly higher expression levels of ADAM10/17 compared to HEK293T wild-type cells. However, we used this system in a proof-of-concept approach to demonstrate the biochemical capability of certain proteases to cleave the IL-11R and IL-6R. Indeed, we observed shedding of the IL-6R and the IL-11R by several metalloproteases, which resulted in receptor fragments of different sizes, indicating cleavage events in different regions within the receptors. For the IL-6R, two different shedding sites within the stalk region have already been discussed [13]. However, two additional fragments appeared upon cotransfection of the IL-6R with ADAM9, meprin β, MT1-MMP, or ADAM17. We hypothesize that those fragments represent cleavage in the linker regions in between the ectodomains D3 and D2, as well as D2 and D1. Meprin α seemed to solely target the stalk region of the IL-6R, resulting in a fragment of about 50 kDa, whereas ADAM10 seemed to cleave within the stalk region as well as within the linker region between domains D2 and D1.

A comparable band pattern occurred in the cotransfection with the IL-11R. The bands at 30 and 35 kDa could represent cleavage events in the linker region between extracellular domains D1 to D3. Analogous to the IL-6R, this band pattern underlines different cleavage specificities for the tested proteases.

Overall, we observed clear differences between the pharmacological and genetically-modified approaches, which has to be considered, particularly with regard to therapeutic manipulation of IL-6/11 signaling.

In sum, the physiological and pathophysiological functions of IL-6 trans-signaling have been studied to a great extent in health and disease [1,9]. In this proof-of-concept study we showed that the IL-11 trans-signaling can be induced by ADAM9, meprin β, MT1-MMP, ADAM10, or ADAM17 in vitro. Future experiments will show if this concept also holds true in vivo. A possibility to address potential pathological impact of some of the proteases analyzed in this study would be the generation of knock-in mouse models to induce protease-specific expression in certain tissues under diseased conditions.

4. Material and Methods

4.1. Chemicals

All chemicals were of analytical grade and obtained from Carl Roth GmbH & Co. KG (Karlsruhe, Germany), Merck KGaA (Darmstadt, Germany), Sigma Aldrich Inc. (Darmstadt, Germany) and Thermo Fisher Scientific Inc. (Waltham, Massachusetts, USA), if not stated otherwise.

4.2. Cells

HEK293T wild-type (DSMZ GmbH, Braunschweig, Germany) and HEK293T ADAM $10/17^{-/-}$ cells (kindly provided by Dr. Björn Rabe [41]) were maintained in Dulbecco's modified eagle medium (DMEM; Gibco^{TM}, Waltham, Massachusetts, USA) with GlutaMAX^{TM} supplemented with 10% (*v*/*v*) FCS, 100 units/L penicillin, 100 μg/mL streptomycin, and 50 μg/mL Gentamycin.

Ba/F3-gp130 cells [42], which have been stably transfected with gp130 cDNA and therefore respond to the IL-6/sIL-6R or IL-11/sIL-11R complexes [43], were maintained in Dulbecco's modified eagle medium (DMEM; Gibco^{TM}) with GlutaMAX^{TM} supplemented with 10% (*v*/*v*) FCS, 100 units/L penicillin, 100 μg/mL streptomycin, 50 μg/mL Gentamycin, and 10 ng/mL Hyper-IL-6. All cells were kept under humidified conditions (5% CO_2) at 37 °C.

4.3. Plasmids

- pcDNA 3.1
- murine ADAM9 in pcDNA 3.1
- human ADAM10 flag-tagged in pcDNA4/TO
- human ADAM17 DDK in pCMV6: purchased on Origene

- human meprin α in pSG5
- human meprin β in pSG5
- human IL-11R N:myc, C:HA in pcDNA3.1
- human IL-11R in pcDNA 3.1 lacking H353-S362
- human MT1-MMP flag-tagged in pcDNA4/TO
- human IL-6R in pcDNA 3.1
- Sequences were confirmed by Sanger-sequencing.

4.4. Shedding Assay

A total of 2.5–3 × 10^6 cells per dish were transiently transfected with an N-terminally myc and C-terminally HA-tagged variant of the IL-11R or different proteases and the IL-11R. pcDNA 3.1 alone was used as mock-control. In the same experimental setting, a N-terminally myc-tagged IL-11R variant lacking amino acids H353-S362 was cotransfected with the proteases. Polyethylenimine (PEI, Polysciences Europe GmbH) was used as transfection-reagent. For cell transfections, plasmids and PEI were mixed in a ratio of 1:3 in serum-free medium. After 30 min of incubation at room temperature, the transfection-mixture was added to the cell culture in a total volume of 5 mL. After incubation for 5 h, an additional 5 mL of medium was added and incubation was continued overnight. Cell culture medium was replaced with serum-free medium for additional 24 h. For the detection of soluble receptors, the supernatant was collected and cleared by centrifugation (500× *g*, 10 min, room temperature). Additionally, the cell culture supernatant was ultracentrifuged at 186,000× *g* for 2 h at 4 °C. Cell culture supernatants were normalized to the amount of protein of the respective cell lysates and analyzed by immunoblotting. For concentration of proteins, respective cell culture supernatants were precipitated with trichloroacetic acid (TCA 10% (*w*/*v*)) following incubation on ice for 60 min. Afterwards proteins were pelleted (15,000× *g*, 15 min, 4 °C), washed with ice cold acetone, and dissolved in sample-buffer.

4.5. ADAM10/17 Stimulation

HEK293T wild-type and HEK293T ADAM10/17$^{-/-}$ cells were transfected (as described above) with the IL-11R. Afterwards, cells were washed twice with sterile PBS and cell culture medium was replaced with serum-free medium. Cells were then stimulated by adding PMA (100 nM) for 2 h or ionomycin (1 μM) for 1 h. Dimethyl sulfoxide (DMSO) served as control. Cells were harvested and lysed (as described below) and cell culture supernatants were analyzed (as described previously) [7].

4.6. Phosphorylation Assay

Ba/F3-gp130 cells were washed four times with sterile PBS and incubated in serum-free medium for 2 h. A total of 10^6 cells per vial were suspended in 300 μL conditioned medium (shedding assay) and either 150 ng recombinant IL-11 or 150 ng Hyper-IL-6 [7,13] was added. Cells were incubated at 37 °C, shaking at 500 rpm, for 15 min before centrifugation at 500× *g* for 3 min at room temperature. The supernatant was discarded, and the cell-pellet was boiled in sample-buffer. SDS-PAGE and immunoblot analysis were performed to analyze protein expression and protein phosphorylation.

4.7. Cell Lysis, SDS-PAGE, and Immunoblot Analysis

Transfected cells were harvested in ice-cold PBS and centrifuged at 1000× *g* for 10 min at 4 °C. Cell pellets were washed three times with PBS prior to resuspension in lysis buffer (cOmplete ™ protease inhibitor cocktail, 1% (*v*/*v*) Triton-X 100, PBS, pH 7.4) before incubation on ice for 45 min. Afterwards, the cell-suspension was centrifuged at 15,000× *g* at 4 °C for 15 min and the protein amount was determined using a BCA protein assay kit (Thermo Fisher Scientific Inc.). Cell culture supernatants were normalized to the protein content of the respective cell lysates to ensure a comparable analysis. Protein samples were boiled in sample buffer containing DTT. After separation by SDS-PAGE proteins were transferred on nitrocellulose membranes by blotting, which were then saturated with 5% dry milk or 3% BSA for 1 h at room temperature. The membranes were then incubated with primary antibody at 4 °C overnight. After washing in TBS, membranes were incubated with horseradish peroxidase-conjugated secondary antibodies (Thermo Fisher Scientific Inc.) in 5% dry milk or TBS for 1 h at room temperature. Chemiluminescence was detected using the Super Signal^{TM} West Femto kit (Thermo Fisher Scientific Inc.) with the LAS-3000 Imaging System (FUJIFILM Europe GmbH). The following antibodies were used for detection: anti-meprin α/meprin β (polyclonal Abs, generated against ectodomains; Pineda); anti-Myc (# 2276, Cell Signaling Technology, Inc., Cambridge, UK); anti-Flag (F1804, clone M2, Sigma-Aldrich); anti-ADAM9 (polyclonal Ab, kindly provided by Carl Blobel); anti-Actin (#097M4883V, Sigma-Aldrich Inc.); anti-IL-6R (4-11 monoclonal, generated against the D1-domain); and anti-pSTAT3 (#9131 (Tyr705), Cell signaling Technology, Inc.); anti-STAT3 (#9139 (124H6), Cell signaling Technology, Inc.).

Author Contributions: M.S., F.P., F.S. and C.B.-P. designed experiments and interpreted the data. M.S. conducted the experiments. J.L., L.W., S.L., C.G., and S.R.-J. provided material. M.S. and C.B.-P. wrote and edited the manuscript.

Funding: This work was supported by the Deutsche Forschungsgemeinschaft (DFG) SFB 877 (Proteolysis as a Regulatory Event in Pathophysiology, Projects A1, A9, A10, A14, and B13).

Conflicts of Interest: The authors declare no conflict of interest. The funders had no role in the design of the study; in the collection, analyses, or interpretation of data; in the writing of the manuscript, or in the decision to publish the results.

References

1. Garbers, C.; Scheller, J. Interleukin-6 and interleukin-11: Same same but different. *Biol. Chem.* **2013**, *394*, 1145–1161. [CrossRef]
2. Agthe, M.; Garbers, Y.; Putoczki, T.; Garbers, C. Interleukin-11 classic but not trans-signaling is essential for fertility in mice. *Placenta* **2017**, *57*, 13–16. [CrossRef]
3. Nieminen, P.; Morgan, N.V.; Fenwick, A.L.; Parmanen, S.; Veistinen, L.; Mikkola, M.L.; van der Spek, P.J.; Giraud, A.; Judd, L.; Arte, S.; et al. Inactivation of IL11 signaling causes craniosynostosis, delayed tooth eruption, and supernumerary teeth. *Am. J. Hum. Genet.* **2011**, *89*, 67–81. [CrossRef]
4. Hilton, D.J.; Hilton, A.A.; Raicevic, A.; Rakar, S.; Harrison-Smith, M.; Gough, N.M.; Begley, C.G.; Metcalf, D.; Nicola, N.A.; Willson, T.A. Cloning of a murine IL-11 receptor alpha-chain; requirement for gp130 for high affinity binding and signal transduction. *EMBO J.* **1994**, *13*, 4765–4775. [CrossRef]
5. Zhang, X.G.; Gu, J.J.; Lu, Z.Y.; Yasukawa, K.; Yancopoulos, G.D.; Turner, K.; Shoyab, M.; Taga, T.; Kishimoto, T.; Bataille, R.; et al. Ciliary neurotropic factor, interleukin 11, leukemia inhibitory factor, and oncostatin M are growth factors for human myeloma cell lines using the interleukin 6 signal transducer gp130. *J. Exp. Med.* **1994**, *179*, 1337–1342. [CrossRef]
6. Garbers, C.; Hermanns, H.M.; Schaper, F.; Muller-Newen, G.; Grotzinger, J.; Rose-John, S.; Scheller, J. Plasticity and cross-talk of interleukin 6-type cytokines. *Cytokine Growth Factor Rev.* **2012**, *23*, 85–97. [CrossRef] [PubMed]
7. Lokau, J.; Nitz, R.; Agthe, M.; Monhasery, N.; Aparicio-Siegmund, S.; Schumacher, N.; Wolf, J.; Moller-Hackbarth, K.; Waetzig, G.H.; Grotzinger, J.; et al. Proteolytic Cleavage Governs Interleukin-11 Trans-signaling. *Cell Rep.* **2016**, *14*, 1761–1773. [CrossRef] [PubMed]
8. Müllberg, J.; Schooltink, H.; Stoyan, T.; Günther, M.; Graeve, L.; Buse, G.; Mackiewicz, A.; Heinrich, P.; Rose-John, S. The soluble interleukin-6 receptor is generated by shedding. *Eur. J. Immunol.* **1993**, *23*, 473–480. [CrossRef] [PubMed]
9. Garbers, C.; Heink, S.; Korn, T.; Rose-John, S. Interleukin-6: Designing specific therapeutics for a complex cytokine. *Nat. Rev. Drug Discov.* **2018**, *17*, 395–412. [CrossRef]
10. Yawata, H.; Yasukawa, K.; Natsuka, S.; Murakami, M.; Yamasaki, K.; Hibi, M.; Taga, T.; Kishimoto, T. Structure-function analysis of human IL-6 receptor: dissociation of amino acid residues required for IL-6-binding and for IL-6 signal transduction through gp130. *EMBO J.* **1993**, *12*, 1705–1712. [CrossRef]

11. Nitz, R.; Lokau, J.; Aparicio-Siegmund, S.; Scheller, J.; Garbers, C. Modular organization of Interleukin-6 and Interleukin-11 alpha-receptors. *Biochimie* **2015**, *119*, 175–182. [CrossRef] [PubMed]
12. Sprecher, C.A.; Grant, F.J.; Baumgartner, J.W.; Presnell, S.R.; Schrader, S.K.; Yamagiwa, T.; Whitmore, T.E.; O'Hara, P.J.; Foster, D.F. Cloning and characterization of a novel class I cytokine receptor. *Biochem. Biophys. Res. Commun.* **1998**, *246*, 82–90. [CrossRef] [PubMed]
13. Arnold, P.; Boll, I.; Rothaug, M.; Schumacher, N.; Schmidt, F.; Wichert, R.; Schneppenheim, J.; Lokau, J.; Pickhinke, U.; Koudelka, T.; et al. Meprin Metalloproteases Generate Biologically Active Soluble Interleukin-6 Receptor to Induce Trans-Signaling. *Sci. Rep.* **2017**, *7*, 44053. [CrossRef] [PubMed]
14. Goth, C.K.; Halim, A.; Khetarpal, S.A.; Rader, D.J.; Clausen, H.; Schjoldager, K.T. A systematic study of modulation of ADAM-mediated ectodomain shedding by site-specific O-glycosylation. *Proc. Natl. Acad. Sci. USA* **2015**, *112*, 14623–14628. [CrossRef] [PubMed]
15. Riethmueller, S.; Somasundaram, P.; Ehlers, J.C.; Hung, C.W.; Flynn, C.M.; Lokau, J.; Agthe, M.; Dusterhoft, S.; Zhu, Y.; Grotzinger, J.; et al. Proteolytic Origin of the Soluble Human IL-6R In Vivo and a Decisive Role of N-Glycosylation. *PLoS Biol.* **2017**, *15*, e2000080. [CrossRef] [PubMed]
16. Qi, H.; Rand, M.D.; Wu, X.; Sestan, N.; Wang, W.; Rakic, P.; Xu, T. Artavanis-Tsakonas, S. Processing of the notch ligand delta by the metalloprotease Kuzbanian. *Science* **1999**, *283*, 91–94. [CrossRef] [PubMed]
17. Matthews, A.L.; Noy, P.J.; Reyat, J.S.; Tomlinson, M.G. Regulation of A disintegrin and metalloproteinase (ADAM) family sheddases ADAM10 and ADAM17: The emerging role of tetraspanins and rhomboids. *Platelets* **2017**, *28*, 333–341. [CrossRef]
18. Janes, P.W.; Saha, N.; Barton, W.A.; Kolev, M.V.; Wimmer-Kleikamp, S.H.; Nievergall, E.; Blobel, C.P.; Himanen, J.P.; Lackmann, M.; Nikolov, D.B. Adam meets Eph: An ADAM substrate recognition module acts as a molecular switch for ephrin cleavage in trans. *Cell* **2005**, *123*, 291–304. [CrossRef]
19. Herzog, C.; Haun, R.S.; Kaushal, G.P. Role of meprin metalloproteinases in cytokine processing and inflammation. *Cytokine* **2019**, *114*, 18–25. [CrossRef]
20. Prox, J.; Arnold, P.; Becker-Pauly, C. Meprin alpha and meprin beta: Procollagen proteinases in health and disease. *Matrix Biol.* **2015**, *44–46*, 7–13. [CrossRef]
21. Broder, C.; Arnold, P.; Vadon-Le Goff, S.; Konerding, M.A.; Bahr, K.; Muller, S.; Overall, C.M.; Bond, J.S.; Koudelka, T.; Tholey, A.; et al. Metalloproteases meprin alpha and meprin beta are C- and N-procollagen proteinases important for collagen assembly and tensile strength. *Proc. Natl. Acad. Sci. USA* **2013**, *110*, 14219–14224. [CrossRef] [PubMed]
22. Broder, C.; Becker-Pauly, C. The metalloproteases meprin alpha and meprin beta: unique enzymes in inflammation, neurodegeneration, cancer and fibrosis. *Biochem. J.* **2013**, *450*, 253–264. [CrossRef] [PubMed]
23. Jackle, F.; Schmidt, F.; Wichert, R.; Arnold, P.; Prox, J.; Mangold, M.; Ohler, A.; Pietrzik, C.U.; Koudelka, T.; Tholey, A.; et al. Metalloprotease meprin beta is activated by transmembrane serine protease matriptase-2 at the cell surface thereby enhancing APP shedding. *Biochem. J.* **2015**, *470*, 91–103. [CrossRef] [PubMed]
24. Wichert, R.; Ermund, A.; Schmidt, S.; Schweinlin, M.; Ksiazek, M.; Arnold, P.; Knittler, K.; Wilkens, F.; Potempa, B.; Rabe, B.; et al. Mucus Detachment by Host Metalloprotease Meprin beta Requires Shedding of Its Inactive Pro-form, which Is Abrogated by the Pathogenic Protease RgpB. *Cell Rep.* **2017**, *21*, 2090–2103. [CrossRef] [PubMed]
25. Johnson, G.D.; Bond, J.S. Activation Mechanism of Meprins, Members of the Astacin Metalloendopeptidase Family. *J. Biol. Chem.* **1997**, *272*, 28126–28132. [CrossRef] [PubMed]
26. Herzog, C.; Haun, R.S.; Ludwig, A.; Shah, S.V.; Kaushal, G.P. ADAM10 is the major sheddase responsible for the release of membrane-associated meprin A. *J. Biol. Chem.* **2014**, *289*, 13308–13322. [CrossRef] [PubMed]
27. Hahn, D.; Pischitzis, A.; Roesmann, S.; Hansen, M.K.; Leuenberger, B.; Luginbuehl, U.; Sterchi, E.E. Phorbol 12-myristate 13-acetate-induced ectodomain shedding and phosphorylation of the human meprinbeta metalloprotease. *J. Biol. Chem.* **2003**, *278*, 42829–42839. [CrossRef] [PubMed]
28. Bedau, T.; Peters, F.; Prox, J.; Arnold, P.; Schmidt, F.; Finkernagel, M.; Kollmann, S.; Wichert, R.; Otte, A.; Ohler, A.; et al. Ectodomain shedding of CD99 within highly conserved regions is mediated by the metalloprotease meprin beta and promotes transendothelial cell migration. *FASEB J.* **2017**, *31*, 1226–1237. [CrossRef] [PubMed]

29. Bedau, T.; Schumacher, N.; Peters, F.; Prox, J.; Arnold, P.; Koudelka, T.; Helm, O.; Schmidt, F.; Rabe, B.; Jentzsch, M.; et al. Cancer-associated mutations in the canonical cleavage site do not influence CD99 shedding by the metalloprotease meprin beta but alter cell migration in vitro. *Oncotarget* **2017**, *8*, 54873–54888. [CrossRef]
30. Schutte, A.; Ermund, A.; Becker-Pauly, C.; Johansson, M.E.; Rodriguez-Pineiro, A.M.; Backhed, F.; Muller, S.; Lottaz, D.; Bond, J.S.; Hansson, G.C. Microbial-induced meprin beta cleavage in MUC2 mucin and a functional CFTR channel are required to release anchored small intestinal mucus. *Proc. Natl. Acad. Sci. USA* **2014**, *111*, 12396–12401. [CrossRef]
31. Keiffer, T.R.; Bond, J.S. Meprin metalloproteases inactivate interleukin 6. *J. Biol. Chem.* **2014**, *289*, 7580–7588. [CrossRef] [PubMed]
32. Chan, K.M.; Wong, H.L.; Jin, G.; Liu, B.; Cao, R.; Cao, Y.; Lehti, K.; Tryggvason, K.; Zhou, Z. MT1-MMP inactivates ADAM9 to regulate FGFR2 signaling and calvarial osteogenesis. *Dev. Cell* **2012**, *22*, 1176–1190. [CrossRef] [PubMed]
33. Itoh, Y. Membrane-type matrix metalloproteinases: Their functions and regulations. *Matrix Biol.* **2015**, *44–46*, 207–223. [CrossRef] [PubMed]
34. Zunke, F.; Rose-John, S. The shedding protease ADAM17: Physiology and pathophysiology. *Biochim. Biophys. Acta Mol. Cell Res.* **2017**, *1864*, 2059–2070. [CrossRef] [PubMed]
35. Edwards, D.R.; Handsley, M.M.; Pennington, C.J. The ADAM metalloproteinases. *Mol. Asp. Med.* **2008**, *29*, 258–289. [CrossRef] [PubMed]
36. Lokau, J.; Wandel, M.; Garbers, C. Enhancing Interleukin-6 and Interleukin-11 receptor cleavage. *Int. J. Biochem. Cell Biol.* **2017**, *85*, 6–14. [CrossRef] [PubMed]
37. Baumann, H.; Schendel, P. Interleukin-11 regulates the hepatic expression of the same plasma protein genes as interleukin-6. *J. Biol. Chem.* **1991**, *266*, 20424–20427. [PubMed]
38. Balic, J.J.; Garbers, C.; Rose-John, S.; Yu, L.; Jenkins, B.J. Interleukin-11-driven gastric tumourigenesis is independent of trans-signalling. *Cytokine* **2017**, *92*, 118–123. [CrossRef] [PubMed]
39. Putoczki, T.L.; Thiem, S.; Loving, A.; Busuttil, R.A.; Wilson, N.J.; Ziegler, P.K.; Nguyen, P.M.; Preaudet, A.; Farid, R.; Edwards, K.M.; et al. Interleukin-11 is the dominant IL-6 family cytokine during gastrointestinal tumorigenesis and can be targeted therapeutically. *Cancer Cell* **2013**, *24*, 257–271. [CrossRef] [PubMed]
40. Shaarawy, M.; Zaki, S.; Sheiba, M.; El-Minawi, A.M. Circulating levels of osteoclast activating cytokines, interleukin-11 and transforming growth factor-beta2, as valuable biomarkers for the assessment of bone turnover in postmenopausal osteoporosis. *Clin. Lab.* **2003**, *49*, 625–636. [PubMed]
41. Riethmueller, S.; Ehlers, J.C.; Lokau, J.; Dusterhoft, S.; Knittler, K.; Dombrowsky, G.; Grotzinger, J.; Rabe, B.; Rose-John, S.; Garbers, C. Cleavage Site Localization Differentially Controls Interleukin-6 Receptor Proteolysis by ADAM10 and ADAM17. *Sci. Rep.* **2016**, *6*, 25550. [CrossRef] [PubMed]
42. Gearing, D.P.; Ziegler, S.F.; Comeau, M.R.; Friend, D.; Thoma, B.; Cosman, D.; Park, L.; Mosley, B. Proliferative responses and binding properties of hematopoietic cells transfected with low-affinity receptors for leukemia inhibitory factor, oncostatin M, and ciliary neurotrophic factor. *Proc. Natl. Acad. Sci. USA* **1994**, *91*, 1119–1123. [CrossRef] [PubMed]
43. Lokau, J.; Agthe, M.; Garbers, C. Generation of Soluble Interleukin-11 and Interleukin-6 Receptors: A Crucial Function for Proteases during Inflammation. *Mediat. Inflamm.* **2016**, *2016*, 1785021. [CrossRef] [PubMed]

MDPI
St. Alban-Anlage 66
4052 Basel
Switzerland
Tel. +41 61 683 77 34
Fax +41 61 302 89 18
www.mdpi.com

International Journal of Molecular Sciences Editorial Office
E-mail: ijms@mdpi.com
www.mdpi.com/journal/ijms

www.ingramcontent.com/pod-product-compliance
Lightning Source LLC
Chambersburg PA
CBHW061847210926
51850CB00006B/988
* 9 7 8 3 0 3 9 3 6 6 4 8 4 *